T0073404

MIND THIEF

MIND THIEF

THE STORY OF ALZHEIMER'S

HAN YU

Columbia University Press
New York

Columbia University Press
Publishers Since 1893
New York   Chichester, West Sussex
cup.columbia.edu

Library of Congress Cataloging-in-Publication Data
Names: Yu, Han, 1980– author.
Title: Mind thief : the story of Alzheimer's / Han Yu.
Description: New York : Columbia University Press, [2021] |
Includes bibliographical references and index.
Identifiers: LCCN 2020019015 (print) | LCCN 2020019016 (ebook) |
ISBN 9780231198707 (hardback) | ISBN 9780231198714 (pbk.) |
ISBN 9780231552769 (ebook)
Subjects: LCSH: Alzheimer's disease. | Alzheimer's disease—History.
Classification: LCC RC523 .Y8 2021 (print) | LCC RC523 (ebook) |
DDC 616.8/311—dc23
LC record available at https://lccn.loc.gov/2020019015
LC ebook record available at https://lccn.loc.gov/2020019016

Cover design: Milenda Nan Ok Lee
Cover art: Spiderplay © iStock

CONTENTS

PREFACE

very 65 seconds in the United States, a person develops
Alzheimer's.[1] Every three seconds in the world, a person
develops dementia, mostly due to Alzheimer's.

In 2019, 5.8 million Americans lived with Alzheimer's. By 2050,
13.8 million will. By the same year, 131.5 million people worldwide
will be living with dementia.

In the United States, death from Alzheimer's increased more than
120 percent over the last twenty years. It is now the sixth-leading
cause of death. Worldwide, Alzheimer's and other dementias are the
fifth-leading cause of death.

In 2019, it cost $350,000 to care for one American with dementia.
That year, the total cost of Alzheimer's and other dementias was
$290 billion in the United States and over $1 trillion in the world.

Still higher is the human cost of Alzheimer's—the loss of friends
and families, memories and purposes, lives and souls.

At present, there is no effective prevention or treatment, let
alone cure, for Alzheimer's.

As a species, we desperately need to do something.

ACKNOWLEDGMENTS

I thank my agent Kate Johnson at MacKenzie Wolf for her steadfast support, superior work ethic, keen eyes, and literary smartness.

The very same goes to my editor Miranda Martin at Columbia University Press.

These two brilliant women recognize the value of lost histories and silenced voices. They appreciate the humanistic in science and the emotional in facts. They look beyond the artificial division between experts and laypeople. They advocate for non-Western, non-masculine authors. Working with them has been an absolute pleasure.

Thanks also go to my colleague Deborah Murray, who introduced me to the charming Meadowlark retirement community; my colleague Elizabeth Dodd, who gave me invaluable insight in conceiving a big picture for the book; and my colleague Katherine Karlin, who patiently read proposal and chapter drafts, encouraged, guided, and helped in every way she could. I am very lucky to have you all in my life.

Thank you, too, to the very capable editorial and production team at Columbia University Press and its partner who saw this book through various stages of production: Brian Smith for tying up various loose ends, Ben Kolstad and Sue McClung for copyediting, Leslie Kriesel for coordinating, and other behind-the-scene folks. Your professionalism inspires me.

And, of course, thank you to my husband, who puts up with my long hours bound to the desk, who never hesitates to challenge me, and who is prouder of me than I am of myself.

MIND THIEF

1

LOST WIVES

In 1864, a sprawling structure was rising on the outskirts of Frankfurt, Germany. Emulating the medieval gothic style, the structure had imposing towers, an arched entryway, and enormous windows. It was by all architectural accounts an impressive building. Yet, as the home of the Institution for the Mentally Ill and Epileptics, it felt more depressive than impressive under the gloomy Frankfurt sky. Among the locals, it was known as the Castle of the Insane.

Stumbling through the door of the castle in the winter of 1901 was a local woman, Auguste Deter, fifty-one years old. Auguste was tall, with long brown hair. One could tell she had been handsome in her youth, and she still was, except that her dark eyes looked confused and helpless. Standing by her side was her husband, Karl. Auguste had no idea where she was, or, really, who she was.

One of four children, Auguste was born and raised in Kassel, Germany. Her father died young, supposedly from a skin infection, and her mother died at sixty-four of pneumonia.[1] As was the custom for working-class girls of her time, Auguste attended primary school, worked for a wage, and then did the only honorable thing a woman could: marry.[2] At age twenty-three, Auguste tied the knot with Karl Deter. Together, they moved to Frankfurt, where Karl worked as a railway clerk. For the next twenty-eight years, Auguste proved herself a hardworking and amiable wife, maintaining the household, tending her husband, and raising a healthy daughter.

But Auguste hadn't been herself lately. She neglected housework and messed up her cooking. She had trouble remembering things. She was paranoid that others were plotting against her. She wandered around paying unwanted visits to neighbors. She even grew suspicious of Karl, accusing him of having an affair with a neighbor.

Humiliated and helpless, Karl brought Auguste to their family doctor. Noting Auguste's memory problem, manic state, and sleeplessness, the doctor decided that she was suffering from general paresis. Common at the time, general paresis was a feared condition that led to mental deterioration, personality change, and ultimately death. Today, it is known as *neurosyphilis*, a disease caused by bacterial infection in the brain and nervous system. With the advent of penicillin as a treatment, it is no longer lethal. But in 1901, there was no penicillin, nor was there a clear understanding of the cause of the disease. The family doctor declared Auguste's case hopeless and ordered that she be sent to a mental institution: the Castle of the Insane.

Despite its fearsome name and severe appearance, what went on inside the Castle of the Insane was more heartening.[3] Established under the direction of Heinrich Hoffmann, a psychiatrist better known to some as the author of the popular children's book *Der Struwwelpeter* (Shockheaded Peter), the hospital operated based on Hoffmann's belief that mental illness is a true disease. Mental patients are not morally corrupt, weak, or possessed. As with other patients, their suffering has a physical origin: a damaged brain, something less visible than a broken limb or gashing wound. If anything, mental patients need the gentlest and most humane care because their illness is of the most devastating type.

Acting on this belief, the institution banned straitjackets and gave patients as much freedom as possible. Coercive measures were avoided, and isolation was employed only as a last resort. Widely used instead was bath therapy. Patients lay in bathtubs immersed in warm water for hours, or even days, to help soothe themselves. They were also encouraged to exercise, garden, and listen to music.

Poor Auguste, however, didn't get to enjoy the amenities much. She was too far along in her confusion and agitation. Upon her arrival at the institution, she was unable to recall her name, her husband's name, or even that she had a husband. If Karl Deter had had any hope before, he realized that day that his wife of twenty-eight years was gone.

That same year, someone else at the institution lost his wife. Thirty-seven years old, senior physician Dr. Alois Alzheimer was a kind colleague, a content father of three, an avid researcher, and indeed, a recognized expert on general paresis, the disease Auguste supposedly had.

In February 1901, Alois's wife of seven years, Cecilie, died of an undetermined illness. Like Auguste and Karl's marriage, Alois and Cecilie's was a happy one—only it started, rather than ended, with mental illness.[4] Back in 1892, Cecilie Alzheimer was still Cecilie Geisenheimer, the wife of Otto Geisenheimer, a wealthy merchant suffering from general paresis. That year, Otto embarked on an expedition to North Africa with Cecilie and his personal physician. During the trip, Otto suffered a sudden turn for the worse. His physician, a friend of Alois Alzheimer, telegraphed him for help.

Hastening to Algeria, Alois met the group and did what he could, but he couldn't save Otto, who passed away on their way back to Germany. In the real-life plot of a modern-day drama-romance, the young widow took a liking to Alois. More remarkably, *she* eventually proposed to *him*. Alois accepted, and the two were married two years after the fateful expedition.

In addition to three children and a family, Cecilie gave Alois an enormous fortune from her previous marriage. Alois Alzheimer never needed to work for a living again.

But work he did, and hard. If anything, financial freedom made him more determined to pursue his interest in medicine and science, an interest he had developed since a young boy. Born in the small German town of Marktbreit, Alois attended the local elementary school and then went to Aschaffenburg for secondary school. Young Alois distinguished himself at Aschaffenburg, showing strong interest and ability in natural science. Upon graduation, he pursued medical study at universities in Berlin, Würzburg, and Tübingen. Among the subjects he studied, two in particular fascinated Alois: psychiatry and histology (the microscopic study of cells and tissues).[5] The combination of these two subjects was a decisive factor in his eventful encounter with Auguste.

By the time Auguste arrived at the Castle of the Insane, Alois Alzheimer had worked there for thirteen years, first as an intern and eventually as chief physician. As part of his hospital routine, Dr. Alzheimer liked to talk with patients to develop a rapport and observe their symptoms. His first interview with Auguste happened the day after she was admitted, during lunch. On the menu that day were cauliflower and pork.

ALOIS: What are you eating?
AUGUSTE: Spinach!
    She chewed the meat.
ALOIS: What are you eating now?
AUGUSTE: First I eat the potatoes and then the horseradish.[6]

No wonder the poor woman couldn't cook, as her husband complained. She couldn't even tell her veggies from her proteins. It's funny how things haven't changed much over the centuries. Today, the inability for women to perform in the kitchen—"forgetting her apple pie recipe"—is frequently what first alarms husbands and children.

For days, months, and years, Dr. Alzheimer and his colleagues examined Auguste. Detailed notes were kept on how she passed time at the hospital, how she interacted with other patients, and how she responded to doctors. We almost didn't find out about all these because Auguste's medical file went missing around 1910. But after many years of searching, by pure luck, the file turned up in 1995, buried in the archives of the University of Frankfurt's psychiatric clinic:

The blue-colored cardboard pocket, still in pristine condition, contains photographs of Auguste D. [one of which is shown in figure 1.1] and samples of her attempts at a signature. There are also several pages of Alzheimer's hand-written notes, in a now-outdated German script, documenting in detail his patient's behavior during the first 5 days of her hospitalization, and other pages by two colleagues describing subsequent changes in her condition.[7]

From these files, we saw snapshots of Auguste that spanned the four years from her admission to her eventual passing at the institution. She was confused about time and place. She had trouble forming sentences and couldn't read or write, not even her own name. With difficulty, she remembered her husband's and daughter's names, but she couldn't say much about them. She hallucinated, was violent with other patients, and lashed out when doctors tried to examine her. She screamed and wailed, wandered aimlessly, and was heard crying out to God for mercy.

Augustine's condition puzzled and intrigued Alois Alzheimer. Although it was well recognized at the time that advanced age can bring on mental deterioration—the so-called senile dementia—Auguste was only fifty-one, hardly senile. Her symptoms also seemed more severe and aggressive than those of senile dementia. The alternative diagnosis of general paresis, which Auguste's family doctor suspected, could explain the age factor, since this illness strikes indiscriminate of age. But if it were general paresis, there were other factors that did not add up. Although Auguste had the mental impairment seen in general paresis, she lacked other telltale signs: her pupils reacted normally to light, her gait was not disturbed, and her knee reflexes were normal.[8]

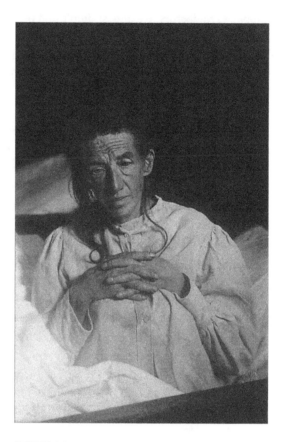

FIGURE 1.1 Auguste Deter

As Dr. Alzheimer struggled to make a diagnosis, Auguste continued to deteriorate. Toward the end, she "was lying in bed in a fetal position completely pathetic, incontinent."[9] In April 1906, she died. Following an autopsy, the causes of death recorded in her file included blood infection due to bedsores, hardening of small brain blood vessels, fluid buildup in the brain, brain atrophy, pneumonia, and kidney inflammation.[10]

Dr. Alzheimer was not the one who performed the autopsy. In fact, he was no longer at the Castle of the Insane. He had left

Frankfurt three years earlier to work for the renowned psychiatrist Emil Kraepelin at the newly constructed Royal Psychiatric Clinic in Munich, where Alzheimer was promised extensive opportunities to conduct neuropathology and microscopy research.[11]

As Emil Kraepelin later wrote in his memoir, Alois Alzheimer was instrumental in the opening of the Munich clinic. Alzheimer assisted him "in the most conscientious and tireless manner" and with "unshakable loyalty and dependability."[12]

With his wealth, Alzheimer agreed to work at the clinic without drawing a salary. As Kraepelin recalled, "Alzheimer entered the service of the clinic without pay because I did not have a position for him and because he wanted to be in charge of how he spent his time."[13] Alzheimer worked as an unpaid "scientific assistant," set up the clinic's anatomy lab, and paid for lab supplies and equipment and the salaries of other personnel. He also helped arrange every single furnishing for the clinic, smoothed out conflicts, and supervised the medical service.[14]

Even though he was busy at his new job, Alzheimer didn't forget about Auguste. His old colleagues at the Castle of the Insane kept him updated on her condition and promptly informed him when she passed away. They even arranged to send Auguste's file and brain to Alzheimer so he could perform a thorough microscopic examination at his Munich lab.

What Alzheimer discovered under the microscope would make history. Inside Auguste's brain, Alzheimer saw neurons in various degrees of disintegration, seemingly destroyed by fibrils, or tiny fibers, growing inside them. In some neurons, one or several fibrils could be seen. In more severe instances, the fibrils were combined into thick bundles. In the hardest-hit areas, neurons had completely disintegrated, and only a tangle of fibrils remained to indicate where neurons once were.[15] In addition to the fibrils, Alois found numerous tiny plaques, or deposits, scattered between neurons. These deposits were made of an unknown substance and could be plainly observed.[16]

Alzheimer's discovery of fibrils was claimed by some scholars to be the very first in medical history. But earlier reports, scant though they were, had identified fibrils in the senile dementia brain.[17] Plaques were not a novel find either. They had been reported some twenty years earlier in the senile dementia brain.[18] However, finding *both* in the same patient, a patient who was *not senile*, was remarkable.

The autopsy results, together with his clinical observation of Auguste at the Castle of the Insane, led Alzheimer to believe that what Auguste had was something unique—something that didn't fit into senile dementia or other mental illnesses known at the time. Still, he wasn't ready to announce that he had found a *new* disease. Instead, he called it "a peculiar disease of the cerebral cortex." Under that title, he published an article describing Auguste's case a year later in the *General Journal for Psychiatry and Psychic-forensic Medicine*.

Naming the illness "Alzheimer's disease" would fall to Alzheimer's boss at the Munich clinic, Emil Kraepelin.

Emil Kraepelin was an influential figure in the history of modern psychiatry—some would say a founder of modern psychiatry itself. Through his best-selling textbook *Foundations of Psychiatry and Neuroscience*, which went into nine editions, Kraepelin made a significant contribution to classifying mental illnesses. His work imposed order on the chaotic array of mental diseases so that systematic diagnoses, research, and treatments could follow.

Right around the time Alzheimer published his paper on Auguste, Emil Kraepelin started revising his textbook for its eighth edition. In that edition, published in 1910, he included, matter-of-factly and without warning or pretext, a new disease that he called "Alzheimer's disease." In that entry, Kraepelin detailed what the disease looked like in clinical symptoms, based on Alzheimer's case observation:

> Over the years the patients gradually regress mentally, become weak in memory, impoverished in thought, confused, and unclear. They can no longer find their way around, fail to recognize people, and give

their things away. Later a certain restlessness develops; the patients chatter a lot, mutter to themselves, sing and laugh, run about, fiddle around, rub, pluck, and become unclean. . . . [T]he patients do not understand any requests or gestures, do not recognize objects and images, do not complete any ordered tasks. . . .

The speech disorders, above all, are most profound. The patients can produce comprehensible individual words or sentences well but usually fall into a meaningless babbling. . . .

Finally the patients fall into complete silence, emitting, only when excited, single comprehensible words or meaningless groups of syllables. Writing is impossible. At the same time, the highest conceivable degree of stupor develops. The patients perhaps look up when one turns to them, occasionally smile, but no longer understand a word or a facial expression, no longer know their next of kin, and answer only to direct physical intervention with gestures of irritation and hastily pronounced, ill-defined syllables. . . .

The final condition represented here can continue with a very gradual worsening or remain seemingly unchanged for many years. Death ensues in the cases I have observed through comorbid illnesses.[19]

Following these clinical symptoms, Kraepelin included a paragraph of brain autopsy findings: the widespread neuronal death and numerous plaques and fibrils. Kraepelin admitted that these autopsy findings overlapped with those of senile dementia, but he emphasized that patients with Alzheimer's disease were much younger and their clinical symptoms were especially severe. "In such cases," Kraepelin concluded, "at least one would therefore have to assume a *presenile* dementia, if we are not in fact dealing with a peculiar disease process that is largely independent of age."[20]

Kraepelin's words being authoritative, Alzheimer's disease was born. It is often referred to simply as *Alzheimer's* or by the acronym *AD*. (Incidentally, those are the initials of its first patient: Auguste Deter. Coincidence, or fate?)

2

CURSED INHERITANCE

Emil Kraepelin knew full well that naming a disease makes it real in popular belief and creates ripple effects beyond just an entry in a textbook. It regulates lives, legitimizes treatments, and sells remedies. *Female hysteria* used to be a catch-all term to condemn women's emotions and urges. Bizarre hypotheses (like the wandering uterus) and condescending treatments (pelvic massages) had been proposed and sold for centuries.

Still, Kraepelin couldn't have imagined the controversy that he stirred up by naming a disease after his loyal employee. Naysayers questioned the medical basis of Alzheimer's disease, arguing that there were few cases supporting its existence and suggesting that naming it was nothing but a publicity stunt.

It is difficult to know how many Alzheimer's cases Kraepelin was aware of. As far as published cases went, only six, including that of Auguste Deter, had made it into the medical literature by the time Kraepelin's book was printed.[1] Moreover, of the six cases, three involved patients who were over sixty years old, the common threshold for senility, and thus presented a contradiction of Kraepelin's proposition of *presenile* dementia.

But there were also unpublished cases, of which Kraepelin knew at least one. It happened at his own Munich clinic and was seen by Alois Alzheimer himself. This was the case of Johann Feigl, Alzheimer's patient number 2.

A fifty-six-year-old, widowed laborer, Johann was admitted to the Munich clinic in late 1907.[2] His symptoms were remarkably similar to those of Auguste's. He was forgetful and disoriented. He couldn't bathe himself, didn't know what to do with a comb, and couldn't write anything except his own name. He could articulate words but had trouble comprehending them. When interviewed, he repeated questions rather than answering them. He fixated on words and ideas, calling a key a "kneecap" after having talked about kneecaps.

Johann's deterioration happened fast. When first admitted, he was able to put a cigar in his mouth, strike a match, light the cigar, and smoke it. Six months later, when given a cigar, he hopelessly rubbed it against the matchbox. As time passed, Johann became restless, resistant, and mute. Toward the end, he was incontinent and completely nonresponsive. Three years after admission, he died.

Because Kraepelin witnessed Johann's symptoms firsthand, this case likely had a strong impact on his decision to name a presenile Alzheimer's disease. What Kraepelin didn't know was that there was more to this case than it seemed.

After Johann passed away, Alois Alzheimer promptly performed a brain autopsy. Like Auguste's, Johann's brain had shrunk and was overtaken by plaques of what seemed the same unknown substance. The plaques were numerous in some parts of the brain, and many had an extraordinary size.[3] However, despite many attempts at sampling different parts of the brain, Alzheimer couldn't find a single neuron invaded by fibrils as he had seen in Auguste's brain.

Until Johann, all the cases that were diagnosed as Alzheimer's disease had abnormal fibrils. It is the combination of fibrils, which grow from inside neurons, and plaques, which are deposited between neurons, that made the microscopic case for a disease called Alzheimer's. Johann's case muddled this picture. Even today, the so-called plaque-only Alzheimer's disease is an uncomfortable topic that, if taken seriously, threatens the pathological basis of an Alzheimer's diagnosis. The good news is that such cases are uncommon.[4] Researchers chalk them up to an early stage of Alzheimer's: had the patients lived longer, as the idea goes, fibrils would eventually have appeared.[5]

Back then, Alzheimer did not have the luxury of comparing a large number of cases. Only a handful existed, and he personally examined just two. One had both plaques and fibrils, and the other had plaques but no fibrils. Given this lack of common pathologies, Alzheimer seemed quite uncomfortable that a new disease bearing his name had now been created. As he wrote in a 1911 paper about Johann's case, "There is then no tenable reason to consider these cases as caused by a specific disease process. *They are senile psychoses, atypical forms of senile dementia.*"[6] The most that he would admit to is that these cases "assume a certain separate position, so that one has to know of their existence . . . in order to avoid misdiagnosis."[7]

Did Alzheimer blame Kraepelin for jumping the gun? That, we don't know, but some historians certainly did.[8] They thought that Kraepelin acted hastily based on scarce scientific evidence and did what he did only for social and financial gain. Supposedly, Kraepelin, by naming a disease after his clinic employee, was trying to outshine a rival clinic in Prague and protect his own clinic's financial stability. At the Prague clinic, a researcher named Oskar Fischer had found plaques in multiple senile dementia brains and was proposing a plaque-based subtype of dementia called presbyophrenia. "Alzheimer's disease" would preempt "Fischer's presbyophrenia." Another theory has Kraepelin competing with Sigmund Freud and his school of psychoanalysis. Freud attributed mental disorders to a conflicted psyche and sexual anxiety, whereas Kraepelin and his school of organic psychiatry attributed mental disorders to biological abnormalities in the brain. An "Alzheimer's disease" based on microscopic brain damage would deal a blow to the idea of the psyche.

Both theories sound attractive, as conspiracy theories go. But neither is adequately supported.[9] There is no evidence that Kraepelin's clinic was in financial crisis. Even if it were, it is doubtful that naming a disease after an employee would have offered immediate relief. The alleged need to trump Freud also seems dubious when the organic school had just scored a big victory: the discovery that general paresis, Auguste's initial diagnosis and a common mental illness of the time, has a biological origin, being caused by bacterial infection in the brain.

Both theories also ignored the fact that Kraepelin was an eminent scholar with a professional reputation who would think twice about proclaiming a disease he did not believe in. Although there weren't many cases in front of him, what Kraepelin read and witnessed must have been compelling. If people in their forties and fifties could deteriorate the way they did, there were probably unique pathologies behind it.

To doubt Kraepelin is also to ignore the achievements of Alois Alzheimer, implying that he went down in history only by virtue of working for an influential boss. On the contrary, Alzheimer was an astute physician, a meticulous researcher, and an outstanding teacher. He was admired for his dual emphasis on clinical and microscopic observation, and recognized for his expertise in multiple conditions, including general paresis, brain arteriosclerosis, and epilepsy. During his lifetime, he published more than fifty research papers,[10] almost all of which he wrote as a single author. He was *not* just in the right place at the right time to catch a rare case.

For his achievements, Alzheimer was appointed full professor and chair of psychiatry at the Silesian Friedrich-Wilhelm University in Breslau in 1912. Unfortunately, three years into the position, he succumbed to a streptococcal infection and rheumatic fever and died at age fifty-one. Walther Spielmeyer, his successor at the Munich clinic, spoke highly of him: "Alzheimer never had to fight for recognition for his research work. The clarity of his oral presentations and writings convinced even a distant observer about the importance of his results. At these times of prolific publishing where everyone believes to have something of importance to say and where many advertise the little things they find over and over again, Alzheimer never took the floor unless he had something truly important to say."[11]

These were the same qualities that held Alzheimer back from believing in Alzheimer's disease. Had he lived longer and encountered more cases, would he have changed his mind? That, I don't know; but had he foreseen the plight of his countrymen the Volga Germans, I'm sure he would have.

For a Russian emperor, Peter III was ill fated. Crowned in 1762, he was overthrown in a mere six months, by none other than his wife, Catherine II. Peter was imprisoned and died mysteriously days later, while Catherine (or Catherine the Great as she came to be known) went on to rule Russia for more than three decades.

One of Catherine's first and most ambitious undertakings was to develop Russia's southeast frontier along the Volga River, a stretch of vast, unpopulated wilderness.[12] It was a behemoth of a task, but Catherine knew just the way to accomplish it: lure European immigrants to settle the land for her. The recruitment campaign was advertised in a 1763 *Manifesto* throughout Europe, which painted a glorious picture of the frontier. The land was rich and fertile, had "an inexhaustible wealth of multifarious precious ores and metals," and was "richly endowed with forests, rivers, seas, and oceans convenient for trade."[13] The *Manifesto* also promised many privileges to those willing to come: travel assistance, daily allowances, equipment and supplies, religious freedom, military service exemptions, tax breaks, and self-governance.

The campaign attracted some 30,000 settlers and had the most success in German's two Hesse states. The Hessians had been plagued by endless wars, barren land, and extreme poverty. Thinking that they had nothing to lose, they plunged into the long journey to Russia. Among them were the Reiswig family: Johannes, the husband, Catherine, the wife, and George, their eight-year-old son.[14] Very little record of the family remains, but we know that they survived the arduous journey (many didn't) and arrived at the Volga frontier.

The end of the journey, to the dismay of the Reiswigs and their fellow travelers, was only the beginning of their hardships. The land that they faced couldn't be more different from what they had been promised: primitive soil, wild vegetation, extreme seasons, and pests of all kinds, including ruthless human outlaws. There was no shelter, and food and medical supplies were minimal. With the Russian winter moving in, the settlers had no choice but to dig pits in the ground to shield themselves from the subzero temperatures. Sickness prevailed, taking the lives of many, including that of Johannes Reiswig. But miraculously, Catherine and little George survived.

Most of the immigrants in Volga were not farmers, and their lack of farming experience was matched by the scarcity of the land and resources. But the people gave the last of their physical and mental strength—now they *really* had nothing to lose. Eventually, against all odds, the crops started to yield, livestock were produced, log huts were erected, and the population grew. The Reiswigs expanded too. Through little George, their descendants grew up to call the Volga frontier home. With other settlers, they turned the once-unproductive wasteland into Russia's agricultural center and industrial hub. All the while, the Volga German immigrants maintained a close-knit community, married within, and kept their language, religion, and customs.

Just as things were turning around, the good started to wane. A prolific people, the Volga Germans ran out of land to sustain their growing population. Anti-German sentiment also flared up in Russia, with talk of forced assimilation and military service in the air. To seek a new land and life, the Volga Germans sent investigators to the United States to scout out the prairie lands along the Missouri River. The scouts came back with favorable reports: there were enough land, railway jobs, religious freedom, and opportunities for every hardworking individual to amount to something. As brave as their ancestors, the first of the Volga Germans departed for the United States in 1874, and hundreds of families in tow followed over the next two years. Experienced farmers, many of them settled in the states of Kansas and Nebraska.

Among them were little George Reiswig's descendants, a family of six: Jacob, the father, Anna, the mother, and four young boys. Arriving in Kansas in 1878, the family moved to Oklahoma several years later, settling down in a town called Kiel. The oldest boy, Christian, married in his early twenties, producing five children of his own. By Volga German standards, that was a small family. For them, having fewer than eight children was an exception; ten and twelve were not uncommon.

That standard was resumed by Christian's first-born, John, who married in the early 1900s—when Auguste Deter was spending her final days in the Castle of the Insane in Frankfurt. In 1906, Auguste Deter died. That same year, John's oldest daughter was born. Another thirteen boys and girls would follow.

The Dust Bowl also followed. Years of aggressive farming, drought, and dust storms crippled the ecology and agriculture in the Oklahoma and Texas panhandles. The sky turned dark, the land desolate, and pneumonia and malnutrition prevailed. Fortunately, all fourteen of John Reiswig's children survived the elements and grew into healthy adults. But like their ancestors, they survived one journey, only to endure a more relentless plight.

The bad omen started when the patriarch of the family, John, began to show signs of forgetfulness and distraction at age forty-eight. No one thought much of it, attributing it to the burden of raising a large family during hard times. Plus, John was functioning. He was taking care of himself, managing the farm, and driving family members around on errands—until one day, during one of those drives, he mysteriously collided his truck into an oncoming train, killing his wife and injuring one of his sons.

John survived the crash but was devastated and became ever more confused. He gave up the farm and moved in with his adult children. Day and night, he had to be watched and cared for like a little child; day and night, he wore a vacant stare on his face. At age sixty-two, he died. But the nightmare did not die with John. It was just starting. Scarcely a decade would pass before John's sons and daughters, now in their forties and fifties, started to wear that same stare. One daughter had to quit her job as a secretary because she could manage it no longer. Another was divorced because she was unable to maintain the household. One son could no longer put together the farm equipment that he had once known like the back of his hand. Another would get lost if he ventured only a little way away from home. One after another, they embarked on the same journey of memory loss, utter confusion, and personality and behavioral changes. For some, the journey lasted a few years; for others, it took two long decades. In their final days, they became paranoid, fearfully aggressive, unresponsive, and incontinent—just as Auguste Deter had been decades earlier. And one after another, they died in their fifties and sixties. Of the fourteen children, only three were spared.[15]

For a long time, the Reiswigs didn't know what was plaguing them. But in the 1960s, after scores of doctor's visits and years of terror, they finally had a name: Alzheimer's disease (more precisely, what's known today as early-onset Alzheimer's disease). There is no hard and fast rule on how early qualifies for "early-onset," but younger than sixty to sixty-five years of age is the rule of thumb. The Reiswigs also learned that they were not the only ones. As of 1992, eighteen families, all of Volga German descent, had been reported to suffer the same condition.[16]

With these families, Emil Kraepelin is vindicated.

Kraepelin's proposition of a presenile dementia dovetails with today's early-onset Alzheimer's, a subset of Alzheimer's disease. What Kraepelin, and Alois Alzheimer himself, didn't imagine was that the disease could be inherited—even though a case was right in front of them: Johann Feigl, Dr. Alzheimer's Munich patient.

Genealogical studies a century later showed that Johann Feigl was not the only person in his family to be demented.[17] Johann's maternal great-grandfather and grandfather, as well as his mother, all died under circumstances that suggest early-onset Alzheimer's. Johann's brother Jakob was mentally incapacitated for several years before his death at sixty-six. Two of Johann's other siblings likewise died suffering from mental illness. Altogether, the disease appeared to have affected sixteen of Johann's maternal relatives and two of his paternal relatives.

As for Auguste Deter, very little of her family record exists, but it is possible that she too was related to the Volga German disease line, since she was from Germany's Hesse region.[18] Supporting this theory, a family with early-onset Alzheimer's has been found in present-day Hesse to harbor the same genetic mutation as afflicted Volga Germans (for more about genetic mutations, see chapter 4).[19] Conceivably, Auguste, this contemporary Hessian family, the Reiswigs, and other afflicted Volga Germans in the United States all share a common curse.

3

LEARNING TO WALK

If something already called Alzheimer's disease was identified and catalogued by the authoritative Emil Kraepelin in 1910, why did it take generations of Reiswigs dying to get a diagnosis? What happened to Alzheimer's in the good part of the twentieth century?

Well, nothing—or very little—according to the U.S. National Library of Medicine. The world's largest biomedical library, it houses an enormous store of research data and impressive electronic tools. One search engine in particular, PubMed (available at https://pubmed.ncbi.nlm.nih.gov), combs through thirty million publications to churn out information about biomedicine and health publications.

Through PubMed, we can get a bird's-eye view of modern Alzheimer's research. From 1910 to 1960, 62 publications with the word "Alzheimer's" in the title or abstract are found in PubMed. That's an average of about 1 publication per year. By comparison, in 2019 *alone*, 10,747 publications are found. The changes that happened in between are shown in figure 3.1, and there you have it: modern Alzheimer's research started at a crawl. It crawled through

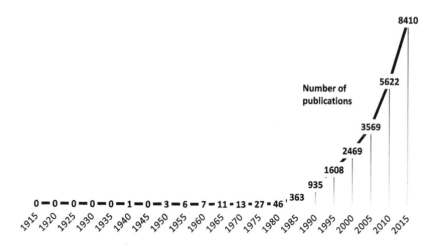

Number of publications

0 ▬ 0 ▬ 0 ▬ 0 ▬ 0 ▬ 1 ▬ 0 ▬ 3 ▬ 6 ▬ 7 ▬ 11 ▪ 13 ▪ 27 ▪ 46 ' 363

1915 1920 1925 1930 1935 1940 1945 1950 1955 1960 1965 1970 1975 1980 1985 1990 1995 2000 2005 2010 2015

935
1608
2469
3569
5622
8410

FIGURE 3.1 The exponential growth of Alzheimer's research in the past 100 years

the first half of the century, started to wobble and walk in the 1970s and 1980s, broke into a trot in the 1990s, and only started running (and then sprinting) over the last fifteen years or so.

Why was so little attention paid to Alzheimer's in most of the twentieth century, then followed by a sudden explosion of interest? Was there a dramatic outbreak of the disease? Not really. Behind the crawl-to-sprint transformation were reasons political and cultural—and then scientific.

Prior to the 1970s, it was believed that Alzheimer's disease, epitomized by the fifty-odd-year-old Auguste Deter, was rare.[1] Rare diseases don't exactly attract research attention—or funding. When patients and cases are limited in number, it takes tremendous time and effort to accumulate knowledge about a disease. Even if research succeeds and a drug is discovered, it sells to only a limited number of people, with meager financial payback. Research into

rare diseases is thus considered unworthy of public-sector investment and unduly risky for pharmaceutical companies.[2]

But what about *senile* dementia, you ask—dementia that happens
to people in their sixties and beyond? That wasn't rare back then, was
it? Indeed it wasn't, and there *was* some suggestion that Auguste's
rare, presenile dementia and the common, senile dementias were
related. After all, the plaques and fibrils found in Auguste's brain had
long been seen in the brains of patients with senile dementia.[3]

But the problem was that we didn't know what to think of senile
dementia back then. Is it a pathological condition, a disease, or is it
simply part of getting old, a matter of "benign memory loss"?[4] If it is
part of aging, then it is inevitable; and if it is inevitable, then there's
no point in studying it. As one researcher recalled, until the 1960s,
"[i]f you raised the subject of senile dementia with doctors, not lay-
men, but doctors, they didn't want to know about it. . . . They'd say, 'I
don't want to discuss it. Look, we all grow old, we can't help it, it's a
sheer, total, waste of time working on it'."[5] With this attitude, studying late-life brain diseases was considered career suicide.[6]

The tide started to turn only in 1968. That year, researchers from
Newcastle University, England, did a "counting" study to see if the
number of brain plaques is related to dementia.[7] Using interviews
and cognitive tests, they first assessed the existence of dementia
among elderly participants. When these participants passed away,
brain autopsies were performed so the number of plaques in their
brains could be counted. Armed with numbers from sixty participants, researchers showed that there is indeed a consistent, if
somewhat imprecise, connection: the more plaques, the worse the
dementia. The demented brain contained up to four times more
plaques than the nondemented brain. The implication? Senile
dementia corresponds with brain anomalies, and the anomalies
are not inevitable in old age—at least not quantitatively. Ergo, senile
dementia is *not* inevitable in old age.

Following this conclusion, pioneer researchers started to argue
in the 1970s that the so-called presenile Alzheimer's disease "is
essentially identical" to senile dementia.[8] They share the same brain

pathologies: plaques and fibrils. They present the same clinical symptoms: memory loss, confusion, and personality and behavioral changes. Except for the age of onset, they have no real differences. It is time, researchers advocated, "to drop the arbitrary age distinction and adopt the single designation, Alzheimer disease."[9]

This proposition has profound implications. If presenile dementia were the same as senile dementia, and both were simply Alzheimer's disease, then none of it can be normal, so studying it wouldn't be a waste of time.[10] Also, now that Alzheimer's was wearing two hats, an early-onset and a late-onset variety, it was no longer rare. Estimates at the time, based on data from England and other countries, were that 4.4 percent of people over sixty-five had dementia, and about 65 percent of that was due to Alzheimer's.[11] By this estimate, 600,000 Alzheimer's cases existed in the United States in the 1970s and caused 100,000 deaths annually.[12] The once rare disease became "a major killer," the fourth- or fifth-leading cause of death in the United States.[13]

Of course, it is naive to think that a few scientists, whose influence rarely reached beyond obscure research journals, could singlehandedly transfer Alzheimer's into a national health concern. In the United States, that transformation involved a host of other people: celebrities, politicians, patients and their families, and businesspeople.

To many Americans at the time, the person who made Alzheimer's "real" was Rita Hayworth: dancer, movie star, and the great American love goddess. Wildly successful as a sexy femme fatale, Hayworth starred in more than forty films and was a beloved pinup girl among American servicemen during World War II. Sadly, the love goddess started to show signs of mental distress and memory loss in her fifties. She drank heavily, couldn't remember her lines, and was consumed by emotional outbursts. For years, people thought her demon was alcohol, but then an Alzheimer's diagnosis was made in 1979.[14] Hayworth died eight years later, at age sixty-eight, trapped in her Central Park West apartment in Manhattan in "a state of utter helplessness."[15]

As Hayworth's life was unraveling, someone else took the stage and played an instrumental role in putting Alzheimer's on the map: the National Institute on Aging (NIA). Established in 1974, NIA is one division, or institute, within the collective U.S. National Institutes of Health. NIA is a bit of an odd division. Unlike other divisions that focus on specific diseases or organs—such as the National Cancer Institute and the National Heart, Lung, and Blood Institute—NIA had a broad charge: to tackle any and all "problems and diseases of the aged."[16]

A broad charge, contrary to what one may think, closes rather than opens doors. As a former National Institutes of Health officer put it, "when a new institute starts . . . particularly one devoted not to a disease or diseases but to a life stage, it is very hard, given the political realities of how budgets are fixed, to get money for studies of aging. You need to find something that is going to attract the interest of the political forces."[17]

That *something*, NIA's first director, Robert Butler, decided, was going to be Alzheimer's. "The way we were going to get resources for the Institute was not going to be the way a scientist might wish it," Butler revealed decades later, "because people don't perceive themselves as dying of molecular biology or basic science. They do perceive themselves as afflicted by whatever. I thought it was justified, since I was already interested in dementia in the broadest sense and Alzheimer's disease in particular, and since Alzheimer's was a common disease. It wasn't as though I was picking telangiectasia, progeria, or something."[18]

With a disease picked, NIA then "primed the pump," so to speak, actively recruiting researchers from related fields to study Alzheimer's and helping them identify research opportunities.[19] The goal wasn't necessarily to find a cure for Alzheimer's right then and there, but rather to generate compelling scientific stories to present to Congress for funding.

In addition, NIA waged a publicity campaign, reaching out to the press whenever an interesting finding about Alzheimer's was made.[20] More cleverly, it cultivated a social group whose members could speak to the human sufferings of Alzheimer's and catch the attention of Congress. At the time, several grassroots Alzheimer's groups already existed, formed mostly by patients' families looking for support and information. NIA put these groups in touch with each other and urged them to form a national organization. In 1980, that organization was born: the Alzheimer's Disease and Related Disorders Association, later renamed, simply, the Alzheimer's Association.

The association performed beyond NIA's wildest dreams. In a mere six years, it grew to 125 chapters and affiliates that covered forty-four states, with up to 35,000 volunteers.[21] It became a strong voice in outreach and disease advocacy. Using private donations and government funds, it lobbied and advocated for patient care, public education, favorable research activities, and support policies. Today, the association remains the largest and best-known not-for-profit organization on Alzheimer's, boasting a budget of $380 million, a staff of 2,600, and a volunteer pool of 62,000.[22]

What is less well known is how the association rose to sudden stardom, which involves another celebrity, Pauline Phillips.

Under the pen name Abigail Van Buren, Pauline wrote the popular, syndicated newspaper column "Dear Abby" from 1956 to 2000. In it, she advised readers on topics in their lives ranging from marriage to children to jobs. On October 23, 1980, the following exchange of letters appeared in her column:

DEAR ABBY: About two years ago I began to notice a change in my husband. He became increasingly forgetful and easily confused even though he was only 50. He had a physical checkup and was found to be in excellent health, but his memory got so bad it wasn't safe to let him drive anymore. Then he had to quit work.

We saw several doctors before one finally seemed familiar with my husband's condition. He told us he had Alzheimer's disease, for which there is no known cure. Alzheimer's disease occurs in people as young as 40 and 50 as well as in some older people.

Abby, my husband is too young for a nursing home, and besides, he is completely healthy otherwise—only his mind is affected.

I fear for his safety and have to watch him every minute. For a while he seems perfectly normal, then he becomes dependent and forgetful. Have you ever heard of Alzheimer's disease? I feel so helpless. How do others cope with this affliction?

DESPERATE IN N.Y.[23]

In reply, "Dear Abby" wrote:

DEAR DESPERATE: You are not alone. Approximately 1 million people in U.S. suffer from Alzheimer's disease. There are now groups of concerned friends and relatives who have banded together to provide support, develop and disseminate helpful information, and encourage much needed research on Alzheimer's disease.

Send a stamped, self-addressed, long envelope to Alzheimer's Disease and Related Disorders Association, 32 Broadway, New York, N.Y. 10004 for up-to-date information. It's free.[24]

We don't know if Desperate in N.Y. acted on the advice, but 25,000 people sure did.[25] Twice a day, two huge sacks full of letters requesting information would arrive in the association's office. Some people also sent in checks. "I was very much interested in the checks," said Jerome Stone, a Chicago businessman and founding chairman of the association, whose wife had Alzheimer's. "I knew then that there were a lot of people who were interested in subsidizing this new organization, and that was very heartening."[26] As the association scrambled to respond to the letters, Stone scrambled to raise more funds. "I went to my friends and anybody who was involved in Alzheimer's. As I say, people who sent even $50 or $100, I called to see if we could get better contributions from them. Anyways, we raised over $200,000 the first year, which was encouraging."[27]

It was, indeed, very heartening and encouraging—except that the "Dear Abby" letter was, um, staged. As Stone let on later, a friend of his who was a friend of Pauline Phillips arranged to get that letter printed. "Desperate in N.Y." was a publicity stunt.

Does the end justify the means? I don't want to judge. As someone who had seen the ugly disease up close and personal, I'm biased. I mean, 25,000 people (and possibly more) found the letter relatable, and $200,000 did find its way to charitable causes. "Desperate in N.Y." may not have been a real person, but her plight seemed to resonate with tens of thousands. Evidently, the advice columnist Pauline Phillips herself didn't judge and was in on it—it is poignant to think that coincidentally, she too would be diagnosed with Alzheimer's twenty years later.

But Peter Whitehouse would have judged, had he known this well-kept little secret. A neurology professor at Case Western Reserve University, Whitehouse is the author of *The Myth of Alzheimer's: What You Aren't Being Told About Today's Most Dreaded Diagnosis*. In it, he argued that Alzheimer's is *not* a definable disease, but rather an extension of normal brain aging. The "disease," according to him, is a myth—a label created by politicians, researchers, and other stakeholders for their own benefit and gain.

"We must remember," Whitehouse wrote in his book, "that multimillion-dollar organizations like the Alzheimer's Association and other private foundations, as well as the Alzheimer's programs at the National Institute on Aging and in universities, which collectively house hundreds and hundreds of employees, would not be able to sustain their funding if Alzheimer's disease did not exist as a cause."[28] "While this is not a conspiracy," Whitehouse concluded, "it's a clever business and politically motivated strategy executed by organizations that know how to tell compelling stories with vivid language that will help sustain themselves."[29]

When asked what they think of the book, some of Whitehouse's colleagues said, "Peter has lost his marbles."[30] While I wouldn't go that far and indeed appreciate Whitehouse's point about money and politics, I did find his book sparse on novel scientific evidence, while exuberant with personal conviction. It implies that the public is clueless and can be manipulated into believing in a colossal myth. It suggests that without the "Alzheimer's" label, we would cease to stigmatize cognitive decline and start to celebrate old age.

I'm not sure this is true. Average men and women are plenty capable of recognizing their own pain and suffering, regardless of what doctors or the media say. Heck, we are *better* than doctors at recognizing our own pain and suffering. Yes, Alzheimer's is a scary diagnosis, but without this label, don't we just double down on stigmas? Wasn't Rita Hayworth, for all her charm and beauty, once deemed a pitiful alcoholic? Weren't millions of nameless Ritas, upon losing their minds, deemed incapable, irresponsible, or simply crazy? Musing the (yes, somewhat arbitrary) difference between "normal" and "diseased" is a good philosophical exercise (for more about this, see chapter 20), but asking the elderly to get on with their lives when the life they know no longer exists is more than a little impractical.

NIA's efforts started to pay off by the mid-1980s. In 1976, NIA's budget for Alzheimer's research was a measly $800,000; by 1985, it had grown to $28.9 million.[31] Not all this money went to build the political-economic empire that Whitehouse imagined. In fact, we have something very good to show for this early investment: the first modern hypothesis of what caused Alzheimer's—and from it, the first treatment approved by the U.S. Food and Drug Administration (FDA).

This early work approached Alzheimer's from the very basics: what is memory? Memory, of course, amounts to the things, people, and events that we remember. However, from a neuronal perspective, memory is essentially a pattern: a pattern in which certain neurons are fired, or activated, together. When neurons are activated together, the connections among them are strengthened. If we repeat the connections enough times, an automatic and long-lasting pattern forms. A face will immediately conjure up a name, personal attributes, past experiences with that person, and so on. That's when we say we remember that person.

For neurons to fire and connect with each other, we need something called *neurotransmitters*, which are chemical messengers that transfer impulses from neuron to neuron. Multiple types of neurotransmitters exist, each responsible for certain actions. One that caught the attention of early researchers is acetylcholine, or ACh for short.

In the late 1970s, autopsy studies showed that in the brain of an Alzheimer's patient, ACh production is significantly reduced.[32] The cause for this reduction was soon revealed—by none other than Peter Whitehouse. At the time, Whitehouse was studying at Johns Hopkins and found that with Alzheimer's disease (he had no problem calling it a disease back then), the brain suffered significant neuronal loss in a lower-front region called the *substantia innominate*.[33] Remarkably, these lost neurons would be the primary source of ACh production, responsible for supplying ACh to other brain regions important for cognition and memory formation.

These findings ushered in the cholinergic hypothesis—the word *cholinergic* means using ACh as a neurotransmitter. According to this hypothesis, when ACh production runs low, the relay of messages between neurons stalls, which causes memory to break down and cognition to falter. Accordingly, to cure Alzheimer's, we must increase ACh production.

A main ingredient in ACh production is choline, which is a common dietary nutrient. Animal-based food items such as meat, egg, and fish are rich in choline, whereas most fruits and vegetables are low in it.[34] In rat studies, the administration of choline raised brain ACh.[35] Dietary intervention worked too. A choline-rich diet increased ACh in the rat brain, while a choline-poor diet reduced it.[36]

In human experiments, however, choline administration failed to improve cognitive performance, despite repeated attempts.[37] Ingesting dietary sources of choline similarly had no effect.[38] Not ready to give up, researchers came up with an alternative idea: if boosting ACh production doesn't work, maybe we can prevent its degradation? When ACh is released into the brain, it is quickly broken down by an enzyme called *acetylcholinesterase (AChE)*. If we inhibit AChE, ACh can stick around long enough to do its job. Drugs that inhibit AChE are called *AChE inhibitors*.

The first AChE inhibitor that went into human trials is tacrine (trade name Cognex). In multiple trials with Alzheimer's patients, tacrine was found to improve cognitive performance, as well as over-all behavior and function.[39] The amount of improvement, however, was small: while untreated patients would deteriorate 0.5–1.0 point on a standard cognitive test during a twelve-week period, tacrine-treated patients ranged from deteriorating 0.38 point to improving 0.12 point.[40] The drug also had serious side effects, including liver toxicity, and about 25 percent of patients simply couldn't tolerate it at all. Among patients who could, about 40–50 percent showed short-term improvement.[41] Despite tacrine's side effects and less-than stellar therapeutic performance, in 1993, it was approved by the FDA as an Alzheimer's treatment—the first ever.

After tacrine, three other AChE inhibitors emerged, demon-strated similar benefits with fewer side effects,[42] and were approved by the FDA in quick succession between 1996 and 2001. Among them, donepezil (trade name Aricept) was approved for treating all stages of Alzheimer's, from mild to moderate to severe. The others—rivastigmine (trade name Exelon) and galantamine (trade name Razadyne)—were approved for treating mild-to-moderate Alzheim-er's. Safer than tacrine and comparable in effect, the new AChE inhibitors essentially pushed tacrine out of the market.

In addition to ACh, early researchers looked at another neurotrans-mitter, glutamate. Like ACh, glutamate facilitates neuron connection and memory formation. It is, in fact, one of the most abundant and powerful neurotransmitters out there, capable of exciting neurons even at low concentrations.[43] Precisely because it is so powerful, glutamate is a double-edged sword. If they receive too much of it, neurons will be overexcited to the point of exhaustion and death—a phenomenon known as *excitotoxicity*.

Because the Alzheimer's brain suffers damage in certain neurons and neuronal structures that interact intimately with glutamate,[44] researchers suspected excitotoxicity and experimented with a drug that blocks glutamate: memantine (trade name Namenda). It was a quick success. In mid-to-late-stage Alzheimer's patients, meman-tine preserved basic abilities such as bathing and toilet use, reduced caretaker dependence, and improved behavioral issues such as agitation.[45] On account of these effects, in 2003, the FDA approved memantine for treating moderate-to-severe Alzheimer's.

Today, AChE inhibitors and memantine remain the go-to drugs for Alzheimer's. Their success is impressive, especially considering that they came out of pioneering research performed before the field even took off. Since then, no new Alzheimer's drug has been approved.[46] It is, however, important to note that existing neurotransmitter treatments are all palliative. That is, they do *not* cure Alzheimer's, nor do they stop the progression of the disease. They only reduce symptoms—marginally and temporarily. They are, at best, a bandage.[47]

A bandage offers no explanation for where the wound came from. It presents no explanation for why Auguste Deter fell ill when scarcely fifty, while Rita Hayworth bought herself a bit more time. It gives no explanation for what runs in the Volga German bloodline and haunts generations of the Reiswigs. To answer these questions, we turn to the crown jewel of modern Alzheimer's research—genetics, starting with a refresher, or crash course, on chromosomes.

4

SEARCHING FOR THE ALZHEIMER'S GENE

Chromosomes live inside the nucleus of a cell. Under the microscope, they look like little rods made of coiled cotton threads. The threads are long deoxyribonucleic acid (DNA) strands, so chromosomes are, simply stated, tightly packed DNA. A full definition of DNA can run for pages. For now, it suffices to say that DNA is a molecule that directs protein production and thereby determines the way that life forms and functions.

Each person has twenty-three pairs of chromosomes—*pairs*, so a total of forty-six. Of the twenty-three pairs, one pair consists of the sex chromosomes: X and Y. This pair of chromosomes determines our sex: XX for a female; XY for a male.[1] Early-onset Alzheimer's does not discriminate between males and females, which suggests that the "cursed" inheritance is not conveyed by the sex chromosomes.

The remaining twenty-two pairs of chromosomes are called *autosomes*, which are the same in males and females. They are numbered 1 through 22. Chromosome 1 is the largest and has the longest DNA strand; the rest follow in descending order.[2]

A child inherits half of its chromosomes from the mother and half from the father. So, for example, with the chromosome 1 pair, one copy comes from the mother and the other from the father. If a defect in either inherited copy is sufficient to cause a disease, the disease is called *autosomal dominant*; if both copies must be defective to cause a disease, the disease is called *autosomal recessive.*

Given its robust inheritance pattern, early-onset Alzheimer's is autosomal dominant. This is a cruel reality for children born within an afflicted family because their chance of getting the disease is a steadfast 50 percent. Suppose that the disease resides on chromosome 21, and a mother carries one defective and one healthy chromosome 21 as her own pair. Because a child *must* inherit one chromosome 21 from the mother, there's a 50 percent chance that the child will get the defective copy and inherit the disease. The math is the same if the father is the affected parent.

Having more children doesn't change the math either. Each time a child is conceived, the coin is tossed for the same 50 percent odds. This is a new game each time it is played, and playing more rounds doesn't improve the luck. Among the fourteen Reiswig children, eleven lost—a string of bad, bad luck. To make matters worse, with early-onset Alzheimer's, inheritance is fully penetrant, which means when a child inherits the defect, there is no escaping the disease.[3] Their fate is sealed at birth. For these reasons, a single patient in a prolific family can create profound ripple effects on generations of descendants. And if these descendants marry within a close-knit community, as the Volga Germans do, they give rise to large clans that suffer from the disease.

Sadly, despite all their suffering, the Volga Germans aren't the worst hit by Alzheimer's. On the outskirts of Medellín, the capital of Colombia's Antioquia province, live the Paisas (from *paisano*, meaning "countrymen" in Spanish), another close-knit, isolated community. In the Paisas' mountainous villages, researchers located twenty-five exceedingly large families with 5,000 members suffering from early-onset Alzheimer's. Their disease line is suspected to come from a single European colonist a few centuries earlier. Before modern science reached the villages, the Paisas called their disease *La Bobera* ("the foolishness"). They thought it was caused by the curse of a witch, by the revenge of a priest, or by touching a certain tree.

On top of the large number of cases, the Paisas face a disease onset that is shockingly early, even by the standard of "early-onset." Many start experiencing symptoms in their forties or early fifties. Elderly parents, if they are not themselves ill, are condemned to care for their demented, adult children. At age eighty-two, Mrs. Cuartas looks after three. One son, Oderis, fifty, has failing memory but denies it. The other, Darío, fifty-five, "babbles incoherently, shreds his socks and diapers, and squirms so vigorously he is sometimes tied to a chair with baggy blue shorts."[4] A daughter, María Elsy, sixty-one, showed symptoms at forty-eight and is now "a human shell, mute, fed by nose tube."[5]

"To see your children like this. . . . It's horrible, horrible. I wouldn't wish this on a rabid dog. It is the most terrifying illness on the face of the earth," cried Mrs. Cuartas.[6]

Recognizing the abyss of their suffering, we should be thoroughly grateful to the Paisas, Volga Germans, and others like them because their misfortune opened an invaluable door to understanding Alzheimer's disease. Individuals from these communities willingly take part in research studies, undertaking cognitive tests, blood panels, brain scans, lumbar punctures, and, eventually, autopsies in order to yield data for modern science. Many know in their heart of hearts that what they are doing is too late for themselves, but they hope that they are giving something valuable to their children, their children's children, and other ill-fated souls.

And they are. With their shared ancestry and living environments, these large clans provide data that contain little of the variation and noise that researchers usually must account for when calibrating results. By comparing the genetic profiles of healthy and diseased clan members, researchers can more effectively locate the defective chromosome and then zero in on the stretch of DNA on that chromosome that causes the patients' suffering.

That said, the process is a lot more complex in practice. DNA can be written as a string of A, T, C, and G letters, which stand for the different nitrogen bases—adenine (A), thymine (T), cytosine (C), and guanine (G)—that randomly line up to form a DNA strand. Human DNA is three billion bases long, and no two people have exactly the same base sequences, so researchers have to rule out many innocent differences before they can pinpoint the real defect. To make the task feasible, they need to start with a likely suspect rather than trying their luck by going from chromosome 1 all the way to 22.

Of the twenty-two pairs of autosomes, chromosome 21 was the favorite suspect. This chromosome is implicated in Down syndrome, a genetic disorder featuring intellectual disabilities and developmental delays that affects 1 in 700 babies in the United States. Down syndrome is caused by being born with one extra copy of chromosome 21: three rather than the usual two. This is relevant to Alzheimer's because children with Down syndrome almost always develop Alzheimer's-like brain pathologies (namely, the plaques and fibrils) when they reach middle age. Eventually, many also develop outward Alzheimer's-like symptoms, including changes in function and personality. So, it seems possible that certain defects on chromosome 21 are also the culprit in Alzheimer's. In the 1980s, this theory interested researchers around the world, who dissected this little chromosome in search of patterns of inheritance.[7]

To appreciate what they were looking for, we need to know a little something about sex. During sex reproduction, our body naturally seeks to maximize its genetic diversity. Before a parent passes his or her chromosomes to a child, some mixing and shuffling happens. The mother's paired chromosomes (for example, the two copies of chromosome 1), which she separately inherited from *her* parents, would physically cross over and exchange pieces of DNA. After this mixing, a copy gets passed on to her child, so the child essentially inherits from both *the maternal grandmother and grandfather*. The same happens on the father's side.

Imagine this genetic shuffling as card shuffling. When we shuffle and then cut a deck of cards, cards that were very close to each other beforehand are more likely to end up in the same stack. The same is true of DNA. DNA segments that sat close to each other on a chromosome are more likely to stay together during the shuffling and be inherited together. Given this mechanism, researchers can search for unknown, defective DNA on a chromosome by using known DNA as markers. If a marker is consistently inherited by those who are diseased, then the defective DNA probably sits close by. This method of study is called *linkage analysis*.

Although sound, linkage analysis is also a fishing expedition—a random search in the hope of finding something, as opposed to a targeted search for a specified thing. Even though chromosome 21 is a small autosome, it still contains 48 million DNA bases with numerous known markers. If you don't know what you are looking for, it's going to take a while to find it. It's a good thing, then, that as chromosome 21 tied up many researchers, others quietly made progress in different ways.

When Alois Alzheimer first saw plaques in Auguste Deter's brain, he didn't know what he was looking at; he dubbed them "depositions of a special substance."[8] Twenty years later, the Belgian psychiatrist Paul Divry figured out what the substance was: a protein called *amyloid*.[9] Truth be told, it is not all that special. Amyloid was found in plants in the 1830s, and then in the human nervous system in the 1850s.[10] The name comes from the Latin *amylum*, which means "starch," because like starch, amyloid has a waxy, gelatinous appearance. It also stains the same color when exposed to certain dyes, and that was how Paul Divry identified it.

Despite its starchy, innocent look, amyloid is quite the troublemaker. It can build up and cause damage in various organs, including the heart, kidney, liver, and spleen. In the heart, amyloid deposits stiffen the heart muscle and reduce its ability to beat, with potentially fatal consequences. Given amyloid's shady character, finding it in the Alzheimer's brain was an "aha" moment. Presumably, this troublemaker was damaging normal brain tissues, which could conceivably cause Alzheimer's.

Of course, we need more evidence than amyloid's bad reputation, because to be clear, amyloid is a general term that refers to a kind of pleated protein structure, not a specific protein. Many proteins can form the amyloid structure, so to get closer to the nature of Alzheimer's plaques, we need to determine their specific identity—a task that took another half-century. You see, amyloid is quite stubborn. It doesn't easily dissolve and it's hard to purify, so studying it hasn't been easy. It wasn't until 1984 that two researchers from the University of California, San Diego finally succeeded.

George Glenner and Caine Wong's success had a lot to do with their plan of attack: rather than targeting the plaques that scatter around neurons in the Alzheimer's brain, they went after the ones inside brain blood vessels, which are more soluble and easier to analyze.[11] From these plaques, they isolated two proteins: the first they named *alpha;* the second, *beta.* Alpha turned out to be an unrelated substance, so henceforth, the infamous substance that accumulates in the Alzheimer's brain would be known as *beta-amyloid.*

It should be noted that beta-amyloid is not a foreign intruder. It is produced naturally in the brain, playing functions that we do not currently understand. Normally, it is degraded by enzymes or moved out of the brain into the circulation. In Alzheimer's disease, though, it has somehow stayed and clumped into plaques.

Having isolated beta-amyloid, Glenner and Wong moved to the next step: determining its building blocks, the amino acids. Amino acids are like tiny beads that can be strung together to form a chain. When the chain gets long (usually having more than fifty beads) and is stable, it becomes a protein. Our body uses twenty common amino acids, strung together in different ways, to produce the numerous proteins it needs.

Figuring out beta-amyloid's amino acid makeup is important because amino acids are generated—coded, so to speak—by unique

DNA bases. Once the amino acids are determined, we can deduce the DNA sequence that codes for beta-amyloid. This stretch of functional DNA is what's known as a *gene*. With a gene in hand, we can then test whether it is mutated in Alzheimer's patients. If it is, the mutation is likely the cause of beta-amyloid deposits in their brains—and the root of their disease!

What Glenner and Wong found, however, was a bit of a letdown: beta-amyloid turned out to be a rather short chain, consisting only of twenty-four amino acids, not long enough to qualify as a protein.[12] The DNA that codes for it, then, won't be a "real" functional gene. But this discovery made researchers realize that beta-amyloid could be a protein *fragment* cut from a longer, *precursor* protein, as pictured in figure 4.1.

Using Glenner and Wong's finding as a clue, several research groups quickly teased out the full precursor protein, which is about 700 amino acids long.[13] Now *that* is a real protein. Its gene, the amyloid precursor protein (APP) gene, has about 2,100 coding DNA bases and sits on chromosome 21.[14] As mentioned before, this little chromosome was a favorite suspect in Alzheimer's because of its involvement in Down syndrome. Things are falling into place! Now, we just

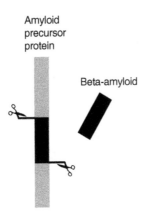

Amyloid
precursor
protein

Beta-amyloid

FIGURE 4.1 Beta-amyloid protein fragment being cut out of a longer amyloid precursor protein by enzymes

need to test Alzheimer's patients to see what mutation was happening in their APP gene, and we would find the cause of their suffering. But so close to victory, things ground to a halt. For years, researchers tested the Alzheimer's patients they had access to, but one after another, their APP genes all turned out fine. Despite everything that's going for it, maybe this gene wasn't the answer?!

It's a good thing that scientists are a persistent bunch. As they persisted, breakthroughs finally came. In February 1991, John Hardy and his team from St Mary's Hospital Medical School, London, found an APP mutation in two early-onset Alzheimer's families.[15] The mutation was tiny: a single DNA base change, from C to T, among the thousands of bases in the APP gene. Once this mutation was found and researchers knew where to look, several more mutations were spotted in the same or nearby DNA bases in other diseased families.

But not everyone was celebrating. Most other researchers remained frustrated and confused. Working with their patient samples, they still couldn't turn up any mutations. Pressured and paranoid, they commiserated at conferences and wondered what they were doing wrong.[16]

Then, the revelation came: if so many labs couldn't find mutations in the APP gene, if so many early-onset patients don't seem to have an APP mutation, then maybe APP is not the *only* gene that can trigger the disease. For all we know, there is another gene out there waiting to be discovered, waiting for someone else to make history. With that thought, hopelessness and paranoia turned back into frenzy, and teams rushed to conduct more linkage analyses. This time, chromosome 21 was no longer the singular favorite. This little chromosome has been searched over and over with a fine-toothed comb; not much was left to mine. Where else to look in the other twenty autosomes was anyone's guess. It's a huge body of water, but it was surmised that if we kept fishing, someone would get lucky.

And someone did. In October 1992, after years of fishing, Gerald Schellenberg from the University of Washington spotted a suspicious marker region on chromosome 14.[17] Schellenberg had not pinpointed a gene yet, but he was able to link nine early-onset

Alzheimer's families to this region. Because APP mutations were found in only a handful of families, Schellenberg's discovery promised something bigger and better.

Scarcely a month after Schellenberg's discovery, *three* other teams came forward to publish their findings linking early-onset Alzheimer's to the same chromosome 14 region. The three articles appeared in the same issue of *Nature Genetics*—back to back, no less. This replication of findings was no doubt a good sign that the discovery was solid, but it was also quite curious that multiple teams should, in supposedly independent fishing expeditions, chance upon the same fish at the same time.

Behind closed doors, rumors were spread about leaks, piggybacking, and even sabotage.[18] Caught in the thick of it was Peter St George-Hyslop from the University of Toronto, the leader of one of the three teams. In its *Nature Genetics* article, St George-Hyslop's team linked several early-onset Alzheimer's families to chromosome 14,[19] but what's interesting about their work is that five years earlier, they had linked some of these same families to chromosome 21.[20] That earlier finding, in retrospect, was the result of a mistake in statistical calculation.[21] But St George-Hyslop couldn't bring himself to admit that. In its new paper, the team repeated the earlier conclusion about chromosome 21 . . . and then asserted that they now had more robust evidence to connect the families to chromosome 14. The way they saw it, that was not so much a mistake as evidence that the genetics of Alzheimer's are complex.

St George-Hyslop first submitted the new paper to *Nature*. But the reviewers at *Nature* turned it down—on the ground that one cannot link the same families to two different chromosomes. Although there are likely to be multiple genetic mutations behind Alzheimer's; there is likely only one mutation going on within a single family, which cannot simultaneously sit on two chromosomes. Giving up on *Nature*, Hyslop turned around and submitted the paper to *Nature Genetics*, where it was eventually published. But that detour cost him precious time and the opportunity to preempt others. Now his paper is only one of three.

Rejections by journal reviewers are common. As gatekeepers, reviewers are supposed to help journals uphold high academic standards. But what is scandalous in this case is that the reviewers were on one of the other two teams who also published in *Nature Genetics* about chromosome 14. In fact, these reviewers were none other than John Hardy, who first discovered the APP gene mutation, and his lab partner. Small world.

Rumor had it that St George-Hyslop felt that Hardy purposefully rejected his paper in order to buy time and finish his own paper. Hardy, on the other hand, maintained that *his* team had been on to chromosome 14 long before he saw anything published about it.

Remember that this neck-and-neck race concerned only the suspect region on chromosome 14; the exact gene and mutation that reside there hasn't been caught. Catching them, rather than wallowing in allegation and self-pity, would be the best revenge. St George-Hyslop did just that two and a half years later; this time, his paper *was* accepted in *Nature*, announcing both the gene and its mutation—and not just one, but five different mutations.[22] The gene has since been titled PSEN1, as in presenile dementia.

As with APP mutations, once researchers knew where to look, additional PSEN1 mutations soon surfaced. To date, more than 150 have been discovered, and one of them is responsible for the suffering of Paisas in Colombia. Compared with APP mutations, PSEN1 mutations are far more prevalent. They account for up to 70 percent of early-onset Alzheimer's, whereas APP mutations account for only about 10 percent to 15 percent.[23] PSEN1 mutations are also particularly vicious, as endured by the Paisas: they accelerate disease progression, causing onset in one's forties and fifties, and sometimes as early as the thirties.

In and of itself, the PSEN1 discovery was a victory, but what's more encouraging is the subsequent realization that this gene is closely related to what we already knew about beta-amyloid. Recall what we said earlier—beta-amyloid is a fragment cut from the longer, amyloid precursor protein. This cutting doesn't happen on its own; it requires enzymes that physically snip the long precursor.

And that's where the PSEN1 gene comes in: it codes for part of a cutting enzyme (called *gamma-secretase*). When PSEN1 mutates, it derails gamma-secretase. As a result, improper cuts are made on the precursor protein, creating harmful or excessive beta-amyloid fragments that accumulate into Alzheimer's plaques. The puzzle pieces are fitting together nicely.

But one piece is still missing: the Reiswigs and their Volga German kin turned out not to have an APP mutation or a PSEN1 mutation. Apparently, another gene is still at large.

This gene, the last one known to cause early-onset Alzheimer's, was found *very* quickly, less than two months after the PSEN1 discovery. The person who found it was someone we've seen before: Gerald Schellenberg, who discovered the chromosome 14 suspect region but didn't spot the PSEN1 gene on it. This time, Schellenberg went all the way, connecting the Volga Germans to yet another chromosome, chromosome 1, and zeroing in on the culprit gene.

There is a reason for Schellenberg's extraordinary speed this time around.[24] When the PSEN1 gene was found, two researchers involved in that study, Rudolph Tanzi and Wilma Wasco, searched the international gene depository GeneBank to see if they could find anything similar to PSEN1. Much to their surprise, they did. A computer query led them to an unidentified human gene that was 80 percent similar to PSEN1 in its DNA sequence. Tanzi and Wasco immediately suspected that this unidentified gene was a sister gene to PSEN1 and, very likely, the missing Volga German gene.

According to Tanzi, he was in possession of a Volga German brain and could easily have used it to test his theory, but he didn't because he didn't want to scoop his good friend Schellenberg, who had been chasing the Volga German gene for years and was getting close. So Tanzi phoned Schellenberg and let him in on the news about the sister gene. Schellenberg confirmed that this unidentified gene sat squarely in the chromosome 1 region that he had been examining.

The two then went on to test their own Volga German brain samples to see if they could spot a mutation.

Good friends or not, when Tanzi and Schellenberg completed their separate testing, they didn't trust each other to just blurt out the results in a phone call. Suppose that one person revealed his findings and the other confirmed them, how was the first person to know that the second person *really* had the same findings and wasn't just repeating what he heard?

The two solved the conundrum by Schellenberg agreeing to share his findings on the phone while Tanzi agreed to FedEx a copy of his data to Schellenberg as evidence of his independent findings. With that arrangement, Schellenberg talked: he found a single-base mutation in the sister gene among the Volga Germans. Later that day, Tanzi's FedEx was duly sent and showed the same results. A month later, their findings were published in *Science*, with both Schellenberg and Tanzi as senior authors.[25] Their gene was later named PSEN2, a sister of PSEN1.

PSEN2 is similar to PSEN1 not only in DNA sequence, but also in biological function: it codes for part of the gamma-secretase enzyme that cuts beta-amyloid out of its precursor protein. Compared to PSEN1 mutations, PSEN2 mutations are rare, accounting for no more than 5 percent of early-onset Alzheimer's cases.[26] Still, the discovery was huge. It solved the Volga German mystery, solidified beta-amyloid's role in the disease, and was another sign that researchers were on the right track.

As for Dr. Tanzi, despite his proclaimed loyalty to his friend, it must have been hard to share the discovery, to not be the one and only. More than a decade later, Tanzi spoke at a U.S. Senate hearing about Alzheimer's disease. In the audience was Gary Reiswig, a survivor of the Reiswig family, who were among the human subjects used by Schellenberg and Tanzi to find the PSEN2 mutation.

For decades, Gary Reiswig had been fighting the war of Alzheimer's on the personal front, losing his father, uncles, aunts, and cousins. For decades, he didn't know where he himself stood with the mysterious disease, but he rallied the family to contribute to

Alzheimer's research. Naturally, Gary felt warm toward Tanzi, a fellow fighter who helped solve the mystery. After the hearing, he approached Tanzi to introduce himself, expecting some kind of acknowledgment, empathy, or comradery. Tanzi, not particularly interested, interrupted Gary and dashed off with some important parting words: "I found that gene, you know. . . . I did!"[27]

Good for you, Dr. Tanzi! And congrats and thanks to all the researchers who helped fish the three early-onset Alzheimer's genes out of the water! Theirs are impressive achievements of modern science—made possible through their dedication and assiduous labor. But as their work creates bragging rights, media headlines, and springboards for research funds, the achievements of their human subjects are rarely celebrated. We see them as patients, victims, samples—we don't acknowledge (not enough, at least) their heroic efforts to make modern science possible. Without their bravery and willingness to give up their lives and bodies for something that's too late to benefit themselves, we would never have gotten to where we are today.

5

LATE-ONSET ALZHEIMER'S

My uncle didn't participate in any research experiment. Twenty years ago, Alzheimer's research was rare in China. It still isn't widely prevalent today. Alzheimer's may be the same human tragedy everywhere, but cutting-edge research, in many ways, is the privilege of a few. At the forefront are countries like the United States, Japan, Britain, and Germany.

These countries have the requisite infrastructure: equipment and supplies, databases and protocols, doctors and researchers. And they have sizable financial and social support for that infrastructure. In the United States, the National Alzheimer's Project Act, signed into law in 2011, makes it a national priority to tackle Alzheimer's, develop treatment, and improve diagnosis and care. Research funding has been steadily increasing, with the 2020 estimated federal budget reaching $2.6 billion, a nearly threefold increase from 2016.[1] Social advocacy, through organizations such as the Alzheimer's Association, drives public engagement and private donations.

China, by comparison, still lacks widespread social awareness of the disease, let alone research infrastructure. Labs, personnel, protocols, and long-term strategic plans will take years, even decades, to develop. In the meanwhile, the toll of Alzheimer's has already caught up: as of 2010, 5.69 million Chinese were estimated to have Alzheimer's, and with China's large population base, that number will surge in the decades ahead.[2]

If one has to find a silver lining in these statistics, it is that the vast majority of these Alzheimer's patients, in China and elsewhere, will *not* have their minds robbed in their thirties, forties, or fifties, the way that Volga Germans and Colombian Paisas do. For all the work done in the 1980s and 1990s fishing for the early-onset Alzheimer's genes, their mutations explain a mere 5 percent of Alzheimer's cases. By far, the majority of people we hear about, know of, or are close to have so-called late-onset Alzheimer's, which doesn't manifest until the patients are in their sixties—as if that's "late" enough.

My uncle, struck at sixty-seven, by definition had late-onset Alzheimer's. Having served in the military, survived the Cultural Revolution, and put in decades of honest work in a local factory, he was supposed to enjoy his retirement years. They were supposed to be golden. They were anything but.

When the diagnosis first came, I was in graduate school outside the country and didn't give the news much thought. "Alzheimer's disease," which was commonly translated into Chinese as something along the line of "elderly dullness/retardation," didn't strike me as something serious. It sounded like something that we will inevitably get a little bit when we get old. And when we do, we become a little more forgetful, disorganized, and stubborn, as old people are. That's the kind of misconception that prevailed then and still exists in China today.

Busy with classes, I didn't get to visit my uncle until two years later. That's when I realized how serious things were. My uncle still lived at home, but he could no longer recognize me, let alone talk to me. During my entire visit, he lay in bed facing the wall, moving very little and saying nothing but a few incoherent syllables. I was stunned that someone could change so much so quickly.

Of course, I wasn't there to witness my uncle's gradual descent. Years later, I would read about it in my cousin's journal. It's strange seeing someone you know turned into a victim. It's like watching an imminent train wreck. It's terrible, but you can't look away. I saw my uncle at his dinner table, surrounded by his wife, children, and grandchildren, and yet he sat and ate as if he were alone, on a planet far far away. I saw him losing interest in mahjong, his favorite game, which he used to play with his mates for hours on end. I recognized the road that he got lost on and then lay down to rest, exhausted, while his family searched frantically for him.

In that moment, I started to remember something—an inkling of change in my uncle that I had seen several years *before* his diagnosis. He was once the chef at family gatherings, in command of the kitchen and proud of his cooking. He was never a big talker, but he did talk, especially after a few sips of his rice wine. Then, I don't remember when exactly, he started to cook less and talk less. At family dinners, he became a stoic face, locked in silence and what seemed mild displeasure. I can't be sure of my memories, but Alzheimer's does start in subtle ways. And once started, it can take decades to play out.

For seven years, my aunt and cousins cared for my uncle at home, but eventually, it became too much, and he was admitted to a hospital care unit. Incontinent, oblivious, fed through a tube, and restrained so he wouldn't pull out the feeding tube, my uncle never left. He passed away in 2016, ten long years after his diagnosis.

Not related to my uncle by blood, I never stopped to think what his illness meant for me. And fortunately for my cousins, there isn't too much for them to worry about either. Genetics have but a small hand to play in the world of late-onset Alzheimer's. There is no grim chance of a 50 percent inheritance, as with early-onset cases. Instead, the number one risk factor is age: the older we get, the more likely we will be ill, simple as that. Among Americans aged sixty-five to seventy-four, 3 percent have Alzheimer's; among those aged seventy-five to eighty-four, 17 percent do; among those eighty-five and older, 32 percent do.[3]

Viewed in this way, late-onset Alzheimer's is a curse of the modern age. Improved living conditions and health care preserve our physical bodies, so now we live long enough to be robbed of our minds. It is as if evolution hasn't quite prepared our brains for when human life expectancy raced forward. In 1900, the average American men and women could expect to live to forty-seven; in 1950, sixty-eight; and in 2010, seventy-nine.[4] A century or so ago, people had no reason to worry about a disease that would strike in their sixties.

Tied to age rather than pedigree, late-onset Alzheimer's is also called *sporadic Alzheimer's*, a random and isolated condition that could strike anyone. But the word *sporadic* is a little misleading, for the disease is not entirely random; it does have *some* genetic origin. That fact we have known since the 1990s.

Back then, when most researchers were busy fishing for the early-onset Alzheimer's genes, Allen Roses from Duke University chose to work with late-onset cases. This was a road less traveled because late-onset Alzheimer's shows no clear pattern of inheritance. To fish for a gene, Roses and his team recruited eighty-seven unrelated patients across different families to see what genetic markers they tended to share. If they shared a marker more frequently than expected by chance, that marker may sit close to the genetic root of their disease.

Through this method, Roses's team first narrowed their search in June 1991 to a suspect region on chromosome 19.[5] Two years later, they pinpointed a gene in that region: ApoE (pronounced *A-PO-EE*).[6] This gene directs the production of the ApoE protein, whose function is to bind with cholesterol and fat and deliver them to different parts of the human body for use.

Unlike the early-onset Alzheimer's genes, ApoE does not mutate in late-onset cases. Instead, it is *polymorphic*—as in having multiple (*poly*) forms (*morph*). What this means is that the gene exists in several versions among the general population: ApoE2, ApoE3, and ApoE4. Of the three, ApoE3 is the most common, found in 60–90 percent of the world's population; ApoE4 is less common, found in

10–20 percent of the world's population; and ApoE2 is rare, found in only 0–20 percent of the population (frequencies vary in different geographical regions).[7] Because we have two sets of chromosomes, we each have two chromosome 19s, and thus two ApoE genes. The two genes may be of the same version (for example, two copies of ApoE4) or different versions (for example, one ApoE2 and one ApoE3).

The three ApoE versions have slightly different DNA sequences, but they are all considered normal. In fact, polymorphism is common in nature. In humans, it results in different blood types like A, B, and O; in other animals, it can cause different colors and markings in the same species. That said, not all polymorphic versions are necessarily created equal, and some may make us more susceptible to certain diseases. That's what Allen Roses found about one version of ApoE: ApoE4.

The finding came down to numbers and statistics. ApoE4 is significantly more prevalent among people with late-onset Alzheimer's than those without.[8] Among diseased individuals, 64–80 percent carry at least one ApoE4; among nondiseased individuals, only 31 percent do.[9] The pattern becomes more telling when we look at different ApoE combinations. At zero copies of ApoE4, our chance of developing late-onset Alzheimer's is about 20 percent; at one copy, that chance increases to 47 percent; and at two copies, the chance jumps to a whopping 91 percent.[10] Because we inherit one ApoE from each parent, if ApoE4 runs in one or both sides of the family, we will have an increased risk for late-onset Alzheimer's.

Not only does ApoE4 increase the disease's risk, Roses found that it hastens its onset. At zero copies of ApoE4, the average age of onset is eighty-four; at one copy, onset comes earlier, at seventy-six; and at two copies, onset moves up to sixty-eight.[11] If we must have Alzheimer's, a later onset is obviously better. It allows us to enjoy more symptom-free years. It is also a practical way to escape the disease altogether. It sounds a little cruel, but a delay of five or ten years means that a number of us will die of old age and other ailments before getting Alzheimer's. Having seen my uncle, I'm convinced those are better ways to go.

Shortly after denouncing ApoE4, Roses and his colleagues sent more encouraging news: another version of ApoE, ApoE2, is significantly more common (16 percent) among people without Alzheimer's than those with (1 percent).[12] In other words, carrying ApoE2 appears to protect us against late-onset Alzheimer's.

Roses's findings were initially met with suspicion—they sounded too simple and uncomplicated, too good to be true. But they were validated in study after study and soon became the standard for estimating late-onset Alzheimer's risk, with these major principles:

- ApoE3, the most common version of ApoE among the general population, is a neutral player and has no effect on disease risk.
- The less common ApoE4 is a troublemaker, increasing disease risk.
- The rare ApoE2 is a do-gooder, reducing disease risk.

Different ApoE combinations and their conferred risks are estimated in Table 5.1.[13] The ApoE3/ApoE3 combination is assumed to have a neutral risk of 1, and other combinations confer higher or lower risk. It is interesting to see that if we have both the good ApoE2 and the bad ApoE4, our risk still increases 3.2 times.

When looking at these numbers, it is important to remember that they are generalized estimates. Depending on race and gender and on individual people, the actual risk will change. Caucasians and Asians, for example, are acutely affected by ApoE4; African Americans and Hispanics, however, are less so.[14] Women in general are also more acutely affected by ApoE4.[15] This may be another reason why more women than men develop Alzheimer's, beyond the fact that women tend to live longer. At the level of individuals, we need to remember that ApoE4 increases our risk, but it does not *cause* late-onset Alzheimer's. Similarly, not carrying an ApoE4 does not mean that a person is home free. Across forty studies of mixed races and genders, 13–42 percent of people with two copies of ApoE4 were

TABLE 5.1 APOE COMBINATIONS AND THEIR ESTIMATED RISK
FOR LATE-ONSET ALZHEIMER'S

APOE COMBINATION	APOE2– APOE2	APOE2– APOE3	APOE3– APOE3	APOE2– APOE4	APOE3– APOE4	APOE4– APOE4
Alzheimer's risk	0.24	0.5	1	3.2	5.5	20.6

free of Alzheimer's, while 10–44 percent of people without a single
ApoE4 still developed the disease.[16]

With these caveats, we return to Roses's simpler conclusion: in the
general scheme of things, what kind of ApoE we have matters. But
why? The short answer is: we aren't sure. Given the breakthrough
that we have made on beta-amyloid, researchers naturally suspect
that ApoE and beta-amyloid are connected.

There is some evidence that they are. The ApoE protein and beta-
amyloid protein fragment can bind to each other, so we know that
these two substances physically interact.[17] Moreover, ApoE4 carriers
develop more plaques in their brains than noncarriers, which seems
to suggest that ApoE status somehow affects beta-amyloid buildup.[18]

But that's where we are stuck. Since Roses's discovery, lots
of studies have tried to uncover how ApoE affects beta-amyloid
buildup, but they yielded different (no, contradictory) findings.[19]
Some suggest that ApoE prohibits beta-amyloid buildup or facilitates its removal from the brain. ApoE4 is a risk factor because it
is an inadequate ApoE—a sluggish worker not very good at its job.
Others, doing a 180-degree turn, suggest that ApoE promotes beta-
amyloid buildup or hinders its removal from the brain. ApoE4 is a
risk factor because it is a super-ApoE, a model worker highly pro-
ficient at its job. Or perhaps both are true, and ApoE has dynamic
functions that exist on a continuum. Depending on the cellular

environment, ApoE may sometimes increase beta-amyloid buildup and sometimes decrease it.

If a verdict ever comes in, Allen Roses won't be around to hear it. He passed away on September 30, 2016, from a heart attack at Kennedy International Airport, en route to a medical conference in Greece.[20] He was seventy-three years old.

The New York Times ran a short piece on this formidable researcher upon his passing.[21] It doesn't say much that I didn't already know about Roses and his accomplishments, but my jaw dropped when I read that in 2009, Roses took out a $500,000 loan on his family house in Durham, North Carolina, to support his research on Alzheimer's. That kind of dedication you do not see everywhere. Rest in peace, Roses, and send us much good luck for the road ahead.

6

THE PARADIGM

Standing in his lab in St Mary's Hospital, assistant professor John Hardy was trying to isolate DNA from a tissue sample. A basic procedure, this task is often delegated to student assistants or low-level technicians. Then again, with his casual outfit and lack of skill at the work, Hardy might well be mistaken for a clueless student.

Hardy had been studying the brain for a long time, but not as a geneticist. While a doctoral student at Imperial College London, he was trained as a neurochemist. What's the difference? A neurochemist works at the level of brain chemicals such as the reward messenger dopamine, not at the level of DNA. Approaching Alzheimer's as a neurochemist, Hardy was examining the changing chemical messengers and other abnormalities in the Alzheimer's brain; as Hardy put it, he was doing forensic investigation after a car crash to find out what caused the crash.[1] Back then, there was no better way.

Then 1983 came, and a better way appeared. That year, a major breakthrough was made in the study of Huntington's, a neurological disease marked by cognitive decline, uncontrolled movements, and emotional outbursts. Using linkage analysis (see chapter 4),

Huntington's researchers had just connected the disease to a genetic marker on chromosome 4.[2] Suddenly, there were more to look at than just a broken windshield and skid marks. Peeking out through the aftermath of the crash is a genetic origin, the promised cause of the disease. As researchers quickly realized, what had worked for Huntington's could work for Alzheimer's too, and thus started the hunt in the 1990s for the Alzheimer's gene (or genes).

Inspired by the new approach, Hardy also joined the search, but as he was not trained as a geneticist, he had to learn molecular neuroscience and genetics from square one. So here he was, starting with extracting DNA. As he continued to struggle, Hardy heard the sound of conversation from down the corridor. He recognized the voice of Bob Williamson, his department head and mentor. Bob was saying enthusiastically to someone, "I'd like to introduce you to somebody I think is going to be the future of British molecular neuroscience. I think he's a really talented person and I think he is going to be the go-to person. So if you ever need to have an editorial written about the future of neuroscience . . . this is who I'd recommend."[3] With that, the door opened. Standing there were Bob Williamson and Peter Newmark, editor of the prestigious scientific journal *Nature*. And Bob said, "Peter, let me introduce you to John Hardy."[4]

Decades later, Hardy would recall the incident with a hearty laugh. "It was such total bullshit. It was such total bullshit," he kept saying.[5] But Bob was dead right. In 1991, Hardy made one of the greatest breakthroughs in modern neuroscience, uncovering the first mutation in the amyloid precursor protein (APP) gene that causes early-onset Alzheimer's. The discovery was published in none other than *Nature*.[6] Today, John Hardy undoubtedly is one of Britain's and the world's most renowned geneticists, and indeed the go-to person for Alzheimer's and other neurological diseases. He is the head of the Molecular Neuroscience Department at University College London and the 2015 winner of the $3 million Breakthrough Prize in Life Sciences.

The APP mutation study that shot Hardy to stardom is the typical scientific research article. It describes, in minute detail, what experiments were conducted and how, as well as what facts emerged.

Namely, a mutation was discovered in two early-onset Alzheimer's families; the mutation is a single DNA base change in the APP gene, which causes the gene to produce a slightly different APP protein right where the protein is cut by enzymes to shed beta-amyloid.

Given these findings, Hardy thought that beta-amyloid buildup must be the cause of Alzheimer's, both the early- and late-onset types. However, he couldn't quite say that out loud in the article. A scientific research article demands facts, not beliefs, and Hardy didn't have all the facts. How exactly does the APP mutation cause beta-amyloid to build up; how does the buildup then cause neuronal damage and cognitive decline; and where do the fibrils that grow inside neurons, the other hallmark of the Alzheimer's brain, come from?

These questions were on Hardy's mind when he visited the United States in late 1991 to give a talk at the National Institute on Aging (NIA). While there, he met and befriended a researcher named Gerald Higgins. Higgins was doing interesting animal studies with the APP gene. He found that by inserting this gene into mice and over-expressing it—and thus overproducing beta-amyloid—he was able to induce plaques, fibrils, and neuronal damage in the mouse brain and turn it into something just like the human Alzheimer's brain.[7]

These animal findings impressed Hardy, and the two agreed that they should write a paper together on beta-amyloid; it would not be a research article, but more of an opinion piece.[8] Higgins contacted a friend who was an editor at *Science*, a high-caliber journal on par with *Nature*, to see if it would be interested in publishing such a piece. It was. Hardy and Higgins went immediately to work, finishing the paper in a mere one week's time.[9]

In April 1992, the paper came out, in a section of *Science* called "Perspective." It was a short paper, barely two pages long, with one image. But it would become one of the best-known articles on Alzheimer's research. In the twenty-eight years since its publication, it has been cited over 3,600 times, while articles from the same time period have been cited, on average, sixty-nine times.[10]

Aptly titled "Alzheimer's Disease: The Amyloid Cascade Hypothesis," the paper officially gave birth to the amyloid cascade hypothesis, the single most influential hypothesis on the cause of Alzheimer's, both the early- and late-onset types.[11] In confident and fast-paced language, the paper explains that the buildup of beta-amyloid is the common first step in a cascade of events that causes Alzheimer's. This cascade is captured in figure 6.1. Hardy and Higgins acknowledged that late-onset Alzheimer's is not directly caused by

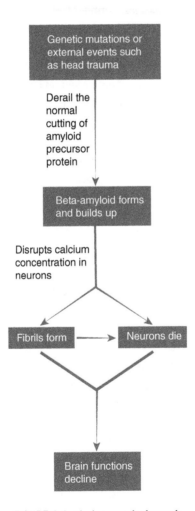

FIGURE 6.1 The beta-amyloid cascade

a beta-amyloid-related genetic mutation, but they suggested that external events such as head trauma or injury could initiate beta-amyloid buildup.

At least two other previous papers had outlined a similar cascade, but they are not nearly as famous.[12] Why? They included finer and more complex details, used more qualified language, and contained more citations, which made the articles several times longer and appear several times *less* confident. Hardy and Higgins's piece, by contrast, was "simple, clear and short. . . . even a venture capitalist or a corporate CEO can read to the end of it."[13]

Between Hardy's 1991 APP mutation article and this 1992 perspective piece, the facts about beta-amyloid had not changed much. Hardy still didn't know what he hadn't known a year earlier, but what he couldn't prove in a research article, he could speculate about in a perspective piece. And the speculation was bold. It made a leap of faith that *association* equated *causation*. Yes, evidence clearly showed that beta-amyloid is associated with (or related to, or involved in) Alzheimer's, but association does not necessarily mean causation. Rather than a cause, beta-amyloid buildup may simply be a pathological feature of the Alzheimer's brain. Rather than a killer, a trigger of neuronal damage, it may be the tombstone, the result of neuronal damage.

Despite its boldness, or rather probably because of it, Hardy and Higgins's hypothesis is mighty attractive. It is the epitome of Occam's razor—given two explanations for an event, the simpler one is more likely to be correct. Their explanation is simple because it follows a straightforward, linear process, as shown in figure 6.1. It is simple because it appeals to everyday logic and common sense:

- The Alzheimer's brain harbors numerous beta-amyloid plaques. These plaques shouldn't be there. Ergo, they can't be good for the brain.
- APP mutations change the amyloid precursor protein right where the protein is cut to shed beta-amyloid. Ergo, the mutations must have led to abnormal beta-amyloid production, and then the plaques. Indeed, shortly after Hardy and Higgins's paper appeared, cell studies

demonstrated that some APP mutations increase beta-amyloid production by six to eight times.[14]

- Early-onset patients carry beta-amyloid genetic mutations, and they develop a severe form of Alzheimer's. Late-onset patients have a milder form of the disease, so they too must have something wrong (but less so) with their beta-amyloid.
- People with Down syndrome have an extra chromosome 21, and thus an extra APP gene, and many of them develop Alzheimer's as they age. Ergo, the extra APP gene must be at fault, probably by overproducing beta-amyloid.

Last but not least, the amyloid cascade hypothesis is attractive because it offers a clear path toward treatment. If beta-amyloid buildup is the cause of Alzheimer's, then all we need to do is remove the buildup or stop it from happening in the first place. That would break the cascade, stop downstream events, and prevent (and maybe even cure) the disease.

In the beginning, all was dark. There was no form or shape. Winds blew from all directions. Water flowed from everywhere. It was chaos. It was confusion.

Before the genesis of the amyloid cascade hypothesis, that's what it felt like in the world of Alzheimer's research. People were pursuing thoroughly different leads for what may cause Alzheimer's: a slow virus, aluminum exposure, accelerated aging.[15] There was no *real* science, at least not according to science philosopher Thomas Kuhn. Kuhn believes that before real science can happen, there must exist a *paradigm*—a scientific model that is sufficiently convincing to attract people away from competing models.[16] A paradigm determines the questions worthy of study and establishes common knowledge and standard methods that allow people to study those questions.

In the amyloid cascade hypothesis, a paradigm was born, allowing what Kuhn calls normal science to happen. During normal science, people focus their attention on the small array of questions that the paradigm deems important. Freed from distraction and guided by shared understanding, they investigate those questions in depths that are not otherwise possible.[17] Discoveries are made and verified, creating still more shared understanding to advance the paradigm.

Inevitably, anomalies—findings that a paradigm and its established knowledge cannot explain—also arise.[18] When they do, the paradigm must revise itself to accommodate the anomalies. Successful revisions make a paradigm stronger because it can now explain more phenomena, and with greater precision. Persistent failure to explain anomalies will stall normal science and eventually crumble a paradigm.

Shortly after the birth of the amyloid cascade hypothesis, one tiny anomaly arose: cell studies showed that not all APP mutations overproduce beta-amyloid. Some do, but most don't.[19] In other words, there is no simple cascade of genetic mutation → excessive beta-amyloid → beta-amyloid buildup → Alzheimer's disease. Instead, most of the APP mutations change the *kind* of beta-amyloid being produced.

Up to this point, we have treated beta-amyloid as if it were a singular substance. But in reality, it is not. Beta-amyloid is cut from the amyloid precursor protein, and depending on the precise location of the cut, beta-amyloid can exist in subtypes of slightly different lengths. The most common subtype is beta-amyloid 40, so called because it is forty amino acids long (recall that amino acids are the building blocks of a protein). Next in line is beta-amyloid 42, which is slightly longer, at forty-two amino acids, and is "stickier" and more prone to buildup.[20] Normally, beta-amyloid 40 accounts for 80–90 percent of the total beta-amyloid in the brain, followed by beta-amyloid 42 at 5–10 percent.[21] This ratio changes when the APP gene mutates: the longer, stickier beta-amyloid 42 increases, at the expense of beta-amyloid 40.[22]

In the face of this finding, the amyloid cascade hypothesis quickly upgraded itself to version 2.0. In that version, the *quality*, not *quantity*, of beta-amyloid was the key. The increase of beta-amyloid 42,

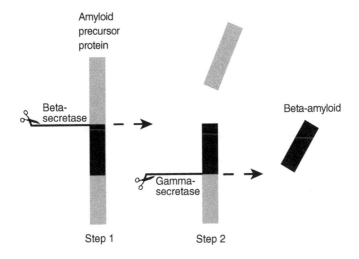

Amyloid
precursor
protein

Beta-
secretase

Beta-amyloid

Gamma-
secretase

Step 1   Step 2

FIGURE 6.2 Two-step process of cutting beta-amyloid

rather than a total increase of beta-amyloid, became the cause of plaque buildup and cognitive decline in Alzheimer's.

Version 2.0 is bolstered by the 1995 discoveries of PSEN1 and PSEN2. These two genes, as discussed in chapter 4, direct the production of an enzyme called *gamma-secretase*, which cuts beta-amyloid out of its precursor protein. To be more precise, the cutting is a two-step process, as illustrated in figure 6.2. Another enzyme, *beta-secretase*, makes the first cut, followed by gamma-secretase.

When PSEN1 and PSEN2 mutate, gamma-secretase malfunctions, and cutting derails at step 2, creating more of the longer and stickier beta-amyloid 42,[23] exactly as version 2.0 predicts.

Normal science continued. But more, and harder, anomalies soon had to be explained.

If the cause of Alzheimer's is excessive beta-amyloid 42, which easily clumps into plaques, then the number of plaques found in the brain ought to correlate with Alzheimer's disease progression. That is, more plaques → greater brain damage → more cognitive decline. This, however, turns out not to be true.[24] In fact, autopsy studies showed that beta-amyloid plaques are quite common in old age and exist in cognitively healthy individuals. According to these findings, 25–30 percent of nondemented older adults have enough plaques in their brains to qualify for a pathological diagnosis of Alzheimer's, but they display no outward symptoms—like Sister Mary.[25]

Sister Mary was a devout Catholic at the School Sisters of Notre Dame convent in Baltimore.[26] Born in a German immigrant, working-class family, Sister Mary was the oldest of eleven siblings. She entered the convent at the age of fourteen and took her religious vows at nineteen. From then until the age of seventy-seven, Sister Mary taught grade school full time. After that, she worked as a part-time math teacher and teacher's aide until she was eighty-four. She then returned to the same convent that she joined as a young girl. There, she remained kind, happy, active, and alert. One day, she "wondered out loud to her doctor if perhaps he was giving her medicine to keep her alive, and after all, her desire was to be with Jesus. Her doctor replied, 'Sister, it's not my medicine that's keeping you alive. It's your attitude!' "[27]

In 1991, when Sister Mary was ninety-nine years old, she and 677 fellow Notre Dame sisters began participating in an aging study known, fittingly, as the Nun Study. Conducted by researchers from the University of Kentucky, the Nun Study examines, among other things, factors that contribute to Alzheimer's. After attending a meeting where the researchers explained the purpose of the study and, especially, the importance of each sister to donate her brain at death for autopsy, Sister Mary was the first to say "Sign me up!"[28]

As part of the study, Sister Mary undertook various tests, which showed that she was cognitively intact—superior, in fact. Her score on the Mini-Mental State Examination, a commonly used cognitive test, was 27, well within the normal range of 24–30. Despite her advanced age, her overall test scores were equal to or higher than the average of the other sisters.

Seven months after her last test, at the age of 101, Sister Mary's wish came true. She joined Jesus as she longed to do, but she left her brain behind for science. When researchers opened up Sister Mary's brain, they were astonished. Weighing only 870 grams (compared to the other sisters' average of 1,120 grams), the brain was riddled with beta-amyloid plaques.

Someone has to explain this.

Bruce Yankner at Harvard Medical School thought he could. Since the late 1980s, he had been convinced of something: beta-amyloid is toxic. When Yankner and his team added high-concentration beta-amyloid to cultured neurons or injected it into rat brains, masses of neurons simply withered and died, as if poisoned.[29] It is a very intriguing theory, and it attracted other researchers trying to replicate Yankner's findings. But not everyone succeeded. Those who did stood by the intriguing discovery, and those who didn't claimed that the theory was far-fetched.

It took a decade for the controversy to resolve itself.[30] It turned out that beta-amyloid's toxicity comes and goes depending on where beta-amyloid is in the process of clumping together. When beta-amyloid is first cut off from its precursor protein and free-floats in the brain, it is not toxic; when it starts to settle and build up, it turns toxic; and when it continues to build up into larger, insoluble plaques, the toxicity disappears again. This is why people had inconsistent results: depending on how or how long they prepared the beta-amyloid for examination, its toxicity may or may not have kicked in or may have already faded.

This elusive toxicity is presumably the answer to the condition of Sister Mary's brain and ushered in the third and most recent version of the amyloid cascade hypothesis. In version 3.0, a new suspect emerges: beta-amyloid oligomers.

The word *oligomer* literally means "a few [*oligo*] parts [*mer*]." It describes a molecular complex composed of multiple small parts. Each small part is the same and is known as a *monomer* (*mono* means "single"). Each beta-amyloid fragment cut from a precursor protein is a monomer. As these monomers stick to each other, they form oligomers—the intermediate stage where toxicity sets in. Oligomers vary in how many monomers they pack in themselves, but generally, that number is not large. For example, an oligomer may be a *dimer* (containing two monomers), a *trimer* (containing three monomers), or a *12-mer* (containing twelve monomers). As beta-amyloid oligomers continue to grow and accumulate, they become plaques.

According to version 3.0, the toxic oligomers, rather than the final plaques, are the culprit in Alzheimer's. In fact, some go so far as to say that plaques are actually "brain pearls."[31] Just as an oyster creates pearls to seal off irritants, a brain builds oligomers into plaques to rid itself of toxicity.

In support of version 3.0, studies showed that beta-amyloid monomers, including the sticky beta-amyloid 42, are not harmful. If anything, at low levels, they are important for normal brain function. When added to cultured neurons, they helped immature neurons grow and protected mature neurons from overstimulation.[32] These findings explain an important anomaly that earlier versions of the hypothesis couldn't: beta-amyloid exists naturally in humans. If it withstood the test of natural selection, it must exist for a reason.

It is only when the natural production of beta-amyloid monomers is disturbed that things go wrong. In early-onset Alzheimer's, genetic mutations increase the ratio of sticky beta-amyloid 42. As a result, monomers more easily clump into and stay as oligomers, exerting prolonged toxicity before growing into plaques.[33] In late-onset Alzheimer's, there is no genetic mutation, but we do have the risk gene ApoE4 (discussed in chapter 5). Although ApoE4's exact function still eludes us, studies suggest that it can stabilize beta-amyloid oligomers and prolong their toxicity.[34] In fact, Alzheimer's patients who are ApoE4 carriers have up to seven times more beta-amyloid oligomers in their brains than noncarriers.[35]

As oligomers form, they are believed to cause havoc in the brain.[36] They disturb neuron receptors, which are tiny structures on the surface of neurons that process biological signals. They snuggle into and damage synapses, which are crevices between neurons where brain signals are exchanged. Oligomers even poke holes in the neuron cell membrane, allowing abnormal ions to flow through and damage neurons. Any one of these events occurs, and our cognitive function goes out the window.

Rat studies support these scenarios. In one study, researchers compared the effect of beta-amyloid monomers versus oligomers in some smart rats.[37] These rats were trained to press two tiny levers a certain number of times *alternately* (for example, pressing one lever six times and then the other six times). If they linger on a lever after the required presses, that's a lingering mistake; if they jump the gun and switch too soon, that's a switching mistake. Only perfect performance is rewarded with food. After the rats had mastered the skill, they were injected with either beta-amyloid monomers or oligomers. On that very day, rats who received oligomers started to flunk at the task. Compared with earlier records, their switching mistakes increased by 173 percent, and lingering mistakes by 215 percent. On the other hand, rats who received beta-amyloid monomers were not fazed.

In humans, ethical guidelines forbid such manipulation, but studies of Alzheimer's patients suggest that brain oligomer levels correlate with the degree of neuronal damage and cognitive decline.[38] Moreover, nondemented older adults generally have low levels of oligomers in their brains.[39]

Once again, normal science continued.

Without a doubt, the amyloid cascade hypothesis faces unanswered questions. Most cripplingly, precisely what we mean by oligomers remains unclear.[40] The formation of beta-amyloid oligomers depends heavily on the external environment. Because of the different chemicals, temperatures, times, and other factors involved in synthesizing or isolating oligomers from brain tissues, we have on our hands a variety of oligomers, from 3-mers to 24-mers, from round structures to ring-shaped structures.

Worse, there is no one way to identify these oligomers, as different researchers have come up with their own names and brands:

type 1 oligomers, type 2 oligomers, beta-amyloid*56, and others. The oligomer used by one lab may overlap with that used by another, but they are called by different names. Alternatively, two labs may think they are using the same thing when they aren't. With these confusions, we have no idea which oligomer is *the* toxic agent to blame in Alzheimer's.[41]

Despite these issues, at present, the amyloid cascade hypothesis is able to explain, and connect, a wide variety of findings about Alzheimer's as no one else could. Through multiple revisions, it has managed to meet anomalies where they arose and hold onto its paradigm status.

Plus, in medical research, perfection is not *that* important. We don't just have disinterested scientists pursuing the ultimate truth; we have profit-driven companies sponsoring research for drug development and millions of patients waiting for those drugs. What is more important is to have drugs that work—or work reasonably well. For that, we don't need to know everything about a disease, just enough to break it—or so the idea goes.

7

ELI LILLY AND MICE

In the heart of Indianapolis, Eli Lilly's headquarters sports a spacious courtyard, beautiful lawns, and a peculiar-looking office complex. The complex's centerpiece is a high-rise block. From the sides of this block grow two small blocks, from those grow two smaller ones, and then two even smaller ones. This expanding building is a fitting metaphor for the company's history. From its humble beginnings as a two-story drugstore in 1876, Eli Lilly undertook scores of mergers, acquisitions, and investments to become today's international pharmaceutical conglomerate.

The company is best known for its antidepressant Prozac, which entered the common lexicon as a synonym for "pick-me-up." Lilly's erectile dysfunction drug Cialis is also famous for being effective for up to thirty-six hours, earning the nickname "the weekend pill." In between these products, Lilly developed drugs for heart disease, diabetes, tumors, and arthritis. Although somewhat faded in the public memory, the company's earlier drug production is equally distinguished. Medications that we take for granted today—insulin, penicillin, and the Salk polio vaccine—were all first mass-produced by Lilly.

With offices across the globe and products marketed in over 100 countries, Lilly raises annual revenues in the tens of billions of dollars. Before Prozac's patent expired and generics appeared, the drug brought in $2.6 billion a year.[1] Cialis, in the company's recent financial reports, racks up $2.5 billion in annual sales.[2] Lilly's antipsychotic medicine Zyprexa, in its patent-protected years, boasted $5 billion in annual global sales; even with generics on the market, the drug's sales amount to around $1 billion.[3] With Alzheimer's, Eli Lilly aspires to earn financial rewards in the same ballpark.

If billions of dollars seems too good to be true for a disease that supposedly affects only "old people," think again. With rising life expectancies throughout the world, we, as a species, are getting old. In 2000, 600 million people worldwide (9.9 percent) were sixty years old or older; in 2015, 900 million (12.3 percent) were; and by 2050, an estimated 2 billion (21.5 percent) will be.[4]

More to the point, in the United States, more than 5 million people currently live with Alzheimer's, a number that is projected to increase about 2.5 times by 2050.[5] Global estimates vary wildly, given different reporting practices among countries. But roughly 30 million people worldwide are living with Alzheimer's today; by 2050, that number is projected to double or perhaps quadruple.[6] This large and fast-growing patient base translates into ready drug consumers. As a chronic condition, Alzheimer's can last years or decades, and so would the need for medication. Imagine two pills a day, 365 days a year, for ten years, taken by 30 million people worldwide; Eli Lilly can expect to do very, very well.

Even more attractive is the state of the Alzheimer's drug market. As mentioned in chapter 3, the current Alzheimer's drugs are all palliative bandages that have only marginal, short-lived effects. Anyone who can come up with a cure would have an instant market monopoly.

Naturally, Eli Lilly is not the only contender here. Other big pharma firms such as Pfizer, GlaxoSmithKline (GSK), Merck, and various biotech startups are eyeing this lucrative market. But if anyone has a chance, Eli Lilly sure does, with its celebrated history and strong foothold in drug development. Over the last decade, Lilly has registered more than forty Alzheimer's drug trials with the U.S. Food and Drug Administration (FDA).

Forty may not seem like a huge number, but it is more than respectable considering how much time and money go into mounting such studies. To start with, companies must pinpoint a chemical product that promises to work, which they do by trying thousands of possible candidates and using their understanding of a particular disease. Once identified, this product is put through preclinical tests to verify its effects (both good and bad). Two types of tests are conducted. First come cell and tissue studies, the so-called in vitro testing—in vitro is Latin for "within the glass," referring to conducting studies in test tubes and other lab equipment. If in vitro results are favorable, the product is then tested in vivo, which is Latin for "within the living." Usually, nonhuman animals are the first to bear the brunt of a drug's potential risks. Guinea pigs were once the animal of choice—hence the use of the expression "guinea pigs" to refer to subjects of an experiment. Today, guinea pigs have all but retired from labs and enjoy lives as family pets; their rodent cousins, mice and rats, are the new guinea pigs.

If animal results are favorable, clinical trials with human subjects finally begin, which is itself a multiphase process regulated by the FDA.[7] In phase 1, a drug is tested in a small group of volunteers (20–100 people) for several months to evaluate its safety. Pending satisfactory phase 1 results, the drug enters phase 2 and is tested in a larger group (several hundred people) to see if it delivers therapeutic effects and continues to be safe. This phase can last from several months to two years. With satisfactory phase 2 results, the drug enters phase 3 and is tested in an even larger group (300–3,000 people) to confirm its efficacy and safety. This phase generally lasts several years. If phase 3 trials are successful, the drug may finally move into the market. Each of these phases, especially the later ones, costs pharmaceutical companies enormous amounts of money for recruiting participants, administering drugs, monitoring responses, and collecting and analyzing data.

Despite all this work and money, more clinical trials fail than succeed: about 70 percent of trials move from phase 1 to phase 2, 30 percent of those move from phase 2 to phase 3, and 25–30 percent of the remainder pass phase 3.[8] All in all, less than 6–7 percent of trials survive the entire ordeal. It is a gamble—in fact, the odds are worse than gambling. Pharmaceutical companies embrace the risk because of the potentially high payoff: with one big success—and an Alzheimer's drug would be more than big—they could recoup all their losses.

Luckily for Eli Lilly, they have an edge in this game.

Because a crucial step in drug development is in vivo animal testing, suitable mouse or rat subjects are a precious commodity. Rodents, however, do not develop Alzheimer's in nature, so we can't use ordinary mice or rats to test the effect of an Alzheimer's drug. What we need are rodents that are genetically modified to develop human-like Alzheimer's brain pathologies and dementia symptoms. By recreating the disease in living, breathing rodents, we can then test the effects of our drugs and hope that the effects will transfer to humans.

But forcing nature to go wrong turns out to be no easier than stopping nature from decaying. Mistakes—and scandals—lined the path toward engineering mice with Alzheimer's.

By the early 1990s, we already knew that mutations in the amyloid precursor protein (APP) gene cause early-onset Alzheimer's (see chapter 4), so that's where we went first to manufacture transgenic Alzheimer's mice. Researchers tried different methods. Some inserted the human APP gene into mice and overexpressed it to produce excess beta-amyloid. Others inserted the gene into mice and modified it to alter the resultant beta-amyloid.

From these efforts, two "successes" emerged. The first was headed by researchers from Molecular Therapeutics Inc. in Connecticut. They inserted a select part of the APP gene into mice and induced beta-amyloid buildups similar to those seen in the human

Alzheimer's brain. Their work was published in a July 1991 issue of *Science*,[9] but eight months later, the researchers realized that they had made a mistake: apparently, the strain of mice they used in the study naturally develop beta-amyloid buildups when they age. Substituting this strain for a different one, the researchers failed to reproduce the buildup.[10]

The second "success" was announced by researchers from Mount Sinai Medical Center and the National Institute on Aging (NIA) in a December 1991 issue of *Nature*.[11] By inserting a select part of the APP gene into mice and overexpressing it, these researchers created transgenic mice that exhibit the whole package of the Alzheimer's brain: beta-amyloid plaques, fibril tangles, neuronal damage—everything.

But no sooner was the paper published than criticism followed. The point of contention was a set of micrographs used in the study that supposedly displayed transgenic mice brains beset by plaques and tangles. But critics thought that the brain tissues shown looked more like human brains than mouse brains. To wit, the paper was accused of passing off actual human Alzheimer's brains as false evidence of having created transgenic Alzheimer's mice!

Of the paper's three authors, the one in charge of analyzing the mouse brains and creating the micrographs was Gerald Higgins, from the NIA. Remember him? He coauthored, with John Hardy, the famous, two-page opinion paper that gave birth to the beta-amyloid hypothesis in the first place (chapter 6). Higgins scrambled to reproduce his mouse tissue results and prove himself innocent. He couldn't, nor could he provide more convincing evidence from the original research. The paper was retracted, a misconduct inquiry was made[12], and Higgins quietly left his job.

Ironically, in deceit (or so it was believed), Higgins played a significant role in contemporary Alzheimer's research. His probable fraud with the mice was an important inspiration behind the seminal paper that he and Hardy wrote to usher in the beta-amyloid paradigm. The paradigm remains today, but no one knows where Higgins is. As Hardy recalled, the last time he saw him, Higgins was interviewing to become an administrator at the National Association for Stock Car Auto Racing (NASCAR) in Florida—a scientist he is no more.[13]

Out of this crumbling mess, Eli Lilly and its business partner, Athena Neuroscience, arose. Athena was a California-based biotech company, and truly a dark horse. It had promising transgenic technology but was short on cash. For its part, Lilly had the deep pockets and drug pipeline. The two entered into an agreement to develop an Alzheimer's mouse model: Athena would hold the rights to the mouse, and Lilly would use the mouse for drug development and share with Athena whatever drug proceeds emerged.[14] In early 1995, their mouse was born and debuted in *Nature*.[15] Over a dozen colorful micrographs were published to show that their mouse, which carries a mutated human APP gene, actually does develop Alzheimer's-like pathologies. Wary of the Higgins scandal, the paper printed, side by side, micrographs of the human Alzheimer's brain and the transgenic mouse brain.

This mouse model, named the *PDAPP mouse*, develops extensive beta-amyloid plaques, which increase as the mouse ages. The plaques are concentrated in the brain regions most severely compromised by Alzheimer's: the hippocampus, the center for memory formation; and the cerebral cortex, which controls language, learning, reasoning, and planning. The mouse also develops other Alzheimer's pathologies and symptoms, including damages in neurites, those slender projections on neurons that receive and send brain signals; loss of synapses, the connections between neurons where signals are exchanged; and memory impairment, as measured by maze and other cognitive tests.

Standing to profit from Eli Lilly's Alzheimer's drugs, Athena was in no hurry to share its mice with others. As researchers elsewhere grumbled, Lilly had swiftly gotten a head start on the development of gamma-secretase inhibitors.

8

INHIBITORS THAT CAN'T INHIBIT

The year 2008 was a big one for Eli Lilly: that year, their Alzheimer's drug semagacestat entered phase 3 human trials. Going this far is an achievement for any drug. It's even more special when the drug is the first gamma-secretase inhibitor to get there.

As the name indicates, gamma-secretase inhibitors inhibit gamma-secretase, which, as mentioned in chapter 6, is one of the two enzymes that cut beta-amyloid out of its precursor protein. By inhibiting gamma-secretase, Lilly hoped to interrupt the cutting, stop beta-amyloid production, and thus terminate Alzheimer's disease.

Two separate phase 3 trials got underway, recruiting more than 2,600 participants with mild-to-moderate Alzheimer's.[1] Lilly duly followed the gold standard of clinical trial protocols, making the trials placebo-controlled, randomized, and double-blind.

A *placebo-controlled trial* assigns participants to either a treatment group or a control group. The treatment group receives the trial drug; the control group receives a placebo that has no therapeutic effect (aka a "sugar pill"). With this setup, if the treatment group got better and the control group didn't, we can be more certain that the effect is caused by the drug rather than some random factors.

A *randomized trial* avoids bias by randomly assigning participants to the treatment or control group. Without random assignment, a company might, for example, put younger participants in the treatment group and older ones in the control group. The treatment group is then likely to perform better regardless of the effect of a drug.

If the participants do not know which group they are assigned to, they are *blind*. Blinding reduces the power of suggestion, which can skew trial results. If someone knows that she is taking an experimental drug, she may *feel* that she is getting better and actually become better. Alternatively, she may feel that she is having side effects and get worse. In a *double-blind* trial, the researchers also do not know which participants are in which group until all the data have been collected—or even analyzed. This reduces researcher bias during drug administration (which may subtly influence participants) and data analysis (which can seriously influence data interpretation).

In semagacestat phase 3, Lilly had multiple treatment groups taking different doses of the drug. The plan was to continue dosing for 1.5 years in order to demonstrate the drug's long-term benefits on cognition and daily living. In 2010, midway through the trials, Lilly undertook a preplanned interim analysis to see how participants were doing.[2] To their astonishment, the treatment groups were in fact declining more than the controls, and those who received the higher doses of semagacestat declined the most. Instead of treating or curbing Alzheimer's, the drug seemed to be worsening and accelerating it. Just as alarming, the treatment groups developed more serious side effects, including skin cancer.

Eli Lilly quickly terminated the trials and ceased all further development of semagacestat. (By the way, the treatment groups saw no reduction in brain beta-amyloid either,[3] so the drug also failed at its very premise: inhibiting gamma-secretase and stopping beta-amyloid production.)

How did a drug that couldn't perform its basic function and actually worsened cognitive symptoms get this far—didn't we say in chapter 7 that in drug trials, a rigorous process is put in place to weed out drugs that don't work? That's a good question.

Eli Lilly did jump through all the hoops in the case of semagacestat. They had cell and animal studies that confirmed the drug's working mechanism.[4] In testing with PDAPP transgenic mice, the drug did reduce beta-amyloid production.[5] The more semagacestat the mice took, the more reduction happened—a typical sign that a drug is doing its job. The drug also passed the phase 1 human trial with flying colors, reducing beta-amyloid production among twenty healthy volunteers.[6]

Then—setback. In an initial phase 2 trial, seventy participants with mild-to-moderate Alzheimer's were randomized to semagacestat or placebo.[7] The semagacestat doses were set at 30 mg for one week and then 40 mg for another five weeks. At these doses, the drug was well tolerated, with few side effects. It also reduced beta-amyloid production. That said, the reduction wasn't statistically significant. It went down, but not by much.

Eli Lilly had a ready answer for this disappointment: the reduction wasn't significant because participants hadn't taken enough of the drug. Accordingly, in a second phase 2 trial, Lilly more than doubled and tripled the dose and the dosing period.[8] Despite these drastic increases, semagacestat still didn't significantly reduce beta-amyloid. Moreover, at higher doses, side effects started to flare up. Particularly noticeable were complaints of rashes and hair color change.

To an outsider, these results might seem pretty bleak. But Lilly was far more positive and had another ready answer: the lack of beta-amyloid reduction was a timing issue. During the trial, beta-amyloid was measured six hours after the patients took the drug, which Lilly now considered too short for the drug to kick in. As for the side effects, the rashes and hair discolorations were reversible once the drug was stopped, and they were, after all, not very serious.

Arguing for the drug's "potential" for therapeutic effect and "arguably acceptable" safety profile, Lilly jumped headlong into phase 3.[9] The doses used in phase 3 were high, comparable to those used in the second phase 2 trial.

This was how Eli Lilly got to where they were. The checkpoints along the way were cleared by assumptions rather than evidence, and faith rather than data drove the process forward. But this still doesn't explain why the drug *worsened* participants' cognitive function. So the drug didn't reduce beta-amyloid. Why would that make things worse?

No one knows for sure, but several explanations were put forward.[10] One is that the trial participants, already frail because of Alzheimer's, were further weakened by semagacestat's side effects, and they underperformed on cognitive tests. Another blames the dosing regimen used in the trial. Participants were dosed once a day at a high dose, but the drug would clear their system in twelve hours, which means the amount of semagacestat in their system fluctuated drastically every day, which is suspected to mess up both beta-amyloid production and cognitive function.

However, the most popular explanation, by far, was one that shook the foundation of the drug. The premise behind semagacestat, like other gamma-secretase inhibitors, was that it suppressed gamma-secretase, which in turn stopped the cutting of beta-amyloid from the amyloid precursor protein. But gamma-secretase doesn't *just* cut beta-amyloid; it has other jobs too. In fact, it processes a long list of proteins involved in vital cell functions. Inhibiting gamma-secretase, then, will give us more than we bargained for.

Among the affected proteins is Notch. Notch was first discovered in fruit flies, and it was named after what it does: a mutated Notch causes the flies to develop notched wings. In humans, Notch regulates skin and hair follicle development. Disrupted Notch, therefore, is a likely cause for the rashes and hair discoloration experienced in the semagacestat phase 2 trial and eventually, as the dosing increased, skin cancer in phase 3 trials. Notch also has important cognitive functions, including regulating how neurons send, receive, and transmit signals.[11] This would explain the worsened cognitive function in the phase 3 participants.

Post-semagacestat, and Notch-scared, big pharma was gloomy about the whole idea of gamma-secretase inhibitors.[12] Some attempts were made to modify these drugs so they could better target the amyloid precursor protein and leave Notch alone, but the results were disappointing. Notably, Bristol-Myers Squibb developed a so-called Notch-sparing gamma-secretase inhibitor called Avagacestat, which supposedly was 190 times more selective toward the amyloid precursor protein than Notch.[13] The drug, however, ended up as a rerun of the semagacestat disaster: at phase 2, the treatment group worsened in cognitive function and developed side effects including rashes and skin cancer.[14] The drug was promptly scrapped, making it doubtful whether a gamma-secretase inhibitor could ever spare Notch.

Lilly didn't waste more time trying to find out. Pulling out of gamma-secretase inhibitor research, it went after another inhibitor. Remember that it takes two enzymes to cut beta-amyloid out of its precursor protein? There's gamma-secretase, and then there's beta-secretase (see chapter 6). Just as we may inhibit gamma-secretase to block beta-amyloid, we can try to inhibit beta-secretase. Naturally, these drugs are called *beta-secretase inhibitors*.

These inhibitors have been difficult to develop. It was relatively easy to make them work in a test tube. But in a living body, they have a hard time reaching the brain, where they are supposed to work. Blocking the way is a literal barrier: the *blood-brain barrier*, which is a wall made of tightly packed cells that separates circulating blood from the brain. This barrier is nature's defense mechanism for the brain. It allows small molecules and elements essential for brain function to enter, but it keeps bacteria, toxins, and other invading pathogens out. Having large molecules, beta-secretase inhibitors can't wiggle through the barrier.

But Eli Lilly succeeded in breaking through using a beta-secretase inhibitor with sufficiently small molecules. Known as LY2811376, it is the first beta-secretase inhibitor that can be taken orally and effectively reach the brain.[15] When fed to PDAPP transgenic mice, LY2811376 was robust in lowering beta-amyloid production. The same was seen in a phase 1 human trial with thirty healthy participants. No side effects were reported except for a few relatively minor complaints, like headaches.

But just when Eli Lilly was about to launch phase 2, bad news came. In parallel with the phase 1 human trial, a study was conducted with rats to test LY2811376's safety for long-term use. Unfortunately for the rats (and Lilly), after three months of dosing, the animals developed eye problems (i.e., enlarged and degraded retinal cells) and neuronal degeneration.[16] The side effects were thought to happen because the drug inhibited not only beta-secretase but other, unknown, off-target elements.[17]

Lilly stopped further testing of LY2811376 but quickly refined it and put forward a second edition, dubbed LY2886721.[18] This revised version was supposed to be better at targeting beta-secretase and sparing other elements in order to avoid side effects. Like its predecessor, the revised LY2886721 lowered beta-amyloid in transgenic PDAPP mice, and the same was seen in a phase 1 trial with healthy participants. Side effects were mostly minor, although two participants had temporary, abnormal liver enzyme levels.

With these results, Lilly advanced to phase 2 in early 2012, making LY2886721 the first beta-secretase inhibitor to get this far. A total of 128 participants with mild or very mild Alzheimer's were recruited, and dosing was supposed to last six months.[19]

Alas, that plan too was jettisoned. During the trial, routine testing revealed several cases of liver toxicity, harking back to the abnormal liver enzyme cases in phase 1. This side effect was again thought to be caused by the drug's interaction with off-target elements. Upon this finding, Lilly was not interested in making any more adjustments. It terminated the trial and stopped further development of LY2886721.

Eli Lilly was not the only one burned by beta-secretase inhibitors. Roche terminated one at phase 1 due to liver toxicity, and Johnson & Johnson pulled the plug at phase 2/3 for the same reason.[20] Toxicity is not the only downfall either. Merck aborted its beta-secretase inhibitor at phase 3 due to interim analysis showing lack of cognitive effects.[21] Similarly, in 2018, five years after LY2886721 failed, Lilly saw another beta-secretase inhibitor go up in smoke. Lanabecestat, jointly developed by Lilly and the UK-based AstraZeneca, was terminated at phase 3 when an interim analysis showed that the drug had no hope of improving cognitive function.[22]

9

POISON OR CURE

An Alzheimer's Vaccine

In some social circles, vaccination is treated with suspicion. By the playgrounds and swimming pools, parents speculate about how vaccination will weaken their children's immune systems and make them sick. Anecdotes are shared, research (real and fake) is cited, and conspiracies are declared. The government is allegedly making vaccine-friendly policy to boost the sales of big pharma, who in turn contribute large sums to political campaigns, and provaccine medical researchers, which are many, are presumably all bought off.

Although provaccine myself, I understand why some parents are concerned. The basic premise of vaccination is to poison our bodies—albeit just a little, so the immune system can be kickstarted. Once prepped, the immune system has a chance to fight a larger quantity or a stronger version of that poison.

The starter poison, in proper terms, is called an *antigen*. It can be derived from bacteria, viruses, and other harmful substances. When introduced into the body, an antigen stimulates the immune system to develop a corresponding protein, which is called an *antibody*. Different antibodies target different antigens and can either isolate them from body cells or tag them for destruction by other immune cells. The beauty of the immune system is that once an antibody has been manufactured, the body "memorizes" it and can produce it in short order when needed to preempt the same antigen strike, making a person immune to a particular disease.

By letting a starter poison into our bodies, we do take a risk. Usually, it is a weakened or inactivated form of bacteria, viruses, or toxins, and it doesn't do much harm—a bit of swelling at the injection site, some muscle aches, for example. But there *is* a possibility that it will go rogue and get out of control. In rare cases, polio vaccines have mutated, traveled to the central nervous system, and caused paralysis, just as a regular polio virus would do.

Such cases are tragic, but they are also very, *very* rare. In medicine, as in life, when things go well, they are taken for granted, but when they go wrong, they commend attention. Global estimates for vaccine-associated polio are, on average, 4.7 cases per million births.[1] Millions of kids who are spared polio by vaccination have nothing to show for *because of* their health, but the few who react badly to the vaccine endure profound suffering that is hard to overlook.

Even more unfortunate, these cases can become fodder for lies and fraud. In 1998, an article appeared in the prestigious British journal *The Lancet*, practically claiming that the measles, mumps, and rubella (MMR) vaccination triggered autism.[2] According to the article, among twelve children who were previously healthy, nine developed autism after vaccination. The popular media had a field day covering the sensational discovery, and fearful parents pulled their children away from vaccination in droves.

But the paper turned out to be a lie—it was fabricated by its lead author, Andrew Wakefield, who was on the payroll of attorneys suing vaccine manufacturers and needing "scientific" evidence. In 2010, *The Lancet* retracted the paper, but by then, twelve long years had passed, the paper had been cited hundreds of times, and far-reaching public damage had been done.

Today, myths continue to circulate about the connection between vaccines and neurological diseases, including autism, epilepsy, multiple sclerosis, and indeed Alzheimer's. On a 2005 episode of *Larry King Live*, comedian, television host, and political commentator Bill Maher blurted out that "if you have a flu shot for more than five years in a row, there's ten times the likelihood that you'll get Alzheimer's disease."[3]

I'm not sure where Maher got his intel; no published study can be found saying that. Such sensational findings would be highly publishable, so I can't imagine any researcher passing up that opportunity. The only clue I can find is Hugh Fudenberg, another discredited researcher whose medical license had been revoked ten years earlier, in 1995. In 1997, Fudenberg spoke at the International Public Conference on Vaccination held in Arlington, Virginia, where he spread the flu shot–Alzheimer's ruse[4] but never produced any study to support his claim.

Fortunately, this nonsense never traveled very far. Had it gained traction in public opinion, Dale Schenk would have never seen the light of day for his idea.

Growing up in Glendora, California, Dale Schenk is the son of a firefighter and a journalist.[5] His sister is an artist and his brother is an automotive shop owner, making him the only scientist in the family. When he was a teenager, Schenk made a career choice between piano, a longtime hobby, and medical research. Besides wanting to help people, he thought, "piano is a hard row to hoe."[6]

After receiving his PhD in pharmacology and physiology from the University of California, San Diego, Schenk went to work for Athena Neuroscience as a staff scientist in 1987. Remember Athena? That was the small biotech company that developed the PDAPP transgenic mouse in partnership with Eli Lilly (discussed in detail in chapter 7). In 1996, Athena was acquired by Elan Pharmaceuticals, a Dublin-based pharmaceutical company. Schenk followed and eventually became Elan's chief scientific officer.

In many ways, Schenk resembled Dr. Alois Alzheimer: a generous colleague, a curious researcher, and someone with a good sense of humor. One of his colleagues at Elan recalled:

> We were on a bus in Kyoto, Japan, setting out for a day of serenity and visiting temples. A bewildering array of buttons with mysterious symbols were embedded in the back of the seat facing us. Out of the corner of my eye, as if in slow motion, a finger floated by, targeting a particularly interesting one of them. I thought "Oh, no. Dale, NO!" Sure enough, the bus screeched to a stop, and all the other, mostly Japanese, passengers stared at us. Dale had, of course, pressed the emergency stop. His mischievous curiosity just couldn't resist the urge to see what would happen. He disarmed everyone with a big, goofy laugh, the whole bus joining in.[7]

Without this mischievous curiosity, the urge to see what would happen, one would never have dreamed up what Schenk did: a vaccine to immunize people against Alzheimer's. The way that Schenk imagined it, he would inject human beta-amyloid into mice. Faced with this antigen, mice might develop an antibody against it. The antibody could then be harvested and used in humans to wipe out beta-amyloid and eradicate Alzheimer's.

When Schenk shared this idea with his Athena/Elan colleagues, it was met with raised eyebrows and chuckles.[8] People thought that it was so far-fetched, it bordered on ridiculous. Vaccines are used to prevent infectious diseases that spread from person to person, such as chickenpox and flu (and polio). The infectious nature of these diseases makes a preventive vaccine valuable. But Alzheimer's is not contagious. The very idea of vaccination, then, seems wrong.

In addition, antibodies are large molecules. Even if some kind of beta-amyloid antibody could be raised, it would have a hard time getting past the blood-brain barrier to reach the brain. And even if it did manage to sneak through the barrier, who knows what damage this little poison might cause in the brain.

Schenk was imaginative enough to ignore this conventional wisdom. He was also fortunate because his employer owned the PDAPP Alzheimer's mice that he needed for his experiments. These mice are genetically engineered to develop beta-amyloid deposits in the brain when they are six to nine months old, which is like middle age for mice. To test his idea, Schenck picked youngsters that were only six weeks old and gave them beta-amyloid injections.[9] Almost miraculously, the mice raised functional antibodies. They grew up free of beta-amyloid deposits and other brain damage, dodging their genetic fate.

Inspired, but unsure what this finding meant to old mice who, as human patients, already had developed beta-amyloid deposits in their brains, Schenck injected beta-amyloid into eleven-month-old PDAPP mice.[10] Amazingly, these animals also raised antibodies, which seemingly slowed—and even reversed—their disease progression. Several months after the injection, the mice saw a more than 96 percent reduction in their brain beta-amyloid. Compared with untreated mice, they also had a 55 percent reduction in neuronal damage. They seemed, for all intents and purposes, cured!

When Schenk's findings were published, they were touted by news media as stunning, groundbreaking, and, above all, a sign of hope. Within the research community, a buzz was going around as researchers jumped on the previously unimaginable vaccine bandwagon. One burning question on everyone's mind was, "So the vaccine reduced beta-amyloid in mice—does that actually help their cognitive function?"

To answer this, we resort to a mouse-sized puzzle: the Morris water maze. Named after British neuroscientist Richard Morris, this maze consists of an indoor circular pool about half full of water. A small platform is placed inside the pool, either above the water or hidden slightly underwater. A mouse is put in the pool and trained to find the platform. Signs can be posted on the walls to help orient the critter visually. Although mice are capable of swimming, they really hate doing it, so they try to get on the platform as soon as possible. How quickly they can find the platform, then, becomes a measurement of their spatial memory and learning.

In one water maze study, several cohorts of mice competed against each other: transgenic Alzheimer's mice that already showed beta-amyloid plaques and cognitive deficit, the same transgenic mice immunized with beta-amyloid, and normal mice.[11] Four rounds of competition were held over the course of 17 weeks. In each round, to add to the challenge, the platform was hidden in a different quadrant of the pool. When the results were tallied, the regular Alzheimer's transgenic mice performed poorly, as expected. But the immunized transgenic mice performed significantly better—and sometimes as well as normal mice that do not carry Alzheimer's mutations.

On the heels of these animal findings, Elan launched human trials. Its vaccine, named AN-1792, is a chemically produced beta-amyloid. In phase 1, Elan recruited eighty mild-to-moderate Alzheimer's patients who were injected either different doses of the vaccine or placebo. Four injections were administered over a six-month period; four more were added over the next twelve months.[12]

Elan had two goals for the trial. First, find out if the vaccine, or a certain dose of it, could raise antibodies in humans as it did in mice. Second, find out if injecting disease-causing beta-amyloid into humans was actually safe. Both goals were realized—at least partially. After the initial four injections, about a quarter of the participants raised sufficient antibodies; after the additional four injections, a little over half did; and among participants who received a higher dose of the vaccine, 77 percent raised sufficient antibodies at some point of the study. Participants did report a long list of side effects ranging from rashes to agitation, but they were generally mild and happened to both the treatment and the placebo groups.

On account of these findings, Elan moved forward to phase 2 in 2001.[13] Phase 2 adopted the higher dose of the vaccine tested in phase 1 and enrolled over 370 mild-to-moderate Alzheimer's patients. Six injections were planned over a twelve-month period. In this trial, Elan wanted to further establish the safety of the vaccine and start assessing if it could clear beta-amyloid and improve patients' cognitive function.

But this time, things fell apart. Four months into the trial, four people in the treatment group developed brain inflammation. Elan stopped the injections, but by then, most participants had received two or more doses. Subsequent follow-ups revealed that another fourteen people also developed brain inflammation with serious symptoms, including seizure and disorientation. Nobody in the placebo group had these side effects.[14] The little starter poison, apparently, was doing a number on the brain.

Despite this safety fiasco, some of the trial's findings were tantalizing. Even with limited dosing, about 20 percent of participants in the treatment group raised adequate antibodies. This cohort also performed better than the placebo group on one cognitive test, though not on other tests.[15] If only we could figure out what caused the brain inflammation, this AN-1792 vaccine would still be very promising.

So, what went wrong? A case of death from the trial lives to tell the tale.

In Catalonia, Spain, an anonymous seventy-six-year-old man signed up for Elan's phase 2 trial.[16] By then, he had lived with Alzheimer's for three years and had moderate symptoms. Upon entering the trial, he was randomly assigned to the higher-dose vaccine group and received two injections within a month.

Six months after the second injection, his speech and gait inexplicably worsened. Another three months later, his movement, communication, and overall cognitive function grew worse still. Brain imaging tests showed that the man had developed brain inflammation. The condition calmed with anti-inflammatory drugs, only to return with a vengeance weeks later. At this point, he had become severely demented and was transferred to a nursing home, where he died two months later.

Upon autopsy, the man's brain was found to contain a surprisingly low number of beta-amyloid plaques. It seems that Elan's vaccine, although it had terrible side effects, did what it was designed to do: clean out beta-amyloid. Also found in the man's brain were wildly activated T-cells. A type of white blood cell, T-cells are the brute force of our immune system. Some T-cells, the so-called Killer T-cells, attack and wipe out infected and damaged cells; others, the Helper T-cells, activate the Killer T-cells and coordinate their attacks. When T-cells are overstimulated, the immune system goes out of control and turns on itself, attacking not only diseased but healthy tissue. This condition is known as *autoimmune responses.*

This scenario, Elan figured, is what happened in their botched phase 2 trial: the injected beta-amyloid activated Helper T-cells, which then mobilized the immune system to attack healthy brain tissue, causing inflammation and rapid cognitive decline.

If so, all is not lost. We just need a vaccine that won't trigger T-cells but will raise beta-amyloid antibodies. Thanks to the highly specialized way in which immune cells work, this is not impossible. T-cells work by zeroing in on certain parts of the beta-amyloid protein, but they are practically blind to the rest of it. If we trim off the parts that T-cells recognize, we can avoid autoimmune responses. Meanwhile, beta-amyloid antibodies can still be raised because antibody production is the job of another type of immune cell: the B-cell.

Pharma, big and small, seized on this concept, jumped in to develop a second-generation vaccine that could bypass T-cells, and quickly emerged with prototypes. In animal testing, these products were safe, raised antibodies, cleared out beta-amyloid, and even reversed the animals' cognitive impairment.[17]

Entering human trials, companies' number one concern was *not* to repeat Elan's mistake and end up with brain inflammation on their hands. To their relief, that didn't happen. The T-cells were kept quiet while patients continued to raise beta-amyloid antibodies.[18]

But other, *desirable* markers were also being quiet and not chang-ing. A vaccine developed by Novartis, a Swiss big pharma company, failed to reduce trial participants' beta-amyloid or slow their cog-nitive decline, even after two years of dosing.[19] Another vaccine by

Pfizer/Johnson & Johnson suffered a similar fate. In fact, some vaccine recipients experienced more brain shrinkage.[20] Facing these setbacks, some companies shelved their products and called it quits. Others hung onto the hope of tweaking a product that was at least safe. But the most bizarre move by far was made by an Austrian biotech company, Affiris.

Affiris had two vaccines named, quite literally, AD01 and AD02, as in "Alzheimer's Disease 01 and 02." Both drugs proved safe in phase 1, but AD02 performed better, so it was selected to move to phase 2.[21] The phase 2 trial enrolled more than 300 early Alzheimer's patients, who were randomized into one of four dosing routines: injections of AD02 at a low dose, injections of AD02 at a high dose, injections of placebo 1, and injections of placebo 2.[22]

After eighteen months of dosing, participants taking AD02, whether the low or the high dose, saw no cognitive benefit. But unexpectedly, one of the placebo groups experienced less cognitive decline and brain shrinkage. Things like this happen sometimes. A "sugar pill" group may have declined more slowly due to chance, or there could have been some statistical errors due to small sample sizes or incorrect analysis. Either way, the logical thing for Affiris to do was to declare the trial a failure.

But that's not what Affiris did—in fact, they did the very opposite. In an astonishing move, the company proudly announced that it had made "nothing less than a breakthrough in Alzheimer's disease therapy."[23] What was the breakthrough? The placebo! Supposedly, this sugar pill had been misunderstood all along. It shouldn't be a placebo. It is, as Affiris's trial proved, a superior Alzheimer's drug.

Affiris promptly dubbed this formerly nameless placebo AD04 and declared its plan to pursue it for clinical development. So what *is* AD04? It is an adjuvant (a term coming from the Latin *adjuvare*, meaning "to help"). Adjuvants are often added to vaccines to help create a stronger immune response by, for example, increasing antibody production. On their own, however, adjuvants do *not* provide immunization. That's why, on their own, they can be used as placebos. What kind of adjuvant is AD04? Affiris wouldn't say. How does it provide cognitive relief? Affiris didn't know.

Six months would pass before Affiris spilled the beans. Their mysterious AD04 was an aluminum salt, which is an aluminum-containing compound and common adjuvant used in vaccines.[24] This revelation threw more irony to the story—because of a lingering suspicion among some that aluminum contributes to Alzheimer's. Indeed, this idea fueled the rumor that flu shots produce Alzheimer's.

Of course, it is not just vaccines that contain aluminum. Aluminum is abundant in the natural environment and found in everyday products ranging from baking soda, foil, and cookware to antacids, antiperspirants, and cosmetics. An aluminum-Alzheimer's connection became known in the 1960s, when a study found that injecting an aluminum salt into rabbit brains caused Alzheimer's-like brain pathologies.[25] Then, in the 1970s, another study found elevated aluminum in human Alzheimer's brains.[26] This research became the foundation of a so-called aluminum hypothesis and inspired a wave of new studies. But rather than supporting the hypothesis, these studies poked holes in it. The abnormalities developed in rabbit brains upon aluminum injection turned out *not* to resemble those found in Alzheimer's.[27] Elevated aluminum in the human Alzheimer's brains was also questioned and attributed to contamination from aluminum-containing chemicals used in the experiment.[28]

To find more solid evidence, researchers looked into the prevalence of Alzheimer's in regions where drinking water contains elevated aluminum, in people with high dietary aluminum intake, and among workers who are exposed to aluminum dust and fumes. The results of these studies are wide-ranging. Depending on whom you listen to, aluminum is either a risk factor for Alzheimer's, has no connection to the disease, or prevents Alzheimer's.[29]

The more decisive evidence came from kidney failure patients.[30] Normally, our kidneys can filter aluminum and remove it from the body via urine. But people with kidney failure lose this ability. If they are chronically exposed to aluminum-containing medications, they can accumulate high levels of aluminum in the brain, which lead to neurological dysfunction and death. But these patients do *not* develop beta-amyloid plaques or other Alzheimer's-like brain pathologies. Their clinical symptoms are also different. They usually suffer from seizures, which are rare in Alzheimer's; their deaths are imminent, about six months after initial symptoms present, whereas Alzheimer's lasts years, or even decades. In other words, excessive aluminum in the brain *is* toxic and *does* cause neurological disease, but what it causes is not Alzheimer's.

Given these findings, the aluminum hypothesis is all but abandoned by researchers today. Still, finding an Alzheimer's *therapy* in aluminum is ironic. We don't know what Affiris thinks about this backstory and its connection to their placebo-turned–aluminum drug. No update has been announced about AD04, and it is not listed among the company's pipeline drugs.

10

ELI LILLY'S THREE EXPEDITIONS

On September 30, 2016, Dale Schenk, the father of the Alzheimer's vaccine, passed away at age fifty-nine of pancreatic cancer. It's unusual that the passing of a so-called industry scientist, which Schenk was, would cast such a large, gloomy shadow. Industry scientists are often lesser-known "brains for hire," tolling away as cogs in corporations big and small. University scientists, who are considered disinterested pursuers of knowledge (but in reality have various economic ties with industries), are the ones more likely to garner social respect. Not in Schenk's case. On his passing, both kinds of researchers came forward to call him a dear friend, a visionary thinker, a brave pioneer, and, above all, a wonderful person.[1]

Schenk had big dreams. Just four years earlier, he had founded Prothena Biosciences, a company that commits to "harnessing the power of the immune system to fight progressive disease." It was developing vaccines for a range of neurological diseases, including Parkinson's and Alzheimer's.

Schenk could no longer help promote these adventures. But he was also spared the details of Eli Lilly's expeditions, the embodiment of his vision for an Alzheimer's vaccine.

August 17, 2010, John Lechleiter's birthday, was not at all a good day for the fifty-eight-year-old chairman and chief executive officer of Eli Lilly. That day, the company announced the phase 3 failure of its gamma-secretase inhibitor semagacestat. Not only did the drug fail to slow cognitive decline, it worsened patients' symptoms and caused skin cancer—likely because the drug disrupted the essential protein Notch (as described in detail in chapter 8).

Despite Lechleiter's statements that "Eli Lily and Company remains financially strong" and "is even more determined" in the quest for Alzheimer's treatments,[2] the failure came at a terrible time. That summer, Lilly was facing patent expiration of its top-selling antipsychotic drug Zyprexa and antidepressant drug Cymbalta. As generics enter the market, a loss of $10 billion in revenue was predicted.[3] It was time, Lechleiter declared, to "find our true grit."[4]

Finding true grit amounted to laying off 5,600 employees during the next two years, shutting down old-time labs and plants, paying high dividends to keep investors happy, and working at top speed to push new patent drugs through the pipeline.[5] From 2005 to 2014, a slew of drugs came out or entered late-stage trials: a low-testosterone drug, cancer drugs, diabetes drugs. Yet what attracted investors more than anything was a wild card, another Alzheimer's drug: solanezumab.

What is solanezumab? It's another vaccine, one that promises to detect and wipe out beta-amyloid and immunize people against Alzheimer's.

But didn't the latest vaccines all go belly-up, either causing deadly brain inflammation or offering no cognitive benefit? That's true. But solanezumab is a different kind of vaccine.

The vaccines that we have seen so far work by injecting various formulations of beta-amyloid, the starter poison, into the body. The hope is to coax the immune system to develop beta-amyloid antibodies, which can then wipe beta-amyloid out of the brain. This is the so-called active vaccination: it is "active" in the sense that the body is raising antibodies and building immunity on its own.

Solanezumab, on the other hand, *is* a beta-amyloid antibody. It is collected from immunized mice and modified for human use. When injected into the body, it is supposed to go straight to helping clear out brain beta-amyloid. This approach is called *passive vaccination:* it is "passive" in the sense that the body is relying on outside help to develop immunity.

From active to passive vaccination, the basic belief in eliminating beta-amyloid and inoculating the body against Alzheimer's didn't change. But in practice, passive vaccination through drugs like solanezumab is thought to be safer and more effective.[6]

It is safer because we no longer slip poisonous beta-amyloid into our bodies, which means that T-cells won't be overstimulated to cause autoimmune responses and brain inflammation. Because the drug *is* the antibody, each patient can be given a consistent dose of it for optimal effect, which is difficult to achieve with active vaccination, where individuals' abilities to raise antibodies differ. By the same token, we can just as easily stop the drug and the influx of antibodies at the first sign of trouble, whereas in active vaccination, there is no easy way to quickly stop our bodies from producing antibodies.

Passive Alzheimer's vaccines all have tongue-twisting names that look similar. Many end in *mab,* which stands for *monoclonal antibodies,* meaning that the antibodies were cloned from an identical cell. Preceding the *mab* are various letters that indicate the source of the antibody. The letter *o,* for example, means that an antibody was harvested from mice, and the letter *u* means that it was harvested from humans. If an antibody was collected from a nonhuman animal and modified to work in humans, the letters become *zu.* This is why Eli Lilly's mouse-derived, modified-for-humans monoclonal antibody is called *solane-zu-mab* (I don't know what *solane* means for Lilly).

In cell and animal studies, solanezumab did splendidly. It functioned as a sinkhole,[7] attracting beta-amyloid and drawing it out of mice's brains into their bloodstream. In transgenic Alzheimer's mice, it also improved the animals' memory.[8]

Mice, like us, are curious animals. Their natural tendency is to explore things that are new. When first given an object such as a small toy, mice take their time playing and interacting with it. Hours later, when given the same toy, they won't be as fascinated by it anymore. If a new toy is handed to them along with the old one, the mice will naturally spend more time checking out the new toy. But there is a caveat to all this: the animal needs to *remember* which toy it has seen and played with. When it comes to aged, transgenic Alzheimer's mice, they can't remember. Having forgotten what they saw hours earlier, they would spend equal time exploring the old and new toys. Remarkably, a six-week treatment with solanezumab changed that. Aged, demented mice started to remember, and they acted almost the same as young, healthy mice.

For good measure, Lilly put the mice through another, more rigorous hole-board test.[9] The test employed a small, 18-inch × 18-inch chamber with a floor insert. The floor had sixteen holes, four of which were baited with food. Through multiple sessions, mice were set loose in the chamber and tested on how quickly and accurately they could find the food. To make sure that the critters were using their brains rather than their sense of smell, testers put food in the other twelve holes too, but made them inaccessible through a screen.

The mice's movements were closely tracked. If they entered a hole that wasn't baited, that was an error; if they missed a baited hole, that was another error; and if they reentered a correct hole after having retrieved food from it, that was an error too. In this test, transgenic Alzheimer's mice treated with solanezumab made significantly fewer errors than untreated ones. More remarkably, they did so after receiving only one dose of the drug. Quite the miracle.

Based on these superb results, human trials ensued. Phase 1 enrolled nineteen patients with mild-to-moderate Alzheimer's, who were given a single injection of solanezumab.[10] The drug was well tolerated, with no signs of autoimmune attacks or brain inflammation. Beta-amyloid in the participants' blood increased sharply, which Lilly interpreted to be a result of the drug drawing beta-amyloid out of the brain.

Swiftly, solanezumab entered phase 2, which enrolled fifty two mild-to-moderate patients, randomized to receive solanezumab or placebo.[11] The dosing regime lasted much longer this time: three months. But the drug continued to be well tolerated, showing no serious side effects. Participants' blood beta-amyloid also continued to rise, supporting Lilly's belief that solanezumab was trapping brain beta-amyloid.

What solanezumab couldn't do, though, was slow the participants' cognitive decline. Lilly quickly explained that this was "not unexpected."[12] Any cognitive improvement, they asserted, wouldn't be noticeable within three months. Over such a short period, the placebo group wouldn't worsen enough to contrast the effect of the drug. Yes, the drug did work wonders in mice after a single dose, but human brains are far more complicated, so it'll take much longer for that kind of improvement to happen—more like a year and a half, the time frame that Lilly set for its phase 3 trials.

Going all out, Lilly started two separate phase 3 trials in 2009.[13] Not above sounding heroic, Lilly named the trials "Expeditions 1 and 2." The two expeditions followed the same protocol. Each enrolled over 1,000 mild-to-moderate patients, who would randomly receive solanezumab or placebo for a year and a half.

For a nice change, there was no premature termination this time. Both expeditions ran their course, and the results came back in 2012. The Expedition 1 data came in first, and they didn't look particularly good at first glance: solanezumab failed to improve the participants' cognitive performance or daily living skills. But a closer look suggested that if we disregarded moderate Alzheimer's patients who were further along in their disease and focused only on mild patients, the result was favorable.

At this time, Expedition 2 was still underway, so Lilly changed its plan to focus that trial's statistical analysis solely on the mild patients. This change didn't make a material difference to the ongoing trial, but it would allow Lilly to announce later that they had reached their "planned" goal of treating mild patients, which sure would sound better to investors.

With this change, things indeed got a little better. In Expedition 2's mild patients, solanezumab improved daily living skills such as balancing checkbooks and preparing meals—as judged by one test, the Alzheimer's Disease Cooperative Study–Activities of Daily Living.[14] However, results from several other tests that measure both living skills and cognitive functions (for example, abilities to remember words and name objects) were still negative.[15]

Pushing the envelope, Lilly decided to pool the mild patients from both expeditions, which would give them a larger sample of about 1,300 participants.[16] Maybe things would look better in this larger pool? Things surely did. This time, four tests came back favorable.[17] According to these tests, solanezumab slowed patients' decline in complex daily living skills (such as taking medication) by 18 percent and slowed their cognitive decline by 34 percent.[18] Three alternative tests that measure overall living and cognitive functions, though, were still negative.[19]

Despite this mixed bag of results, mass media proclaimed Lilly's Expeditions a major, landmark breakthrough. Solanezumab "cut the rate of the dementia's progression by about a third," exclaimed BBC News.[20] The drug was declared as the first to slow the progress of Alzheimer's by tackling its underlying cause; market entry was projected to happen in three years.[21]

But before Lilly could market solanezumab, they needed to collect more evidence and replicate the drug's effect on mild patients. A third expedition was needed.[22] Expedition 3 recruited only patients with mild Alzheimer's. Over 2,000 people were enrolled and randomized to solanezumab or placebo for a year and a half.

When the trial results came back in November 2016, Lilly was sorely disappointed: solanezumab recipients did no better than placebo recipients. Dicing the data no longer, Lilly abandoned the plan to seek market approval for solanezumab and announced no further development steps.[23]

Following this announcement, the mass media took a 180-degree turn. "The failure of Eli Lilly and Co.'s Alzheimer's treatment to slow declines in mental capacity of patients with even mild symptoms cast serious doubt" on its entire drug development approach, Reuters wrote.[24] "Some Alzheimer's experts . . . said they were not surprised by Lilly's results," the *New York Times* chimed in.[25]

Overnight, Lilly's stock plummeted, dropping by 10 percent.[26]

Elsewhere in the pharmaceutical world, other passive vaccines were not doing much better. Getting lots of attention was bapineuzumab, developed by Elan, the employer of the founding father of the Alzheimer's vaccine, Dale Schenk.

A mouse-derived, modified-for-humans, monoclonal antibody like solanezumab (hence its similar name), bapineuzumab worked in a different way. Rather than drawing brain beta-amyloid out into the blood, it entered the brain and directly cleared out beta-amyloid.[27] At least, that was what appeared to happen in mice.

But as usual, things became complicated with the human trials. In phase 1, the drug was mostly safe and well tolerated, but at high dosages, it caused fluid accumulation in the brain.[28] Lowering the dosage, Elan marched on to phase 2.[29] But fluid accumulation still occurred. More vexingly, the drug did not slow the trial participants' cognitive decline.

Like Lilly, Elan examined and re-examined its data, performed additional calculations, and emerged with more triumphant findings: the antibody did work on a subset of the patients—those who were non-ApoE4 carriers. As described in chapter 5, the ApoE gene exists in three possible versions: ApoE2, ApoE3, and ApoE4. ApoE3 is risk neutral, ApoE4 increases the risk of Alzheimer's, and ApoE2 decreases the risk. Non-ApoE4 carriers, then, face a smaller disease risk, but a risk nonetheless.

Among non-ApoE4 carriers, bapineuzumab slowed brain shrinkage and cognitive decline, while ApoE4 carriers saw no benefit. In fact, ApoE4 carriers were also the ones more likely to develop the side effect of brain fluid accumulation. It is not clear how or why ApoE4 status would interfere with the drug's function, but in making connections with this genetic risk factor, the finding is certainly interesting.

Around this time, bapineuzumab was purchased by Johnson & Johnson and Pfizer, who decided that the non-ApoE4 angle was strong enough to pursue further trials. Even if the drug worked only on non-ApoE4 carriers, that would still be a large patient—and consumer—base. Johnson & Johnson and Pfizer organized two phase 3 trials, one recruiting over 1,000 ApoE4 carriers and the other over 1,000 noncarriers.[30] Within each group, the participants were randomized to get either bapineuzumab or placebo. It was double the work—but unfortunately also double the disappointment. After a year and a half, bapineuzumab did not slow either group's decline. Carriers or noncarriers, it made no difference. With these results, the companies terminated further development of bapineuzumab.

In retrospect, and for someone who is not looking for favorable results, the ApoE4 finding from the early bapineuzumab trial was not rock solid. That trial was designed to test bapineuzumab's effect on *all* trial participants, not a subset of them. When the drug didn't work on all the participants, the researchers couldn't very well change their minds after the fact to focus on non-ApoE4 carriers. Why? Because the more we play with data, the more likely that we will find something. If we come up with fifty different ways to dice a set of data into subsets and analyze them fifty different ways, then by *chance*, some of the tests will return favorable results. Dicing also reduces the amount of the original data and hence reduces its statistical power.

As a case in point, completely opposite findings regarding ApoE4 have turned up elsewhere. Tramiprosate, developed by the Canadian biotech company Neurochem, is supposed to treat Alzheimer's by reducing beta-amyloid buildup.[31] At phase 3, the drug failed to show cognitive benefits among more than 2,000 mild-to-moderate Alzheimer's patients, but in further dicing and exploration, a subset of the patients stood out to have benefited: ApoE4 carriers![32] That's right—although bapineuzumab worked only on noncarriers, tramiprosate was the very opposite.

True, the two drugs work differently, so there may well be good reasons why their interactions with ApoE differ. That said, it's almost amusing that pharmaceutical companies can spin any story. If a drug has no effect on ApoE4 carriers, it's because these carriers, given their risk gene, are genetically difficult to treat. If a drug *does* have a favorable effect on ApoE4 carriers, it's because these carriers, given their risk gene, decline more rapidly, which underscores the effect of the drug. Either way, the drug is said to warrant further development.

So, whatever happened to tramiprosate? That itself is an interesting story. After the news broke that the drug failed to benefit all the participants in phase 3, Neurochem saw a more than 40 percent drop in its share price.[33] But the savvy Canadian company rebranded tramiprosate, the Alzheimer's drug, as Vivimind, the "memory-enhancing" nutritional supplement. As a supplement, Vivimind is not subject to stringent regulations and is free to enter the Canadian consumer market. A quick online search finds websites claiming that Vivimind is backed by fifteen years of clinical research, has "followed" (note: not "passed") clinical trials *all* the way to phase 3, and was tested on more than 2,000 patients (never mind the results of that testing). A bottle of Vivimind once sold for more than Can$40, but it is out of stock in most online stores now.

11

TAOISM AND TAU MICE

Taoism is China's homegrown religion, believed to have originated from the teaching of Lao Zi (circa sixth to seventh century BC). Despite having deities and scriptures, to many Chinese, Taoism is less a religion than a school of thought, a way of life.

The central principle of Taoism is *Tao*, which means "the Way." A nebulous concept, *Tao* is said to be the basis of all things, the flow of nature, and the essence of the universe. It is invisible and indescribable and can only be accepted and experienced. It sounds like the Western concept of the Force, made popular by the *Star Wars* film series, but it is different. *Tao* is said to simply exist and be, without effort or intention. To live in harmony with *Tao* is to engage in *nonaction*. A paradoxical concept, nonaction does not mean not acting; it means acting effortlessly, liberated from calculated intentions and material attachments.

If we allow it, we can find provoking parallels between Taoism's nonaction and the gradual loss of purposeful intent and awareness typical of Alzheimer's. As some caretakers believe, when patients decline, they do not lose the ability to be content; they just lose the obsession to seek contentment.[1] If so, when our conscious mind drifts, we are not so much losing something that we are entitled to, as returning to the way we were when we first came into this world. Piece by piece, we give back the cognitive functions that we have gained since being newborn.

But what does this have to do with the cause of Alzheimer's? Not much, except as a moniker. In fact, the moniker is not even spelled Taoism; it is *Tauism*.

In 1906, when Alois Alzheimer looked into the microscope at Auguste Deter's brain, he saw numerous plaques scattered between neurons, which we know today are beta-amyloid deposits. He also saw fibrils, or small fibers, growing inside the neurons. In some cases, only a few fibrils peeped out from a neuron; in other cases, they had completely overtaken a neuron, turning it into a hazy tangle. These tangles have since been observed in most Alzheimer's brains. Together with beta-amyloid plaques, they are the two pathological hallmarks of Alzheimer's.

Like beta-amyloid plaques, these tangles resisted study. Difficult to isolate and dissolve, they took seventy years after Alois Alzheimer's observation to figure out.[2] Interestingly, just like beta-amyloid plaques, tangles are abnormal protein buildups. In this case, the protein is called *tau*, whose name came from the nineteenth letter in the Greek alphabet, τ.

Tauism, then, is the hypothesis that tau, rather than beta-amyloid, is at the bottom of Alzheimer's. The *Tauism* moniker was coined, half-jokingly, to rival *Baptism*, where *Bap* is an acronym for beta-amyloid protein. If the beta-amyloid hypothesis should have a sacred moniker, it is only fair that its biggest rival, the tau hypothesis, should as well, and what could be more contrary to Western baptism than Eastern Taoism?

This rivalry is fueled by the string of failed beta-amyloid trials discussed in earlier chapters. According to Tauists, these failures of Baptists are to be expected: their drugs are designed to stop Alzheimer's by reducing, removing, or otherwise getting rid of beta-amyloid, but that's futile because beta-amyloid does not cause the disease. It is the tau protein that does this.

Unlike beta-amyloid, whose functions we aren't very sure of, tau's function is well established: it binds to and stabilizes microtubules in the brain.[3]

What is a microtubule? It is exactly what its name says: a tiny (micro), tubelike structure. Rigid and hollow, microtubules form the skeletal structure that holds up a cell and gives it shape. In neurons, microtubules support axons, the long, slender antenna on a neuron that sends signals to other neurons.

If we imagine axons as paved highways where brain signals travel, then microtubules are the underlying metal reinforcements that support those highways. If tau malfunctions, microtubules will be weakened, axons will collapse, and brain signals will stall, causing cognitive decline. This, the tau hypothesis posits, is the real cause of Alzheimer's.

What makes tau malfunction? The common belief is that tau acts up when too much of it is phosphorylated. A protein is said to be *phosphorylated* when it takes on a phosphate group, which is one phosphorus atom plus four oxygen atoms.

In the normal brain, phosphorylated and unphosphorylated tau is kept in balance. This is crucial because only the latter can bind to and stabilize microtubules. Once phosphorylated, tau loses that ability. *Hyperphosphorylation* thus weakens microtubules.[4] Worse still, phosphorylated tau can direct unphosphorylated tau away from microtubules so the latter can't perform its job either. Unbound, tau begins to clump together and form tangles that destroy neurons.[5] Indeed, as studies show, the Alzheimer's brain features an excess of phosphorylated tau, which is the main ingredient of tangles.[6]

Tau doesn't just act up in Alzheimer's. It happens in various other neurodegenerative conditions. Best known among them is probably chronic traumatic encephalopathy (CTE), featured in the 2015 film *Concussion*, starring Will Smith, about the effects of concussion on professional football players. The condition also frequently afflicts boxers and other athletes who play contact sports, who suffer repeated head injuries.

In many ways, then, tau is like beta-amyloid: both are protein in origin, and both build up into obnoxious clumps. If anything, tau's damage seems more obvious in the Alzheimer's brain: it literally breaks down neurons, whereas beta-amyloid plaques just sit between neurons. Moreover, tau tangles do not spread haphazardly in the Alzheimer's brain. They stick to a consistent itinerary that, remarkably, corresponds with patients' deterioration.[7] The first stops of tau tangles include the navigation and memory hub of the brain (the entorhinal cortex and the hippocampus). Damage in these areas coincides with the early symptoms of Alzheimer's: disorientation and memory loss. Tangles next attack the brain's emotion center (the amygdala and other limbic structures), which parallels patients' personality and behavioral changes. Finally, tangles target regions of the brain responsible for higher-order functions (the isocortical areas), and with that, the late-stage Alzheimer's symptoms manifest: severe impairment of language, reasoning, and overall cognition. This perfect correlation seems bona fide evidence that tau tangles directly cause Alzheimer's.

Given these points, why should Baptism and not Tauism be the research paradigm? What does beta-amyloid have that tau doesn't?

For a while, the answer was genes. As mentioned in previous chapters, from 1991 to 1995, several genetic mutations were found involving beta-amyloid that cause early-onset Alzheimer's. There is the amyloid precursor protein (APP) gene, which codes for the full-length protein from which beta-amyloid is cut. There are the PSEN1 and PSEN2 genes, which code for enzymes that cut beta-amyloid out of the precursor protein.

Tauists had no genetic mutations to speak of and were chaffed by Baptists for suffering from "mutation envy."[8] Then, in 1998, things looked up. That year, tau mutations were discovered in an inherited neurological disease called *frontotemporal dementia with Parkinsonism*.[9] In this condition, neurons are damaged in two areas of the brain: the frontal and temporal lobes. As a result, patients suffer from cognitive impairment and Parkinson's disease–like movement disorders.

Causing this disease are mutations in the MAPT gene (aka the *microtubule-associated protein tau gene*). This is the gene that directs the production of the tau protein. When it mutates, it begets a version of tau that is prone to phosphorylate, to build up, or to detach from microtubules.[10]

This discovery came as a welcome boost to Tauism, providing evidence that tau can cause cognitive impairment all on its own, without any help from beta-amyloid. In other words, abnormal beta-amyloid does not have to be the first, initiating step toward Alzheimer's, as Baptists believe. However, there is an important caveat: in Alzheimer's, the MAPT gene does *not* mutate. Although tau clearly goes awry and forms tangles in the brains of Alzheimer's patients, their MAPT gene is just fine.

To try and explain this conundrum, Tauists need their own transgenic mice.

Gregor Mendel (1822–1884), the posthumously recognized father of modern genetics, was a little-known Austrian monk during his lifetime. In his monastery garden, he planted peas, examined their traits, and established Mendel's Law of Inheritance. Through his work with peas, Mendel demonstrated that genetic traits are separately and randomly inherited, and their dominant or recessive state determines the appearance of offspring.

What people may not know is that Mendel's initial research subject was not peas. It was mice. Mendel bred them in his monastic room, observing how their coat colors were inherited. But his superior put a stop to that, ruling it indecent to have smelly creatures mating in a monastic cell.[11] Mendel compliantly switched to peas, and in 1866, he published his findings in the botanic journal *Experiments on Plant Hybridization*. Because the journal was obscure, his work remained little known until it was discovered at the start of the twentieth century. Shortly afterward, the French biologist Lucien Cuénot confirmed that Mendel's Law applied equally well to the inheritance of coat colors in mice.[12]

But neither scientist, nor any trained scientists for that matter, can be credited for introducing mice into the modern science lab. Mouse fanciers did that. Mouse fanciers are people who keep and breed pet mice for their unique coat colors and behavioral traits. The hobby originated in Asia, going back to at least seventeenth-century Japan, and later spread to Britain and the United States.[13] Long before scientists showed any interest in mice, fanciers had figured out how to breed them in captivity and control their genetic profiles, which made the animals convenient subjects for research experiments.[14] Most of the lab mice used today originated from fancy mice bred in the late nineteenth to early twentieth centuries.[15]

One American breeder, Abbie Lathrop (1868–1918), from Granby, Massachusetts, was especially noteworthy.[16] A certified teacher, Lathrop quit teaching because of health problems. After leaving the classroom, she first attempted a poultry business, which failed. She then turned to breeding fancy mice and rats, initially selling them to pet owners, and then in large quantities to researchers. Her small farmhouse was known to keep more than 11,000 mice at a time.

Nicknamed the "mouse woman of Granby," Lathrop was portrayed in the media as an eccentric, a woman who, unlike most, did not fear mice and indeed made a living by breeding them.[17] What most didn't know, or appreciate, was that Lathrop was a self-made scientist. She collaborated with the renowned pathologist Leo Loeb to perform cancer experiments on mice, and the two coauthored multiple research articles.

Apart from breeders like Lathrop, mice's inherent qualities make them ideal lab animals. They are not picky eaters and can survive on a cheap diet. They breed year round and give birth to as many as a dozen in a litter. They tolerate inbreeding quite well, allowing precise selection of genetic traits. They grow quickly, reaching sexual maturity at one month and entering old age at one year. This last feature makes them especially useful for Alzheimer's research. If we had to wait years (or decades) for an animal to grow old, that alone would grind Alzheimer's research to a halt.

It also turns out that human and mouse genomes are surprisingly similar. Despite how highly we think of ourselves as an intelligent species, we are not very different from rodents genetically. We have essentially the same set of genes, and the individual genes are 60 percent to 99 percent identical.[18] This may be ego-busting, but it is also good news for our health. We can manipulate mouse genomes to create human disease conditions and then use these transgenic mice to test for cures. This is the single most important use of mice in Alzheimer's research.

The first transgenic Alzheimer's mouse, the PDAPP mouse, was introduced in chapter 7. Since then, a string of others had followed, incorporating various beta-amyloid mutations. With the discovery of tau mutations, transgenic tau mice were also born. As much hope as Tauists put on their *own* mice, the tau mice actually complicated rather than clarified the tenets of Tausim.

Initially, Tauists thought that tau was essential for supporting microtubules and the brain information highway, but tau knockout mice—mice whose tau has been genetically removed—not only survive, but appear relatively normal. Apparently, if tau isn't there to support microtubules, other proteins will step up and do the job.[19] Therefore, damaged tau isn't a big deal in and of itself.

Of course, this doesn't exonerate tau malfunction. A case in point is hyperphosphorylation. Phosphorylated tau can attract proteins (tau or otherwise) away from microtubules so *no one* can support the information highway.[20] In other words, tau may cause disease not because it lost its normal function, but because it gained an abnormal function.

Researchers studying mice that have a mutated tau engineered to phosphorylate would agree.[21] At four months old, these mice develop hyperphosphorylation, notably in the amygdala, an area of the brain responsible for detecting fear. Consequently, the animals start to behave oddly, as if they no longer experience fear and anxiety. Normally, mice feel safe in dark, enclosed areas and fear heights and open spaces. But these transgenic tau mice appeared to lose this natural tendency. When placed in an elevated structure with both open and enclosed ledges—like a tall building with both open and enclosed balconies—they would choose to hang out on the open ledges.

Then again, not everyone blames phosphorylation. Mice with a mutated tau that is engineered to build up act as if what they got is more devastating. Normally, mice, like any animals, want to avoid pain. But these transgenic tau mice act like they don't care in the shock chamber test. This test uses a specially designed chamber with two compartments. One compartment is safe. The other has an electric current running through its floor. If mice step into this compartment, they receive a shock to their feet. Once shocked, mice should learn to stay out of this compartment—unless they can't remember the painful experience. Sure enough, mice that were prone to have tau buildup couldn't remember and persisted in going back to the rigged compartment.[22]

Still other studies with tau mice remind us that tau buildup is not one uniform thing. As with beta-amyloid buildup (discussed in chapter 6), first we have tau monomers, which clump together into oligomers, which continue to grow and finally form tangles. Ever since Alois Alzheimer's vivid description of tangles choking out neurons in Auguste Deter's brain, we took "tangles, the neuron killer" for granted, but tau mice suggest something different.

In one study, mice that carried a mutated tau duly developed tangles, lost neurons, and suffered memory loss.[23] When researchers turned off the mutation, the neuron counts stabilized and memory recovered. However, by this time, abundant tangles have already formed in the mice's brains and continued to form—apparently causing no harm. Given this finding, some Tauists believe that what is harmful about tau buildups is not the end stage (the tangles), but the intermediate stage (the oligomers), which are toxic.[24] Tangles, rather than a killer, may be the brain's protective mechanism: by further growing oligomers, tangles try to reduce toxicity and protect neurons—but they failed. As neurons wither away, tangles are left behind to mark the failed rescue missions.

Here, we see striking parallels between the two rival hypotheses. While beta-amyloid plaques were once thought to be a killer, they are now considered tombstones that mark the killing of toxic beta-amyloid oligomers, or even "brain pearls" that protect neurons from oligomers. It is funny how much we can agree when we disagree.

Transgenic Alzheimer's mice, whether beta-amyloid or tau mice, are not perfect research models. Although mice and humans have highly similar genes, the way those genes are regulated to fulfill biochemical functions differs. More to the point, normal mice do not develop Alzheimer's: when they age, they don't lose memory, personality, or self-awareness. Even Alzheimer's genetic mutations don't hurt them as much as they do us.

In humans, a single APP, PSEN1, or PESEN2 mutation causes early-onset Alzheimer's. But when we put one such deadly mutation into mice, the creatures are relatively unfazed. Yes, they develop beta-amyloid plaques and cognitive impairment, but they don't show tau tangles, or the kind of massive neuron loss seen in the human brain.[25] Mice with a mutated tau are the same—they develop tau tangles, neuron loss, and cognitive impairment, but they never grow the beta-amyloid plaques that beset the human brain.[26]

In order for mice to have the full package of Alzheimer's pathologies, we have to put in them not one, but two or more mutations. Double transgenic mice, for example, carry both a beta-amyloid mutation and a tau mutation. It is questionable, though, whether these mice still have the human equivalent of Alzheimer's, because multiple mutations do not naturally coexist in an Alzheimer's patient.

With this backdrop, what mice tell us about Tauism must be taken with a grain of salt. What they do confirm is that Tauism, like Baptism, has its own share of contradictions and unanswered questions. But these won't stop pharma from trying their hand at tau drugs. Who knows—when beta-amyloid drugs continue to fail, fixing the other hallmark of the Alzheimer's brain may just take care of the whole problem.

12

APPLES, OYSTERS, AND UNDERDOGS

On July 28, 2016, Chris Smyth, science editor at the London *Times*, published an inspiring article headlined "Scientists create the first drug to halt Alzheimer's." In it, Smyth announced that a "drug has stopped brain deterioration in Alzheimer's patients for the first time," and mental decline "was halted for 18 months in some patients."[1] Soon, Smyth wrote, this twice-a-day tablet could "become the first medicine given to Alzheimer's patients to keep the disease at bay for as long as possible."

Readers of the piece were overjoyed and left online comments of relief, praise, and delight. Judging by these comments, many readers were older adults living in fear of Alzheimer's or were families and friends of Alzheimer's patients. When a few commenters dared to express doubt, others bitterly chastised them for dashing their hopes.

Meanwhile, Alzheimer's researchers were surprised. How could something this fantastic reach the daily newspaper before they heard about it at conferences or through publications? Usually, that's where a breakthrough is announced, to a peer audience, in order to stake a claim to the discovery. But as researchers realized what drug and study the reporter was referring to, their surprise turned to disbelief.

The story began properly a day earlier in Toronto, where the 2016 Alzheimer's Association International Conference was in session. Scheduled to present at 4:30 that afternoon was Serge Gauthier, chair of the scientific advisory board at TauRx Pharmaceuticals. This Scotland/Singapore-based pharmaceutical company has an obvious business focus: tau drugs.

Gauthier's presentation was eagerly awaited. He was about to reveal phase 3 trial results of LMTM, TauRx's prize Alzheimer's drug. Late-phase trials are always a big deal because of how close they are to market entry. In this case, the stakes were even higher because LMTM was the first tau Alzheimer's drug to make it to phase 3. Amid a sea of beta-amyloid drugs, it is the crown jewel for Tauists.

News reporters were there as well, ready to broadcast the trial results to the world. But they didn't have to wait all day. Before Gauthier's formal presentation, he held a press briefing in the morning. That was where the *Times* reporter Chris Smyth obtained the material for his article proclaiming victory. Smyth must not have stuck around for the formal presentation. Had he stayed, he might have dialed the victory back—way back.

What is LMTM? It is a tau aggregation inhibitor. As the name suggests, it inhibits and prevents tau from clumping together. In animal studies, LMTM was found to reduce tau buildup by 35 percent.[2] Exactly how LMTM does this is murky. Some believe that the drug modifies the tau protein structure in a way that keeps individual tau monomers separate and nonsticking.[3] Others think quite differently, believing that the drug stimulates the cells' natural housekeeping mechanism to degrade tau.[4] TauRx Pharmaceuticals wasn't interested in doing basic research to get to the bottom of the drug's mechanism. They were in a hurry to jump into human trials.

In fact, they were in a hurry to jump *through* human trials. No published data can be found on any LMTM phase 1 trials. A record of one phase 2 trial does exist, but it enrolled only nine people and was terminated early "for administrative reasons."[5] The company also published the results of another phase 2 trial, but that trial wasn't about LMTM as it currently existed. It tested an earlier version of LMTM, which contains the same main ingredient but uses a different formula. According to that trial, the drug was generally safe, and at a dose of 138 mg, it was able to slow Alzheimer's patients' cognitive decline.[6]

Despite having so little to show for its efforts, TauRx Pharmaceuticals marched into phase 3. Two separate phase 3 trials were undertaken. The first enrolled about 900 mild-to-moderate Alzheimer's patients from sixteen countries, who were randomly assigned to one of three treatments: placebo, 75 mg of LMTM, or 125 mg of LMTM.[7] After fifteen months of dosing, the results came in, which were the focus of Gauthier's presentation and Smyth's *Times* article.

During Gauthier's formal presentation, the first thing that he announced was that LMTM didn't work: whether the 75-mg or the 125-mg dose, the drug didn't slow the patients' cognitive decline.[8] This was disappointing but hardly unusual—other Alzheimer's drugs with much stronger earlier-phase data have failed at phase 3.

But Gauthier wasn't finished. He went on to say that the drug *did* work on a subset of the participants. Dicing participants into subgroups in order to achieve positive results is nothing new, as we have seen before. Still, the *way* that TauRx Pharmaceuticals arrived at their positive findings was something else. As Gauthier explained, most of the trial participants took standard palliative drugs on their own during the trial. These drugs, as mentioned in chapter 3, do not treat Alzheimer's but offer minor symptom relief. When these participants were randomized to receive LMTM in the trial, they would, in a sense, be taking LMTM as a so-called add-on drug. A small percentage (about 15 percent) of the trial participants did not take any palliative drugs. These participants, when randomized to LMTM, would be taking LMTM as a so-called mono drug.

As Gauthier then stated, if we compare the LMTM mono-drug takers with the placebo group, the former benefited, showing less cognitive decline and brain shrinkage. These benefits held true for both LMTM doses tested. It disappeared only when participants took LMTM as an add-on drug. These findings, Gauthier concluded, proved that LMTM does work—it just needs to be taken on its own. That's the conclusion that the *Times* reporter ran with.

At first glance, Gauthier's story seems plausible. But, if we slow down and really think about the way that he compared the various subgroups, the story is off—way off. Indeed, as one researcher exclaimed, it's "like comparing apples to oysters."[9]

The key is the placebo group. Gauthier treated it as a single group to compare to the mono-drug group and the add-on drug group. But in reality, the placebo group was itself a mix: it included people who took palliative drugs and those who didn't. Thus, to truly prove the effect of LMTM as a mono-drug, we need to compare people who took *only* LMTM and *only* placebo. This would be an "apples-to-apples" comparison.

Insisting on an apples-to-apples comparison is not just a formality. There are practical reasons for doing so. Palliative drugs are standard prescriptions in Alzheimer's cases. If patients do not take them, it is often because they have stable symptoms or slow declines.[10] In other words, the LMTM mono-drug participants might have declined less on their own, with no help from LMTM. Without an apples-to-apples comparison, we can't rule out this possibility.

Aside from the apples-to-oysters comparison, the so-called mono-drug effect was probably skewed by another factor. At the conference, Gauthier was asked if there was anything unique about his mono-drug participants. Gauthier was evasive in his response.[11] At one point, he said that there was nothing unique about them. At another point, he admitted that many of them lived in Eastern Europe and Malaysia. In a later publication, Gauthier revealed that precisely 56 percent of the participants came from Russia, Poland, Croatia, and Malaysia.[12]

This geographical pattern is significant. People from these regions have limited access to dementia care. As a result, when they are given the opportunity to enter a clinical trial, they do so with high expectations for the trial drug. Reacting to that optimism, their bodies and brains actually performed better—the classic placebo effect. Similar cases have happened before, in which a trial drug would appear to work in Russia but fail in the United States and elsewhere.[13]

So there you have it—the first-ever phase 3 trial of an anti-tau drug. There was more drama than science. And like many good dramas, a sequel awaits, for TauRx Pharmaceuticals still has that second phase 3 trial.

This second trial has roughly the same design as the first. It enrolled 800 mild Alzheimer's patients, tested a different LMTM dose (100 mg), and lasted eighteen months.[14] The trial results were to be presented in December 2016, in the balmy coastal city of San Diego, at the Ninth Clinical Trials on Alzheimer's Disease (CTAD) conference.[15]

Highly anticipated, the presentation was given the first spot in the oral presentation session—where it sure succeeded at getting people talking. With a sense of déjà vu, attendants sat and heard the company representative admit that, compared with placebo, LMTM failed to slow cognitive decline. But, *once again*, it succeeded as a mono-drug!

You would think that in this second go-round, TauRx had refined its statistical analysis and done a true apples-to-apples comparison. Nope: it was the same apples-to-oysters nonsense, comparing LMTM mono-drug takers to the *entire* placebo group. If the company had made an honest mistake before, they had no excuse this time. Attendees at the presentation "were horrified at how the data had been sliced."[16] Many said emphatically that the so-called mono-drug effect "was at best based on a wrong comparison and, at worst, bogus."[17]

But TauRx wasn't done. A bigger surprise was to come: participants who received placebo and did not take palliative drugs also had less cognitive decline. You heard that right—another placebo had turned therapeutic, after Austrian company Affiris's aluminum salt (chapter 9).

What was in *this* placebo? A tiny amount (4 mg) of the same LMTM drug. This is, of course, highly unusual. Ordinarily, a placebo contains no therapeutic ingredients, let alone the same drug being tested. But that arrangement won't work for LMTM because, when ingested, LMTM turns a person's urine blue. By simply visiting the bathroom, participants would know if they had received LMTM or placebo, which would screw up blinding and skew the trial results.

To mask the drug, TauRx Pharmaceuticals came up with the clever idea of adding 4 mg of LMTM to its placebo. At this tiny dose, LMTM is capable of turning urine blue but is not supposed to have any therapeutic effects.[18]

On that point alone, the company admitted that it made a mistake. Apparently, 4 mg is all it takes—4 mg of LMTM was just as potent as 100 mg and slowed cognitive decline just as well.[19] Normally, if a drug works, a higher dose would get a stronger response, so the finding that 4 mg and 100 mg of LMTM were equally effective is itself a red flag.

TauRx Pharmaceuticals was not bothered by these gaping holes. Because 4 mg of LMTM worked equally well and, being a tiny dose, is safer, the company decided that they would ditch the higher doses and pursue a further trial with 4 mg of LMTM. The trial, dubbed LUCIDITY, started in 2018 and is ongoing as of this writing.

The decision makes no sense. Nothing about the LMTM trials did—except that a company was grasping at straws to stay afloat. To pursue a placebo is something to hook investors and bargain a potential buyout with, whereas to admit failure is the end of the road, as underdogs of the pharmaceutical world well know.

Noscira was founded in 2000. A small, private company based in Madrid, its four executives and five board members are either medical doctors or academics holding PhD degrees. It is the typical biotech startup, which relies on intellectual expertise for financial survival. Noscira's focus is on central nervous system disorders, including Alzheimer's. Its target is tau—more precisely, tau phosphorylation.

Phosphorylation, as mentioned earlier, is believed by some Tauists to be the real culprit behind Alzheimer's. Excessive phosphorylation makes tau (and other proteins) incapable of supporting microtubules. As a result, axon highways collapse and brain signals stall. Meanwhile, phosphorylated tau builds up and damages neurons. Given this scenario, if we keep tau from phosphorylating, we can stop the relentless march of Alzheimer's. This is the premise behind Noscira's drug, Tideglusib, which is a phosphorylation inhibitor and a major type of Alzheimer's tau drug.

Phosphorylation being a complex process, many enzymes are involved. Some directly attach phosphate to tau; others modify tau's structure to facilitate phosphate attachment. Disrupting any of these enzymes might stop the process and arrest Alzheimer's.

Naturally, no one knows which enzyme or enzymes should be targeted, so companies are cherry-picking whatever they think could be most promising. Noscira chose an enzyme called GSK-3β, which is a popular choice because it has a pretty straightforward function: it helps attach phosphate to tau. As we age, GSK-3β becomes more active, which explains the fact that age is the number one risk factor for late-onset Alzheimer's.[20] Also, in cell and animal studies, inhibiting GSK-3β successfully reversed tau phosphorylation and cognitive impairment.[21]

With this evidence, Noscira went into human trials. In an initial, six-week trial with thirty people, Tideglusib was proved safe and well tolerated.[22] In phase 2, Noscira expanded and lengthened the trial to 300 patients and six months, hoping to demonstrate the drug's cognitive effects.[23] That didn't happen. Patients receiving Tideglusib and placebo scored similarly in cognitive tests. In fact, their brains showed similar amounts of phosphorylated tau, which cast doubt on the drug's ability to inhibit phosphorylation.

Two months after that trial failed, Noscira went into liquidation.

Allon Therapeutics was founded in 2001. Except for its different geographic location (Vancouver), this company might well be another Noscira. A small biotech, Allon focused its business on neurological disorders and drew investments from hospitals, universities, and people with titles of "Dr." or "Professor."

In tackling Alzheimer's disease, Allon also pursued tau drugs. The company assumed that when tau goes awry, the main consequence is unsupported microtubules, which in turn collapse axon highways, disrupt brain signals, and usher in Alzheimer's. Therefore, rather than going after tau itself, Allon sought to simply protect microtubules with its drug, davunetide, which is a microtubule stabilizer and another major type of tau drug.

In cell and animal studies, davunetide maintained microtubule integrity and prevented axon failure.[24] In human testing, it also got off to a good start: phase 1 trials showed the drug to be safe, and in one phase 2 trial, participants with mild cognitive impairment (mild enough to not qualify for an Alzheimer's diagnosis) saw improved memory after three months of dosing.[25] Pleased by these results, in 2008, Allon made plans to conduct another phase 2 trial, this time recruiting Alzheimer's patients.

Meanwhile, Allon also tested the drug in a phase 2/3 trial on progressive supranuclear palsy, a rare brain disorder that affects movement, behavior, and cognition. This disorder is genetically linked to tau mutation and, like Alzheimer's, is marked by abnormal tau buildups in the brain. Now that davunetide had proved itself promising in treating Alzheimer's, Allon hoped that it would cross over into treating progressive supranuclear palsy as well.

They hoped wrong. In December 2012, the progressive supranuclear palsy trial failed. Patients receiving davunetide experienced no benefits in either cognition, daily activities, or motor skills.[26] The drug, it appeared, had no effect.

In a news release, Allon shared the disappointing results and announced that it "will not allocate any additional capital to research and development activities for davunetide." In fact, it must "take immediate action to reduce its ongoing operating expenses, including a reduction in staff."[27]

Six months later, Allon filed for bankruptcy.

AXON Neuroscience was founded in 1999. Based in Bratislava, the capital of Slovakia, AXON is interested in all diseases related to tau, especially Alzheimer's. Its Alzheimer's drug represents the last major class of tau drugs: vaccines.

These vaccines are not unlike the beta-amyloid vaccines that have been trying (and failing, as discussed in chapters 9 and 10) to cure Alzheimer's. In fact, the underlying principle is exactly the same: use antibodies to attack and destroy defective tau in the hope that this will stop downstream brain damage and Alzheimer's symptoms.

In practice, there is more than one "defective tau" to vaccinate against, thanks to uncertainties in the tau hypothesis. The favored choices include the two suspects we have encountered: phosphorylated tau and tau buildup. And then there is a third choice: tau fragments cut by enzymes from full-length tau that are more likely to phosphorylate, build up, or otherwise malfunction.[28] This last one is AXON's choice.

In fact, the company believes it has discovered *the* tau fragment that causes neuronal damage in Alzheimer's. When rats express this fragment, microtubules in their brains screw up, and Alzheimer's-like tau tangles form.[29] AXON named its vaccine to target this fragment AADvac1.

In rat testing, AADvac1 proved safe.[30] It targeted the abnormal tau fragment and left healthy tau alone. After vaccination, tau buildup and phosphorylation decreased, as did the number of tau tangles in the animals' brains.

In 2013, AADvac1 moved into phase 1 human trials.[31] The vaccine was well tolerated after six months of dosing, and participants raised adequate antibodies. Encouraged, AXON promptly issued a news release, calling its product a "revolutionary vaccine offering new hope for millions of Alzheimer's disease patients."[32] Although claims of being revolutionary are a bit difficult to take seriously when previous vaccine trials have all failed, this *was* the first tau vaccine to enter human trials.

In 2016, AADvac1 started a phase 2 trial dubbed ADAMANT. About 200 patients were recruited to test if the drug remained safe after two years of dosing and offered any cognitive benefits. The trial results came back in September 2019. According to a company news release, AADvac1 continued to enjoy an "exceptional safety profile," but its assertions on cognitive effect read fuzzy.[33] Reportedly, the drug had "positive *signals* for cognitive endpoints . . . among *younger populations*,"[34] which is usually a euphemism for "the drug showed no statistically significant effect among *all* trial participants." Still, Axon called the results "an important milestone for Axon, and for the entire population of the world that suffers from this devastating disease." The company is currently seeking a partner with the necessary means and resources to help it carry out the next phase of clinical trials.

⁓

Despite going under, the small startups Noscira and Allon took tau drugs further than most companies did. Being underdogs, they went all in—they couldn't afford not to. Meanwhile, big pharma has the deep pockets to diversify and study dozens of drugs at once, to test the waters and back out at the first sign of trouble.

Just like Noscira, Merck, in collaboration with Canadian company Alectos Therapeutics, worked on a phosphorylation inhibitor. The inhibitor targets an enzyme that prepares tau for phosphate attachment. According to a press release in 2014, a phase 1 trial was imminent.[35] But no update or publication ever came forward, and the drug is not listed among Merck's pipeline drugs.

Like Allon, Bristol-Myers Squibb tried its hand at microtubule stabilizers. In mice tests, its stabilizer strengthened microtubules, saved neurons, and improved memory.[36] A phase 1 human trial was registered on Clinicaltrials.gov and was recorded as complete in 2013.[37] But no results were ever announced or published—a bad sign.

The same happened to Hoffmann-La Roche's tau vaccine, which reduced tau abnormality in mice, entered a phase 1 human trial in 2015, but has not been heard from again.[38]

With no trial data to consult, we don't know what happened to these vanished drugs: Were they unsafe? Did they fail at performing basic functions? Were they so miserably inadequate that companies had to bury the trial results? Currently, several other tau vaccines are in early-phase trials.[39] If history is any indication, though, the chances are slim that they will come forward and save the day any time soon.

To move forward with Alzheimer's research and drug development, maybe we should look beyond the two obvious brain pathologies, beta-amyloid and tau. Maybe both are tombstones rather than killers. Given our growing understanding of the disease, are there any other clues to solving the mystery of Alzheimer's?

13

TYPE 3 DIABETES

Saturday night. An urban landscape: skyline, shops, restaurants, neon lights, the hustle and bustle of traffic. Young man and women are lining outside popular clubs: the *Heart Throb*, the *Brain Bar*, even the slightly rundown *Toe Jam*.[1] Inside, lights sparkle, music blares, and the crowds are having fun.

The bouncers run a tight ship: couples only; no singles. Young gals without a date are left grumbling at the door, waiting for their luck to change. As more single gals arrive, the lines keep growing, overflowing into the streets and blocking traffic. Meanwhile, people inside the bars are starting to leave, and the places are getting more deserted by the minute. The lights start to flicker, and the music is dying out. Still, single gals are not allowed in and are left in the street.

This odd nightlife scene is quite an accurate portrayal of what happens in diabetes, with the clubs as our organs and tissues, the single gals as blood sugar, and their dates, insulin.

Blood sugar, also known as *blood glucose*, comes from the food that we eat and is absorbed through the gastrointestinal (GI) tract into the bloodstream. It provides the energy that our organs and tissues run on. Insulin, on the other hand, is a hormone produced by the pancreas. It doesn't create energy itself; its job is to escort glucose into body cells to be used as energy. Insulin does this by communicating with the cells and signaling them to open a channel to let the glucose in.

When there is a lack of insulin, glucose is stuck and accumulates in the bloodstream, while our organs and tissues run on empty. Worse, as blood glucose rises, it hardens blood vessels, blocks blood flow, and interrupts the transportation of oxygen and nutrients. Over time, this causes serious damage to the body, including vision loss, kidney failure, stroke, and heart attack. From *Heart Throb* to *Toe Jam*, everyone is affected.

In the conventional wisdom, chronic diabetes comes in two forms: type 1 and type 2.[2] Type 1 diabetes is less common. It usually develops in children and young adults whose immune system, for unknown reasons, attacks their pancreas and shuts down insulin production. As a result, very little or no insulin circulates in the body to escort glucose.

Type 2 diabetes is more common. Mostly affecting adults, it accounts for about 90 percent of all diabetes cases.[3] With type 2 diabetes, the pancreas and insulin production start off fine, but the body grows insensitive to insulin—becoming *insulin resistant*, as doctors call it. Although plenty of insulin is produced, the body just doesn't seem to see it and can't use it to escort glucose. Frustrated, the pancreas kicks into higher gear to produce even more insulin. Elevated insulin forces the body to tune in to it, and all's fine for a while. But eventually, the pancreas burns out and can no longer churn out enough insulin to escort glucose.

Multiple factors cause type 2 diabetes. Genetically, about forty genes have been found to increase the disease risk.[4] Unhealthy lifestyle choices are also risk factors. A high-calorie diet and lack of physical activity increase body fat, and excessive body fat leads to insulin resistance.[5]

In the conventional wisdom, diabetes, whether type 1 or type 2, is not a brain disease. Yes, it can afflict the *Brain Bar*, but it also afflicts many other organs and tissues. In addition, diabetes primarily affects brain's hardware. To wit, high levels of blood glucose can damage and clot blood vessels in the brain, cutting off blood supply and causing a stroke. As for the software, the functioning of our mind, diabetes should have no direct effect. At least, that's what we thought.

Then came 2005, when a group of researchers from Brown University and Rhode Island Hospital announced that they had found a third type of diabetes, which behaves differently.[6] According to these researchers, what we have known for centuries as Alzheimer's really is a unique type of diabetes: type 3 diabetes. Type 3 diabetes has the same features as types 1 and 2—energy crises in body cells, high blood glucose, and insulin resistance—but its wrath is felt primarily by the brain's software.

To call this idea intriguing is an understatement. We already have a pretty good understanding of diabetes, and we have effective medicine and lifestyle adjustments to cope with the disease. If Alzheimer's really is a form of diabetes, then our current methods of treating and managing diabetes may work just as well for Alzheimer's.

But what evidence do we have for this so-called type 3 diabetes hypothesis?

One apparent similarity between Alzheimer's and diabetes lies in proteins. In Alzheimer's, we have beta-amyloid; in diabetes, we have amylin.

Amylin comes from the pancreas. When the pancreas produces insulin, amylin is generated as a by-product. As a sidekick of insulin, amylin prevents glucose spikes by slowing down the action of the GI tract and creating a sense of "fullness" so that we stop eating.

In type 2 diabetes, because the body becomes insensitive to insulin, the pancreas works overtime to produce more insulin. As a result, more amylin is also produced. Excessive amylin then behaves quite similarly to beta-amyloid: it clumps together, first into toxic oligomers and then plaques, damaging pancreatic cells.[7] Autopsies of diabetic patients reveal amylin deposits in the pancreas, just like beta-amyloid deposits lurking in the Alzheimer's brain. Even more intriguingly, amylin and beta-amyloid are about 90 percent similar in protein structure.[8]

In addition to this beta-amyloid–amylin connection, there is a beta-amyloid–insulin connection. In the human body, insulin is degraded by an enzyme with the quite appropriate name of *insulin-degrading enzyme*. Coincidentally (or not), this enzyme also degrades beta-amyloid. If our bodies are resistant to insulin and force the pancreas to overproduce it, excess insulin may tie up the degrading enzyme, leaving little that can degrade beta-amyloid, which then accumulates into Alzheimer's brain plaques.[9]

This scenario has been re-created in mice. When mice were fed sugary water to induce insulin resistance, their blood insulin rose, as did beta-amyloid buildups in their brains.[10] That said, human data paint a different picture: in autopsies, diabetic patients' brains did *not* show increased beta-amyloid plaques, as the hypothesis would predict.[11]

To make a more convincing case, the hypothesis needs to determine whether diabetes and Alzheimer's actually cross paths. That is, are diabetic patients more likely to develop Alzheimer's, and vice versa?

To find that out, we travel to Rochester, Minnesota.

Sitting at a crossroads near the Zumbro River, Rochester was once a stagecoach stop connecting St. Paul, Minnesota, and Dubuque, Iowa. The earliest settlers came in 1854.[12] Migrating from Rochester, New York, they named the new home after their old hometown. Drawn by the region's cheap and fertile land, more settlers soon followed. By the time of the Civil War a decade later, Rochester's population had grown to over 1,400.[13]

The war also brought to town Dr. William Mayo, who served as the examining surgeon for the federal draftees. After the war ended, Mayo decided to stay and open a medical practice. By then, the railway had arrived, making Rochester a regional hub, tripling its population, and giving Dr. Mayo plenty of patients. Mayo and his wife raised three daughters and two sons, both sons taking up medicine as their father did. The Mayos led a productive, if somewhat uneventful, life—until the night of August 21, 1883.

Around 7 p.m., a violent tornado with winds over 200 miles per hour touched down in Rochester. It gushed through northern Rochester, destroying over 135 homes and damaging another 200.[14] At least thirty-seven people were killed, and hundreds were injured.[15] Without a local hospital, the people of Rochester turned lodge rooms and a dance hall into temporary emergency rooms. In these makeshift rooms, Dr. Mayo and his sons cared for the wounded, and the local Sisters of St. Francis congregation assisted. People did the best they could with what they had.

In the aftermath of the tornado, the Sisters of St. Francis were convinced that Rochester needed its own medical facility. They approached Dr. Mayo, promising to raise funds and build one if Mayo and his sons would serve as the resident physicians. The Mayos agreed, and in six short years, Saint Mary's Hospital, more commonly known as the "Mayos' clinic," was born. In the decades that followed, more doctors joined the practice and additional buildings were erected. In 1914, the facility became known officially as the Mayo Clinic.[16]

Thanks to their ancestors' vision and labor, today's Rochester residents have the country's top hospital on their doorstep. For more than 100 years, they have received essentially all their medical care at the Mayo Clinic, with their complete medical histories stored and updated in the same computer filing system.[17] This vast and well-maintained record provided the ideal setting to search for a connection between diabetes and Alzheimer's.

Tapping into this resource, a federally funded Rochester Epidemiology Project got underway in 1966. Not to be confused with *dermatology* (the study of skin diseases), *epidemiology* studies the occurrence and distribution of diseases within a population, also known as *population studies*.

As the Rochester Project revealed, between 1970 and 1984, 1,455 local residents (aged forty-five to ninety-nine) had or developed type

2 diabetes.[18] During this time, and among these diabetic patients, seventy-seven (5.3 percent) also developed Alzheimer's. This rate, compared to that among the general Rochester population, was high. Specifically, if someone was diabetic and male, he was 2.3 times more likely to have Alzheimer's. Diabetic women likewise had a higher risk for Alzheimer's (1.4 times), but this increase was not statistically significant.

These results are corroborated by findings across the Atlantic, from Rotterdam, in the Netherlands. Rotterdam has a more interesting origin for its name than Rochester. In 1270, a dam was built on the Rotte River, attracting earlier settlers to create a small fishing village around the dam. As its population grew, a city was born, and it seemed only fitting to name it Rotter-dam. Today, Rotterdam is the second-largest city in the Netherlands and the largest port in Europe, earning it the nickname "Gateway to Europe."

In the suburb of Rotterdam, a large epidemiology project involving 15,000 middle-aged and elderly residents has been underway since 1989.[19] These residents are interviewed and undergo extensive health checkups every three or four years. Subsets of these residents and their records are recruited for studies of various diseases, including heart conditions, stroke, dementia, and depression.

Starting from 1990, over 6,000 of these residents were followed to detect a possible connection between Alzheimer's and diabetes.[20] At the start of the study, all the participants were free of Alzheimer's; about 700 were diabetic. About two years later, 89 people had developed Alzheimer's. Mapping these cases, researchers found that they were twice as likely to happen to the 700 diabetic participants.

Apparently, from Rochester to Rotterdam, the paths of Alzheimer's and diabetes do cross. The problem is that, elsewhere in the world, their tracks have a different appearance.

Poised at the northern tip of Japan's Kyushu island, directly across the sea from South Korea, is a small, rural town called Hisayama. While much of Japan has embraced the modern life and its (in)conveniences, the people of Hisayama have stuck to their traditional agricultural roots. The town has a stable population: about 6,800 in the

1960s and about 8,500 now.[21] It has a similar male-to-female ratio and age distribution as the rest of Japan.[22] Its people are trustworthy and dependable. It is an ideal place for a population study.

Begun in 1961, this is the Hisayama study, a remarkable population study with extremely low dropout (1 percent) and high autopsy (80 percent) rates,[23] the latter being especially valuable in the study of Alzheimer's. As part of the Hisayama study, from 1985 on, more than 800 elderly residents were meticulously followed.[24] None of the participants were initially demented, but seventy had diabetes. During the next seven years, forty-two cases of Alzheimer's were diagnosed, but there was no connection between these cases and participants' diabetic condition.

More conflicting data came from obesity studies conducted in Finland and elsewhere.[25] Because excessive body fat contributes to insulin resistance and the onset of diabetes, researchers looked for a possible link between obesity and Alzheimer's. Strangely, if we are obese during midlife (around the fifties) or early old age (the sixties and early seventies), we are more likely to develop Alzheimer's. But if we have a higher body mass index in old-old age (late seventies and older), we seem to be protected against Alzheimer's.

Similar talks of protection were heard elsewhere.[26] Among 348 patients at a New York clinic, those with Alzheimer's were found five times *less* likely to also have diabetes. This "startling" revelation, researchers argued, demonstrates that diabetes actually has a protective effect. Maybe high blood glucose, the hallmark of diabetes, can increase glucose supply to the brain, thereby generating more energy for the *Brain Bar* and boosting its performance.

So the results of population studies are a toss-up. That didn't stop believers from looking elsewhere for evidence to connect diabetes and Alzheimer's. In diabetes, it takes two to tango: glucose and insulin. If one or both of these players go awry in Alzheimer's, that's still good evidence for the type 3 diabetes hypothesis. Better yet, if one or both of these players can be adjusted to stop Alzheimer's, then we've got nothing else to prove.

14

KETONES
The Brain Fuel

And when they had come to the multitude, a man came to Him, kneeling down to Him and saying, "Lord, have mercy on my son, for he is an epileptic and suffers severely; for he often falls into the fire and often into the water. So I brought him to Your disciples, but they could not cure him."

Then Jesus answered and said, "O faithless and perverse generation, how long shall I be with you? How long shall I bear with you? Bring him here to Me." And Jesus rebuked the demon, and it came out of him; and the child was cured from that very hour.

Then the disciples came to Jesus privately and said, "Why could we not cast it out?"

So Jesus said to them, "Because of your unbelief; for assuredly, I say to you, if you have faith as a mustard seed, you will say to this mountain, 'Move from here to there,' and it will move; and nothing will be impossible for you. However, this kind does not go out except by prayer and fasting."

MATTHEW 17:14–21

In modern medical literature, it's unusual to encounter biblical references, but Matthew 17:14–21 is something of an exception. In writings—not about Alzheimer's directly but another neurological disorder, epilepsy—this passage is frequently alluded to, thanks to that last line: "this kind does not go out except by prayer and fasting."

The "prayer" part, researchers have no interest in. That, they figure, is pure religion. But the "fasting" part, they believe, is evidence of fasting being used as a neurological treatment in biblical times. Aside from the Bible, the ancient Greek physician Hippocrates, the father of modern medicine, is also said to have treated seizures with "complete abstinence from food and water."[1]

While interesting, these historical references seem dubious evidence of medical practices. Many a Bible verse invokes fasting. Moses was with the Lord for forty days and nights without bread or water. David, too, chose to fast to save his sick child. Rather than medical treatment, fasting seems more like a manifest of religious devotion, just like praying. And if one actually reads Hippocrates, one will see that he condemned dietary treatments as nonsense prescribed by charlatans who pretend that epilepsy is punishment from the gods and must be cured by purification[2]. To Hippocrates, a neurological disorder is not sacred and may be treated no more by fasting than by eating.

A more likely origin of the fasting therapy, then, is anecdotal. Perhaps a child was too sick to eat, or perhaps her religious parents forced her to fast. And behold, seizures subsided. Whatever the origin, by the early twentieth century, fasting had become quite a buzzword. In the United States, the most well known practitioner was Huge Conklin, a man of alternative medicine and faith healing. His method was known as *water treatment*, so called because patients fasted for twenty to twenty-five days, consuming nothing but water.[3] The method reportedly cured many cases of epilepsy, convincing Conklin that toxins released from the intestine were a neurological culprit, so resting the intestine would save the brain.

Reports of other fasting regimens followed. In one case,[4] a ten-year old named Gwendolyn B. was having seizures every ten minutes, and she was admitted to the Children's Memorial Hospital in Montreal. Upon admission, she was put to bed and given nothing but water and a little clear broth for ten days. The treatment was quite a success: during her first twenty-four hours in the hospital, Gwendolyn had sixty seizures; in the next twenty-four hours, she had six; over the next three days, she had one seizure a day; and after the fifth day, she had no more seizures.

As cases and experiments increased, doctors and researchers started to realize that fasting healed the brain not by removing intestine toxins, as Conklin thought, but by changing patients' metabolism: in the absence of food (and thus glucose), patients burned body fat for energy.[5] So, short of starvation, a diet that is very low in carbohydrates (which quickly turn into glucose in the human body) and very high in fat should perform the same trick, forcing the body to burn fat.[6]

Why using fat for energy would stop seizures is not yet clear. But together with the type 3 diabetes hypothesis (discussed in chapter 13), a theory has emerged on how it might help Alzheimer's patients.

The theory starts with neurons, the Energizer Bunnies in the brain. All day long, neurons buzz, fire, transmit signals, and handle information. In that process, they burn a lot of energy. The brain, despite its minor weight (2 percent of adult body weight), consumes 20–23 percent of the body's energy needs.[7] On top of this, neurons cannot store glucose as muscles do, so they require a constant glucose supply. If that supply falters, neurons starve, shutting down the *Brain Bar* and impairing memory and other cognitive functions.

This scenario, the theory claims, illustrates what happens to cause Alzheimer's. By analyzing the blood that flows in and out of the brain, we can tell how much glucose the brain has used. Compared with normal brains, the Alzheimer's brain suffers from reduced glucose use. The amount of reduction varies in different studies, from about 10 percent all the way to about 60 percent, with about 25 percent being the commonly agreed average.[8] Interestingly, many

Alzheimer's patients crave sweet, carbohydrate-rich foods,[9] as if trying to forage more glucose for their starving brains.

Why is glucose reduced in the Alzheimer's brain? We aren't sure about that. Some people blame it on age: as we get older, our overall metabolism goes down, and brain metabolism goes down too. Others blame diabetes, which reduces the brain's ability to access and use glucose. Neither explanation is rock solid: glucose consumption declines in some, but not all, older adults and some, but not all, diabetic patients.[10]

The glucose theory also faces a bit of a chicken-and-egg situation. That is, reduced glucose can simply be a result of Alzheimer's, rather than a cause. As neurons are damaged or lost during the course of Alzheimer's, these once energetic bunnies consume less energy, so as a result, the brain will use less glucose. In response to this critique, advocates cite studies showing that brain metabolism drops *before* Alzheimer's symptoms appear. In one study, mutation carriers doomed to get early-onset Alzheimer's had reduced brain glucose metabolism years before their cognitive symptoms manifested.[11] In another, people carrying the late-onset Alzheimer's risk gene ApoE4 showed reduced brain glucose metabolism decades before the possible disease onset.[12]

Moreover, experiments show that glucose intake seems to act as a remedy for Alzheimer's and enhance cognitive performance: for example, in one study, Alzheimer's patients were given a sugary beverage or a sugar-free beverage, and those who drank the sugary beverage performed better in cognitive tests.[13] They recalled more words, remembered more story details, and recognized more photographs of faces. The benefits applied to both healthy adults and demented patients.

If you are clapping with joy that you can now indulge in sweets as "brain food," sorry. An acute increase in blood glucose from one sugary drink is different from chronically elevated blood glucose from, say, having ten such drinks every day. The former has cognitive benefits; the latter has the opposite effect. When excessive glucose sits around in the blood, it starts to interact with fat and protein.

This process is called *glycation*, which actually promotes the hallmarks of the Alzheimer's brain: beta-amyloid plaques and tau tangles.[14]

Thus, if the Alzheimer's brain *is* a starving brain, we can't constantly feed it with glucose; we have to find other fuels. This is where the epilepsy treatment, the low-carb-high-fat diet, comes in. When glucose runs low in our bodies, the liver can use fatty acids (the building blocks of fat) to produce a small molecule called *ketones*. Ketones can then travel through the bloodstream to the brain and be burned as an alternative fuel. To induce this metabolic shift though, we must keep our carbohydrate (and hence glucose) intake very low. In a strict ketone diet, 90 percent of daily calories are derived from fat, no more than 2–3 percent come from carbohydrates, and the rest are from protein.[15]

Such a diet makes some evolutionary sense. Fat has been an organic and crucial source of nutrition since the dawn of humanity. Indeed, an increased consumption of animal food and fat two million years ago is thought to have helped early humans develop a larger and more sophisticated brain.[16] By contrast, foods rich in carbohydrates are relatively new additions to the human diet. Refined sugars and cereals were scarce in the seventeenth and eighteenth centuries and became available in the mass market only some 200 years ago, after the Industrial Revolution.[17]

That said, not all fats are created equal. Depending on what kind of fatty acid is in the fat, ketone production varies. As shown in figure 14.1, all fatty acids have two basic parts: a carboxyl group and a hydrocarbon chain.

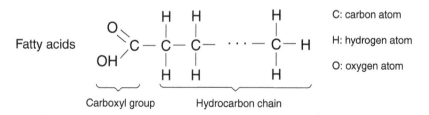

FIGURE 14.1   The two parts of a fatty acid: carboxyl group and hydrocarbon chain

The carboxyl group is the same in all fatty acids, but the length of their hydrocarbon chains varies. The longer the chain, the more hydrogen-carbon links, and thus the more carbon atoms. Fatty acids with more than fourteen carbon atoms are called *long-chain fatty acids;* those with eight to fourteen carbon atoms are called *medium-chain fatty acids* (figure 14.2).

The fats that we typically eat, derived from meat, fish, and vegetable oil, contain long-chain fatty acids, which are not an effective raw material for ketones. They take considerable time getting to the liver and must undergo extra chemical processing before producing ketones.[18]

By comparison, medium-chain fatty acids are far superior ketone producers. These fatty acids are abundant in breast milk, creating extra ketone fuel for the developing infant brain. However, in the typical Western diet, medium-chain fatty acids are generally missing. Their one significant dietary source is the so-called tropical oils: coconut oil and palm kernel oil. Alternatively, one can resort to concentrated supplements known as *medium-chain triglycerides,* with a "triglyceride" being a combination of three fatty acids.

Anecdotal evidence vouches for the miraculous, brain-boosting effect of these oils. In her book *Alzheimer's Disease: What If There Was a Cure? The Story of Ketones,* pediatrician Mary Newport tells the moving story of her husband, Steve.[19] Plagued by Alzheimer's in his early fifties, Steve fell into a mental abyss, unable to cope with daily activities, much less a job. Intrigued by ketones and desperate for a cure,

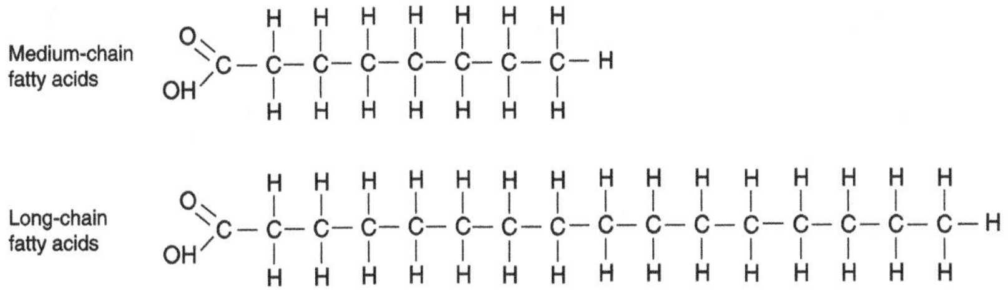

FIGURE 14.2 Medium- and long-chain fatty acids

Mary started Steve first on coconut oil, and then medium-chain tri-glyceride oil.

The effect was instantaneous. On day 1, after having two table-spoons of coconut oil in his morning oatmeal, Steve improved on his cognitive test score. On day 3, he woke up alert, smiling, and talk-ative, having no trouble eating with utensils. On day 6, he was able to clean the pool and vacuum the house. On day 11, he used the dish-washer with minimal help, and on day 24, he did a complete load of laundry by himself. On the rare occasions when Steve missed his oil, he suffered terrible relapses, but a quick dose would have him smiling again in thirty minutes. In the four years before Steve took the oils, his brain underwent significant shrinkage. In the two years afterward, it stayed stable.

In her book, Newport included letters and emails from other caregivers who had given coconut and medium-chain triglycer-ide oils to demented family members. These correspondences are equally glowing, reporting either immediate or gradual improve-ments in memory, cognition, social interaction, and quality of life.

Overnight, we seem to have found the cure we longed for.

For better or worse, anecdotal evidence like Steve's, no matter how heartwarming, carries little weight in the world of empirical science and modern medicine.

From Steve to other supporters, their ages differ, symptoms dif-fer, oil consumptions differ, improvements differ, and their caregiv-ers' qualities of observation differ. Their stories are not carefully quantified and cannot be generalized to other patients. In fact, these printed tales may not represent the whole truth either. Just as big pharma has a stake in Alzheimer's drug development, so does Mary Newport. Her website endorses coconut and medium-chain tri-glyceride oils that she urges people to buy. Apparently, she herself designed one of these oils. For this reason alone, she wouldn't be very motivated to detail in her book anecdotes where the oils didn't work.

This is not to say anecdotal stories have no value. Liberated from the baggage of established science, they hold promise for truly outside-the-box thinking. Fasting as an epilepsy treatment, as I speculated earlier, may have emerged from anecdotes. For some people, personal stories are also more trustworthy than abstract, experimental data that, as we have seen, are subject to dicing and manipulation.

Still, formal experiments provide us with protocols, measurements, and records that we can use to judge their quality in ways that anecdotes don't. To prove that the oils and their ketones work, we need randomized, blind, placebo-controlled clinical trials.

With coconut oil, there haven't been many trials. In the United States, one trial was in the works back in 2013, but it was terminated, reportedly for funding limitations and low enrollment.[20] The product to be tested was Fuel for Thought, a coconut oil beverage manufactured by Connecticut-based Cognate Nutritionals. The beverage supposedly contained a unique blend of coconut oil and other ingredients. In twelve-bottle packs, Fuel for Thought sold for $49.95. But by all appearances, the company and the beverage vanished around 2015, which coincides with the terminated trial.

Outside the United States, one small trial was recently completed in Spain.[21] Forty-four Alzheimer's patients were randomized to a coconut-oil-enriched diet or a placebo diet. After three weeks, those consuming coconut oil improved on several measures of memory and cognition. These results are encouraging, but if history is any indication, one small trial doesn't prove much; the data need to be verified in larger, longer-term trials. As of now, none seem to be planned. The fact that coconut oil is a natural ingredient and can't be easily patented by a pharmaceutical company probably explains, if only partly, this lack of enthusiasm.

For the same reason, ketone diet trials have been limited. These trials also face an extra hurdle: the diet is difficult to enforce. Extremely low in carbs, a true ketone diet prohibits most of the foods we take for granted: bread, pasta, rice; sugar, sweets, honey; and even starchy vegetables, fruits, and seeds like potatoes, beans, corn, and apples. One is allowed less than 20 grams of carbs a day, which is half of a hamburger bun, one small potato, or a few bites of pasta.[22] With these stringent requirements, compliance is a real issue. Moreover, blinding participants to their assigned diet is nearly impossible, so there is no avoiding the placebo effect and potential bias.

In my research, I found only one relevant dietary trial, where twenty-three participants with mild cognitive impairment were randomized to a very-low-carb diet and a high-carb diet for six weeks.[23] Two tests were used to assess cognitive function. One is word recall, measuring participants' ability to remember words. The other is trail-making, which involves more complex cognitive processing. To test this, participants were given a piece of paper with scattered numbers and letters and told to draw lines to connect them in an alternate way (i.e., 1-A-2-B-3-C). At the end of the six weeks, very-low-carb dieters improved on the word recall test but not on the trail-making test. The results are interesting, but again they need to be verified in larger trials.

Fortunately, something else did enjoy more trials: medium-chain triglycerides, the fatty acids directly responsible for generating ketones. These chemical products are easier to administer and potentially patentable, making them a more favorable choice to pharmaceutical companies.

Among current pursuits, the product that came the closest is AC-1202, manufactured by Colorado-based Accera. AC-1202 has a significant advantage over dieting: it is powerful enough to raise ketones even when carbohydrates are consumed, so patients are free to enjoy normal meals.[24]

In a pilot study,[25] twenty patients with mild-to-moderate symptoms randomly received a drink containing AC-1202 or placebo. AC-1202 raised participants' ketone levels, especially for those carrying the Alzheimer's risk gene ApoE4. However, these ApoE4 carriers did not show improved cognitive function; in fact, their cognitive scores worsened. Non-ApoE4 carriers, on the other hand, did show improved cognition. Perhaps, researchers speculated, ApoE4 carriers were not very good at utilizing the ketones raised in their bodies.

Following this pilot study, a phase 1 trial was conducted, but no results were published.[26] Phase 2 quickly followed, assigning 150 mild-to-moderate Alzheimer's patients to daily AC-1202 or placebo for three months.[27] Once again, AC-1202 significantly raised participants' ketone levels. In terms of cognitive effect, the published report, authored by Accera employees, makes a *very* convoluted case and presents a dozen different ways to look at the data. But the bottom line was that AC-1202 had no benefit in terms of cognitive performance or functional behavior, as judged by three tests. Non-ApoE4 carriers experienced some benefits according to one test, but not the other two.

In its report, Accera enthusiastically discussed the one improved test, while attributing the negative ones to the "relative insensitivity of these tests, the small number of subjects, and/or the short duration of the trial."[28] The report ended by predicting that AC-1202 could be further refined and offer a novel strategy for treating Alzheimer's.

Given this prediction, what Accera did next was surprising. Instead of pushing forward into phase 3, the company turned around, made a beverage out of AC-1202, named the beverage Axona, and started marketing it as a medical food.

Medical food, according to the U.S. Food and Drug Administration (FDA), is a food specially formulated for patients to meet the dietary needs of a particular disease or condition.[29] It is a special category sitting somewhat between drugs and dietary supplements. Unlike drugs, medical food is *not* subject to the FDA's stringent clinical trial regulations and premarket reviews. And unlike dietary supplements, medical food *can* label and market itself in terms of a specific disease. It is the best of both worlds for Axona.

The Axona drink was labeled "a medical food for the clinical dietary management of the metabolic processes associated with mild-to-moderate Alzheimer's disease."[30] On its company website, Accera claimed that the drink "may enhance memory and cognition in mild to moderate Alzheimer's" and "helps patients with mild to moderate Alzheimer's disease by addressing diminished cerebral glucose metabolism."[31] Given the latest trial results, these claims are, to say the very least, selective exaggerations.

In a published interview, Accera's executive director of research, Sam Henderson, explained that the company "didn't have the monies to do the long-term development process. . . . For the investors, one of the big appeals of our approach to AD [Alzheimer's disease] is that we're basically using a food ingredient. . . . You don't have to go through the normal long drug development process. You can very quickly come to market with something like a medical food."[32] Henderson added that the company never intended to pursue AC-1202 as a drug and was "always planning on something like a medical food," although Accera's chief executive officer, Steve Orndorff, had spoken of different intentions in the past.[33]

In 2013, four years after the beverage entered the market, Accera completed a case study to vouch for its cognitive effects.[34] The study profiled eight mild-to-moderate Alzheimer's patients who had consumed the drink for more than six months. Tracking the patients' cognitive test scores, Accera announced that their drink "was associated with stable disease or improvement for some patients." A closer look at the published report, however, shows something more nuanced: upon drinking Axona, four of the eight patients continued to decline, and four stabilized or saw small improvements; on average, the patients were still declining, and their rate of decline before and after consuming the drink was no different.

Just when Accera thought it was getting away with its medical food, in late 2013—on the day after Christmas, no less—the FDA issued the company a warning, ruling that the Axona beverage does not meet the definition of a medical food because "there are no distinctive nutritional requirements or unique nutrient needs for individuals with mild to moderate Alzheimer's disease."[35] Instead, Axona was ruled to be a "new drug." As a drug, it must establish safety and efficacy and receive FDA approval before entering the market.

Left with no choice, Accera conducted a phase 2/3 trial, recruiting more than 400 patients and assigning them to AC-1204 (version 2 of AC-1202, with a new formula) or placebo.[36] After six months of treatment, results were announced in early 2017; the trial failed.[37] AC-1204 did not improve patients' cognitive performance—not for ApoE4 carriers, and not for noncarriers. Backed into a corner, Accera explained that the new formula somehow lowered drug levels in patients, which contributed to the failure.

Just when things seemed to be falling apart for Accera, another surprising event occurred. In October 2018, Accera rebranded itself as Cerecin and opened a "global headquarters" in Singapore. The rebranding was enabled by investment from a Singapore-based company called Wilmar International.

Wilmar is not a pharmaceutical company; it is a food manufacturer. According to a company news release, Wilmar's investment is to fund "the Asia Pacific launch of Axona®, a medical food for the dietary management of persons living with Alzheimer's disease."[38] Apparently, although the FDA ruled Axona not a medical food and off-limits to sell in the United States, other markets are fair game. "With a rapidly aging population in Asia Pacific, and in particular China," said Wilmar's head of business development, Gurpreet Singh Vohra, "we recognise the increasing importance of dementia and brain health as priorities on healthcare agendas throughout the region."[39]

Not one word, I am sure, will be said to the "population in Asia Pacific, and in particular China," about those iffy and failed AC-1202/AC-1204 trials. Meanwhile, people like my uncle and their families, desperate for *something*, will line up for this Western-imported, science-based, beautifully packaged, and surely expensive medical food. Given this blatant deception, I can only laugh at what the company said about its new name, Cerecin: it's "a combination of the words cerebrum, the Latin word for brain, and medicine, affirming the company's commitment to brain health."[40]

15

INSULIN FIXES

Sam turned forty last week.[1] An up-and-coming federal prosecutor, Sam works twelve-hour days and loves every minute of it. The late nights and fourth meals has made him gain weight, but he feels good and has no health complaints. In fact, for two years in a row, he has skipped physical checkups. Why bother, and who has the time? Sam would gladly skip it again, except that his wife, annoyed by Sam's laissez-faire attitude, made an appointment for him. Once it's on the calendar, it's settled: Sam is not one to ignore a schedule.

Almost everything at the doctor's went quickly and smoothly. But Sam didn't like it when they pricked his finger to produce some blood. He liked it even less when a little device with a tiny screen picked up the blood and produced Exhibit A. The screen on Exhibit A read 159, which, according to Sam's doctor, means offensively high blood glucose. Without much deliberation, the doctor rendered a verdict: type 2 diabetes.

Sam is no stranger to diabetes; his mother has it. But to get it himself, at forty years of age?! The doctor didn't seem all that surprised. Apparently, Sam's stressful job, family history, and, yes, belly fat all make him a high-risk individual for diabetes. Before Sam could recover from the news, the doctor has moved on to talk about treatment plans.

Sam was introduced to dozens of drugs, and the doctor prescribed several. The one whose name Sam still remembered upon leaving the doctor's office was thiazolidinedione (TZD), which is an insulin sensitizer. It makes Sam's body more sensitive to insulin so that glucose can be escorted into the cells for energy rather than piling up in the bloodstream.

The way that TZD works is a bit ironic: it makes the body store fat—and gain weight. The rationale is that when fat is saved and not burned for energy, the body is forced to become more sensitive to insulin in order to burn glucose.[2] The weight gain is chalked up as the drug's side effect, even though excessive body fat, as in Sam's case, has contributed to insulin resistance and diabetes in the first place.

TZD provided relief for a while but eventually, this and other oral medicine became inadequate for Sam, so he was prescribed insulin to inject. In the doctor's office, Sam is shown the syringe and how to use it: grab the fatty part of his abdomen, pinch the skin, insert the needle at a 90-degree angle, and push the plunger down. Sam goes through this routine several times a day to make sure that there's enough insulin in his body to escort glucose.

With all these treatments, Sam manages his diabetes for decades. At age fifty-five, he takes a job in the private sector. Now a criminal defense attorney at a large firm, Sam handles high-profile—and even more stressful—cases and works even longer days and nights.

As his cycle of diabetic life continues, Sam (and millions like him) are thankful for the range of insulin medicines on the market. If Alzheimer's really is type 3 diabetes, millions of other patients will be thankful too.

Insulin, as described in chapter 13, is responsible for opening cell doors to let glucose in. This seemingly simple action is, in reality, a complex feat of biology. First, insulin must interact with insulin receptors, which are proteins that sit on cell membranes that specifically respond to (or "receive") insulin. The interaction between the two mobilizes a series of other enzymes and proteins, which in turn cause a glucose transporter protein called GLUT4 to move to the cell membrane. Once there, GLUT4 proceeds to lodge itself in the membrane, opening up a channel. Through that channel is how glucose gets into cells. This series of maneuvers is called *insulin signaling*.

Back in the 1950s, we thought that insulin signaling did not happen in the brain. When specially labeled insulin was injected into the bloodstream in rats and humans, it quickly showed up in various organs and tissues, but not in the brain.[3] This finding made sense, because the brain does not solely depend on GLUT4 to transport glucose. It employs other transporters, such as GLUT1 and GLUT3, which can move glucose from the blood into cells *without* the help of insulin.

But soon enough, we realized that we were wrong.[4] The detection method used in the 1950s was not sensitive enough to pick up the small, but still significant, amount of insulin in the brain. Multiple insulin injections, rather than a single shot, were required for brain detection. We also found insulin receptors in the brain, particularly in the hippocampus, the part of the brain critical for memory formation. The fact that insulin receptors sit and wait in the brain suggests that insulin does come, and presumably it has important jobs to do. In fact, we now know that the brain can synthesize its own insulin and is not totally dependent on pancreas-produced insulin. This seems bona fide evidence that insulin not only comes, but also is essential to the brain.

But what is this essential function? After all, the brain, as mentioned previously, is not totally dependent on insulin for glucose transportation. What else could insulin contribute? According to the type 3 diabetes hypothesis, the answer is that it centrally regulates metabolism. In both animal and human experiments, administering insulin into the brain reduced food intake and body weight.[5]

Moreover, as a central regulator, insulin directs a large, complicated symphony of proteins and chemicals in the brain; some promote neuron survival, some cause neuron death, some excite neurons, and some inhibit neurons.[6] Insulin keeps everyone in a happy balance, thereby maintaining cognitive function.

Rat experiments support this belief. When rats were given drugs that destroy brain insulin, the animals suffered neuron damage, their brains shrank, and they had difficulty navigating the water maze.[7] When these rats were treated with insulin sensitizers, the brains recovered, and learning and memory improved.

Preliminary human experiments showed the same. Autopsy studies revealed that the Alzheimer's brain has far less insulin and fewer insulin receptors than normal brains do.[8] As Alzheimer's disease progresses, this shortage grows worse.[9] Conversely, when patients were given insulin, their memory performance improved.[10] These findings would suggest that the cause of Alzheimer's is brain insulin shortage.

But why would there be a *shortage* of insulin in the first place? The whole problem with diabetes is that the body becomes insensitive to insulin, which forces the pancreas to pump excessive insulin into the blood. Plus, the brain can synthesize its own insulin. Where did all this insulin go?

This is where the type 3 diabetes hypothesis hits a snag and gets fuzzy. Supposedly, elevated blood insulin works in mysterious ways. When we get a onetime insulin injection, that insulin can travel from the blood into the brain. However, if the blood insulin is chronically spiked, that somehow *prevents* insulin from moving into the brain.[11] Exactly how this happens is unknown. The general sentiment is that insulin transportation is a delicate business: it is closely regulated, temperature sensitive, and easily jammed. Thus, it is not surprising, we are simply told, that long-term, high-level blood insulin would reduce brain insulin.[12]

If we go along with this fuzzy logic, then insulin sensitizers like the TZD drug that Sam was prescribed could also make fine

Alzheimer's drugs. By making the body more sensitive to insulin, they would keep *blood* insulin low and thus *brain* insulin high, restoring the order in the brain symphony.

One commonly prescribed TZD is rosiglitazone (brand name Avandia), manufactured by GlaxoSmithKline (GSK) and approved by the U.S. Food and Drug Administration (FDA) in 1999 as a diabetes drug. By 2006, the drug's annual sales had surpassed $3 billion. If GSK could prove rosiglitazone's effect on Alzheimer's, the profits from this single drug would go through the roof.

Preclinical studies are encouraging. In cell studies, rosiglitazone boosted insulin signaling and protected neuron health.[13] Transgenic Alzheimer's mice that were fed rosiglitazone showed improved cognition, performing almost as well as normal mice in maze tests.[14] More amazingly, in a six-month pilot study, rosiglitazone improved memory and attention among thirty patients with mild cognitive symptoms.[15]

Inspired, GSK moved into formal clinical trials. But at this juncture, the picture-perfect drug unraveled. In 2007, an article appeared in the *New England Journal of Medicine*, accusing rosiglitazone of increasing patients' risk for heart diseases.[16] The article collected evidence from forty-two rosiglitazone diabetes trials. Together, these trials enrolled about 28,000 participants. A little more than half of them received rosiglitazone; the others received placebo or other drugs. When the two groups were compared, those receiving rosiglitazone had a 43 percent increase in heart attacks. Other articles soon followed, pooling and examining different sets of rosiglitazone trial data. Their conclusions are now always the same, but in general they confirm the drug's cardiac side effects.[17] It is suspected that rosiglitazone elevates the low-density lipoprotein (LDL) cholesterol (aka, the bad cholesterol), which is associated with heart diseases.

As the news broke out, over 10,000 lawsuits were filed against GSK. The company settled most of these out of court, reportedly paying $460 million.[18] It was also issued a fine by the FDA. Lawsuits and fines aside, the FDA did not pull rosiglitazone off the market because the drug was deemed reasonably beneficial for treating type 2 diabetes.

Despite this episode, GSK completed its phase 1 trials testing rosiglitazone's safety and absorption as an Alzheimer's drug. Oddly enough, no published data can be found, which is suspicious given the drug's tarnished safety profile. Notably, one of these trials tested the drug's effect on heart health and was terminated without explanation.[19]

Whatever's going on, we do know that two phase 2 trials followed and were duly published, but neither had impressive results. In the first, rosiglitazone didn't reduce brain shrinkage or improve cognitive function among eighty Alzheimer's patients after a year of treatment.[20] In the second, no cognitive benefits emerged among 500 patients after six months of treatment.[21] Facing these disappointments, GSK made the usual move: looking for better results in subsets of the participants. In a familiar plot, we learned that for patients who did not carry the risk gene ApoE4, rosiglitazone brought cognitive improvement.[22]

Grasping at this ray of hope, GSK limped into phase 3 and launched three separate trials, recruiting a total of more than 3,600 patients.[23] In two of the trials, participants were allowed to take standard, palliative Alzheimer's drugs. In the third, they were not. Other than this, the trials followed a similar design. Through genetic testing, participants were put into ApoE4 carrier or noncarrier groups. Within each group, they then randomly received rosiglitazone or placebo.

This elaborate design didn't save GSK. Twelve months later, all three trials failed. The patients on rosiglitazone showed no improvement in cognition or daily function. It didn't matter if they took rosiglitazone alone or with palliative drugs. It didn't matter if they were ApoE4 carriers or noncarriers. The drug simply didn't work. Upon this news, GSK terminated additional plans with rosiglitazone.

As GSK threw in the towel, the Japanese company Takeda Pharmaceutical was trying to hang on. Takeda manufactures the other commonly used insulin sensitizer, pioglitazone (brand name Actos). The company received free advertising when the news broke that GSK's rosiglitazone causes heart diseases. During that controversy, Takeda's pioglitazone was championed as posing significantly less cardiac risk.[24] It also does a better job at reaching the brain, which makes it a more logical Alzheimer's drug.[25]

But good press didn't translate into cognitive effects. Transgenic Alzheimer's mice that were fed a pioglitazone chow showed no memory improvement and struggled in the water maze.[26] As for human trials, results are all over the place. In one, patients with mild Alzheimer's showed improved cognition after six months of treatment.[27] In another, patients with overall mild symptoms improved on some cognitive tests, but not others.[28] In a third trial, mild-to-moderate patients experienced no benefits after 1.5 years of treatment.[29]

Taking these inconsistencies in stride, Takeda formed a strategy to recruit patients at the early stage of the disease, because their most positive results had come from those patients. This strategy led to the recruitment of 3,500 elderly participants who did not yet show cognitive symptoms.[30] Genetic tests were used to determine who among them were at risk for Alzheimer's by carrying the risk genes ApoE4 and TOMM40 (for more about TOMM40, see chapter 16). At-risk participants then randomly received pioglitazone or placebo to see if the drug can slow their imminent cognitive decline.

It didn't. In January 2018, the trial was terminated when an interim analysis showed a lack of effect.

So much for insulin sensitizers.

With our attorney, Sam, oral medicine eventually fell short, and insulin injection remained the only remedy powerful enough to treat his diabetes. Could the same be true for Alzheimer's? If the Alzheimer's brain is suffering from insulin shortage, then giving it insulin seems a more straightforward solution.

In two separate studies, when Alzheimer's patients received insulin through intravenous (IV) administration (i.e., delivered through a tube inserted into the vein), their memory test scores improved.[31] Presumably, insulin traveled up through the bloodstream into their brains. The effect, however, isn't going to last because the added insulin will quickly wear off. Continuous IV boosts are not an option either, because that means putting excessive insulin into the blood,[32] which, as mentioned previously, is thought to jam insulin transportation to the brain. In addition, excessive blood insulin will use up blood glucose, causing low blood sugar, dizziness, and fainting.

If only we can directly inject insulin into the brain, that would solve our problem. Unfortunately, direct brain injection, of anything, is not a well-established procedure. To solve this conundrum, researchers came up with a clever delivery method: the nasal spray. Insulin is sprayed around the nose, is sniffed in, and travels through the nose cavity into the brain, bypassing the blood altogether. Doing this creates a quicker, more direct access to the brain: once inhaled, insulin peaks in the brain in mere thirty minutes.[33]

In pilot studies, insulin nasal sprays helped patients with early signs of Alzheimer's improve memory and attention.[34] Patients became better at recalling story details and processing conflicting information (e.g., when the word *red* is written in a blue color). Caregivers also reported patients doing better in everyday life. The spray proved beneficial even for healthy participants, who demonstrated improved memory and reported "enhanced mood."[35]

Following these initial studies, a series of trials, fittingly called SNIFF (Study of Nasal Insulin to Fight Forgetfulness), were pursued by a group of university researchers. Some of the trials tested regular insulin; others tested so-called long-acting insulin, which remains active for a longer period of time in the body.

The results of these trials are inconsistent. In one, four months of regular insulin helped Alzheimer's patients remain stable, while those on placebo declined.[36] In another, three weeks of long-acting insulin brought no overall benefit.[37] In a third, neither regular nor long-acting insulin showed overall cognitive benefits after four months, although regular insulin boosted memory test scores.[38]

Despite this mixed bag, trial organizers continued to argue for the potential of insulin sprays. In their words, Alzheimer's is "a devastating illness, for which *even small* therapeutic gains" matter.[39] Pursing that hope, a phase 2/3 trial followed, recruiting 289 patients to receive one year of regular insulin sprays or placebo.[40] Trial results came out in June 2020: failure. The sprays brought no cognitive or functional benefits. Upon this, trial organizers are arguing for further investigation into a better intranasal delivery device.

Effective or not, deceptive or not, the insulin and glucose fixes that we have seen to date are gutsy. They present theories like "Increasing brain insulin treats Alzheimer's" or "Burning ketones as an alternative fuel cures Alzheimer's."

By comparison, other diabetes drugs being tested for Alzheimer's are wimpy. Their advocates do not promise anything specific; rather, they talk in general terms about the need to improve insulin signaling, normalize blood glucose, and protect neurons. Behind these lukewarm positions, one doesn't get exactly where and how these drugs might help with Alzheimer's. Instead, one gets the feeling that we are simply throwing all the available drugs at Alzheimer's and hoping that something sticks.

First among these uncommitted drugs are "insulin stimulators," which is not a technically correct name, but an apt descriptor. Their real name is *GLP-1 agonists*. GLP-1 is a hormone produced by the intestine when we eat. Agonists, put simply, are substances that enable biological actions. GLP-1 agonists, then, are drugs that mimic the natural actions of GLP-1.

What does GLP-1 do? It stimulates insulin production.[41] True, the primary cue for the pancreas to produce insulin is the intake of food and the rise of blood glucose. But food also causes the intestine to produce GLP-1, which further stimulates insulin. Natural GLP-1, however, is rapidly degraded in the body, losing effect in minutes. That's why we need GLP-1 agonists, which have a longer shelf life.

Although GLP-1 agonists stimulate insulin, there is no evidence that they are powerful enough to raise brain insulin, so their direct effect on the Alzheimer's brain is fuzzy. Among the various GLP-1 agonists on the market, liraglutide (trade name Victoza), from the Danish pharmaceutical company Nordisk, went the farthest as an experimental Alzheimer's drug. Notwithstanding its being the best in its group, the drug didn't do too well. Animal test data are inconsistent. In one study, liraglutide reduced brain beta-amyloid in transgenic Alzheimer's mice, helping the critters remember their escape route in the water maze and recognize familiar objects.[42] In direct contrast, no effects on beta-amyloid, the water maze, or object recognition were found in other transgenic mice.[43]

Human results are no better. So far, the results are available for one small, six-month trial with thirty-eight patients: liraglutide did not reduce brain beta-amyloid or improve cognitive test scores.[44] The trial organizers argued that these results are inconclusive—because the number of participants is too small and the trial too short.

One wonders: If that is so, why design the trial in this way in the first place? Had the results been positive, would they still be inconclusive? For more conclusive data, we will have to wait for a larger and longer trial (with 200 patients and one year of treatment) that is currently underway.[45]

If GLP-1 agonists could treat Alzheimer's, by reverse logic, so might another type of diabetes drug. Recall that natural GLP-1 degrades rapidly, which is why GLP-1 agonists were created to lengthen its effect. But guess what—we know what degrades GLP-1: an enzyme called DPP-4. If we inhibit DPP-4, GLP-1 can keep on stimulating insulin and supposedly treating Alzheimer's. Drugs that inhibit DPP-4, naturally, are called *DPP-4 inhibitors*.

A number of DPP-4 inhibitors are on the market, and several have been explored as an Alzheimer's drug, such as linagliptin (trade name Tradjenta), vildagliptin (trade name Galvus), and saxagliptin (trade name Onglyza). In cell and animal studies, these drugs decreased beta-amyloid, reduced tau phosphorylation, protected neurons, and improved the cognitive performance of rats and mice.[46]

Human studies, however, are lacking. Some population studies did find that these drugs improved patients' cognitive function, but these were type 2 diabetes patients, not Alzheimer's patients.[47] Plus, population studies are not randomized, placebo-controlled trials, so their value is doubly limited. As of now, I am not aware of Alzheimer's clinical trials testing DPP-4 inhibitors.

Even worse than these drugs that lack compelling evidence are two that are borderline controversial.

The first is amylin—remember this by-product of insulin, a sidekick that helps control glucose (see chapter 13)? Amylin does this in several ways.[48] It slows digestion so that food is converted to glucose gradually, not in overwhelming spikes. Amylin also sends the "full" signal to the brain, cuing us to stop eating and stop the influx of glucose. Last, amylin regulates liver activities. The liver keeps a glucose reserve and releases it into the blood between meals, and amylin stops this process after we are properly fed.

These functions make amylin a useful drug to regulate glucose in diabetes, but what do they have to do with Alzheimer's? Well, Alzheimer's patients, as mentioned in chapter 14, have a reduced brain glucose metabolism, so who knows—amylin, as a natural glucose regulator, might just help. Plus, amylin levels *are* low in people with Alzheimer's, which suggests that there is some kind of deficit.[49]

But here comes the irony. Amylin also has some suspicious qualities. As mentioned in chapter 13, it builds up in the diabetic pancreas and damages pancreatic cells, just as beta-amyloid builds up in the Alzheimer's brain and damages neurons. More alarmingly, amylin doesn't stay put in the pancreas—it can wander into and clump together in the brain. In autopsied Alzheimer's brains, small amylin plaques are seen lying next to or mixed up with larger beta-amyloid plaques.[50] In this scenario, amylin seems like the sidekick of beta-amyloid rather than insulin. Yes, we can create synthetic amylin that doesn't build up easily, but the idea of using amylin to treat Alzheimer's still seems odd. In transgenic Alzheimer's mice, both natural and synthetic amylin improved cognitive performance, but no human trial data are currently available.[51]

The last diabetes drug worth a quick mention is metformin. Metformin is an old drug, having been around since the 1950s. Safe, cheap, and easily available, it is used worldwide as the first-line, go-to medicine for diabetes. Its primary function is to reduce blood glucose. Like amylin, it does so by regulating the liver and reducing its glucose release.[52]

Metformin does not build up in the brain, but it too is controversial. While some cell and animal studies claim that metformin reduces beta-amyloid and tau anomalies and enhances cognitive function,[53] others found the complete opposite.[54]

To settle the debate, researchers worldwide looked into the cognitive state of diabetic patients who regularly take metformin. In Taiwan, 25,000 diabetic patients were tracked for eight years. Among them, those taking metformin had a reduced risk of dementia.[55] This was confirmed in Singapore, where metformin stalled cognitive decline among 365 older adults with diabetes.[56] On the other hand, in the United Kingdom, long-term use of metformin increased the risk for Alzheimer's among 14,000 elderly adults.[57] This finding was repeated in Australia, where among 126 diabetic patients, cognitive performance was worse in those taking metformin.[58]

Despite these conflicting data, clinical trials were recently pursued. In a small study, twenty participants with mild cognitive symptoms were randomized to metformin or placebo. Sixteen weeks later, the metformin recipients' scores improved in one cognitive test, but not others.[59] Just as gloomy was a larger trial with eighty patients.[60] After a full year of treatment, participants on metformin did better in one test, but not six others. Reflecting the drug's noncommitment, trial organizers concluded that a "larger trial *seems* warranted."[61]

16

BACTERIA IN THE BRAIN

In hot and crowded chaos 14 billion years ago, out of nowhere and everywhere, a Big Bang took place. Cosmic debris was flung every which way, creating and expanding the universe. Very slowly, dust, gas, and fragments were pulled together by gravity, creating stars and planets. It took 10 billion years to form the planet we call home: Earth.

Back then, Earth had nothing to offer but scalding hot rocks. Beneath the surface, these rocks moved and collided, creating volcanoes that hissed and roared all day. Up above, meteorites and debris from the Big Bang visited, hitting Earth and leaving behind giant craters. There was no air, water, or atmosphere of any kind.

Slowly, the surface of Earth cooled, allowing liquid water to form. Whether water molecules always existed inside Earth's rocks or they were brought by asteroids is a mystery. Meanwhile, gases from volcanic eruptions (chiefly carbon dioxide and nitrogen) created a protective atmosphere that would retain water.

Then life formed. How life sprang into existence is anyone's guess. Some thought that a bolt of lightning shocked organic molecules out of inorganic elements. Others thought that meteorites from outer space delivered organic molecules. Whatever the case, fossil evidence shows that microorganisms started to exist on Earth some 3.5 billion years ago.

As life formed, it needed to harness energy and sustain itself. Because there was no oxygen on the early Earth, organisms learned to break down glucose without oxygen, inventing *anaerobic* metabolism. The process isn't very efficient: one unit of glucose results in only two units of energy.[1]

Gradually, some bacteria learned to photosynthesize as modern-day plants do, using light, water, and other elements to produce energy. In that process, they released the crucial by-product of oxygen into Earth's atmosphere. As oxygen grew, a new generation of organisms was born that practices *aerobic* metabolism, using oxygen to break down glucose. Far more efficient than anaerobic metabolism, aerobic metabolism can produce thirty-six to thirty-eight units of energy per glucose.[2]

Then one day, about 1.5 billion years ago, something unexpected happened: a larger anaerobic cell ate an aerobic bacterium. Curiously, the bacterium wasn't digested; instead, the two started to cohabit. The engulfed bacterium enjoyed shelter and nutrients while it lived inside the larger cell; in return, it provided superior, oxygen-assisted metabolism to sustain its host. Over time, the arrangement became permanent, and the ancestral cell to modern-day plant and animal cells was born.

To this day, the engulfed bacteria remain in our bodies. We no longer call them bacteria, though. Today, they are known as *mitochondria:* tiny, bean-shaped structures responsible for harnessing energy from glucose. Each cell in the body contains hundreds or thousands of mitochondria. As little powerhouses, they provide energy that allow the muscles and organs, including the brain, to run.

Since being hijacked, mitochondria gave up most of their bacterial DNA. Only a small part remains today. This DNA, known as *mitochondrial DNA,* codes for a dozen or so proteins that allow mitochondria to perform their powerhouse duties.[3] Mitochondrial DNA resides inside the mitochondria, is physically removed from a cell's nuclear DNA, and is inherited separately during reproduction.

Was our ancestral cell lucky in inventing its own powerhouse? By all appearances, yes. Did it get more than it bargained for? According to the mitochondrial hypothesis of Alzheimer's, also yes.

Because of their crucial role as cellular energy providers, mito-chondria are closely monitored by cells: damaged mitochondria are constantly repaired or cleared out and replaced by new ones. But as we age, mitochondria accumulate increasing damage and muta-tions. At some point, they are beyond repair or replacement and simply go out of commission. When that happens, energy produc-tion drops, shutting down cells in the muscles and organs, includ-ing the brain.

Indeed, the situation is especially dire in the brain, which has a higher energy need and yet a smaller number of mitochondria.[4] These mitochondria also have a slow turnover, which means that they have longer lives and accumulate more damage than other mitochondria.[5] In old rats, brain mitochondrial activities are reduced by 35–65 percent—way beyond what neurons can tol-erate in terms of basic energy needs.[6] When mitochondria are experimentally inhibited in mice, the animals develop Alzheimer's-like symptoms, including neuron loss and cognitive decline.[7] In Alzheimer's patients, brain energy production can drop as much as 20 percent at the beginning of the disease and worsen as the dis-ease progresses.[8]

Not only do damaged mitochondria shirk their powerhouse duties, they do active harm to cells. Inside mitochondria, energy production happens in a complex, oxygen-assisted process. As a normal part of this process, some unstable molecules that contain oxygen, known as *reactive oxygen species*, are created. These mol-ecules are "reactive" because they have an unpaired electron, which pushes them to interact with other molecules like DNA and protein to either steal an electron or give away the extra electron. Molecules that are robbed or endowed also become reactive, creating a chain reaction inside cells.

Normally, this reaction is kept in check. Healthy cells are equipped with enzymes and other substances known as *antioxidants*. Antioxidants can donate or take electrons without becoming unstable themselves, so they balance out the unruly reactive species. However, when mitochondria become damaged, they produce more reactive oxygen species, tipping the balance.[9] The result is *oxidative stress*—chain reactions that disturb normal cell function.

Once again, this stress is felt most acutely in the brain. Because of its high energy needs, the brain is involved in more energy production, and therefore is exposed to more reactive oxygen species.[10] Moreover, the brain doesn't come with robust antioxidant activities and cannot mount an effective defense.[11] Already deprived of energy, and now attacked by oxidative stress, the brain caves in, losing its normal function and marking the onset of Alzheimer's. Sure enough, in autopsies, the Alzheimer's brain shows significantly high levels of oxidative damage.[12]

So if aged and damaged mitochondria are responsible for cognitive decline, why do some of us develop Alzheimer's and some don't? According to the mitochondria hypothesis, all of us *will* eventually get the disease if we live long enough—because defunct mitochondria are the *very definition* of aging, and no one can escape aging. But our mitochondria deteriorate at different rates.[13] Each of us is born with a baseline of mitochondrial function determined by genetics (specific genetic factors unknown). During our lifetimes, both genetic and environmental factors (again, specifics are unknown) determine how fast this baseline drops. If we have a low baseline and fast decline, we age faster and develop Alzheimer's in our sixties. If we have a high baseline and slow decline, we age slowly and won't show cognitive impairment until we are eighty or ninety, when we might have, well, died of other causes, Alzheimer's free.

With its lack of specifics, this mitochondria-driven disease process seems a little on the speculative side. But the hypothesis does have something intriguing up its sleeve. Strictly speaking, late-onset Alzheimer's is not inherited, as early-onset Alzheimer's is. There is no disease-causing genetic mutation, only the ApoE risk genes. But back in the 1990s, we noticed a curious pattern: if a patient has a *parent* who also has Alzheimer's, that parent is more likely to be the mother than the father.[14] In other words, mothers seem to contribute more risks than fathers do and are the "carriers" of the disease, so to speak.

In support of this idea, studies show that people with an Alzheimer's mother, even when they appear cognitively normal, harbor Alzheimer's-like symptoms in their brains: reduced metabolism, brain shrinkage, and beta-amyloid deposits.[15] By contrast, people with an Alzheimer's father do just fine, showing little difference from those who have no family history of the disease.

This maternal pattern lends credence to the mitochondrial hypothesis because mitochondria are inherited exclusively from the mother. During reproduction, the typical mammalian sperm contains 50–75 mitochondria, while the typical mammalian egg contains 100,000 to 100 million mitochondria.[16] Hence one is effectively diluted by the other, and what trace amount of sperm mitochondria exists, the fertilized egg actively destroys, so only maternal copies are passed on.[17] If Alzheimer's risk indeed comes from the mother, then that risk very likely lives inside mitochondria, in its separate little stretch of mitochondrial DNA.

The problem with this theory is that no definitive mitochondrial DNA culprit has been found. Some mitochondrial DNA mutations were reported to be more common in Alzheimer's patients (e.g., the total disappearance of about 5,000 bases), but there's plenty of conflicting evidence.[18] Aside from mutations, researchers looked into normal population differences. As our ancestral mothers settled in different regions around the globe, generations of their offspring developed unique changes in their mitochondrial DNA, giving rise to so-called mitochondrial groups and subgroups. For example, the L group is commonly found in Africa, while the M group dominates Asia. Again, there is conflicting evidence regarding which mitochondrial groups face a higher or lower risk of Alzheimer's.[19]

Missing a mitochondrial DNA smoking gun, the hypothesis turned to a powerful ally with his evidence in nuclear DNA.

The name of Allen Roses is known to everyone in the field of Alzheimer's. A neurologist at Duke University, Roses discovered the ApoE gene in the 1990s (see chapter 5). His discovery enabled and inspired decades of studies that use the ApoE4 risk gene, the ApoE2 protection gene, and the ApoE3 neutral gene to explore Alzheimer's causes and cures.

A maverick, Roses didn't think much of the prevailing beta-amyloid paradigm. He thought that beta-amyloid buildup was the result, not the cause, of Alzheimer's. A true maverick, Roses didn't think too highly of his own success either. He knew full well that ApoE didn't explain everything. Plenty of people without the ApoE4 risk gene develop Alzheimer's, and plenty of people with the gene escape it. All in all, ApoE explains only about 50 percent of late-onset Alzheimer's cases, which means there are additional genes out there waiting to be found.[20]

People *have* been looking, everywhere, for those mystery genes. Thanks to advances in genomic sequencing technology, researchers today can efficiently comb through the DNA of tens of thousands of people in search of risk genes. But the technology is a double-edged sword: its ability to compare huge amounts of DNA from diseased and healthy individuals allows us to detect genes that account for very little, if any, disease risk. At present, more than twenty Alzheimer's risk genes have been found. But compared with ApoE, their effect is minimal, not to mention inconsistent across studies. While the bad ApoE gene, ApoE4, can increase the odds of getting Alzheimer's by three to five times, these other genes make at most a 1.2–1.3 times difference.[21] Given their weak predictive power, these genes have no real value in clinical practice.[22]

Allen Roses had been looking too. In 2010, nearly two decades after discovering ApoE, he announced to the world that he had done it again—he had found the missing piece: the TOMM40 gene.[23]

TOMM40, pronounced just as it's written, resides on chromosome 19, the same chromosome where ApoE lives. In fact, the two are close neighbors—hardly a coincidence, considering that we have more than 20,000 genes scattered over twenty-three pairs of chromosomes. The TOMM40 gene instructs cells to produce the TOMM40 protein, and guess what: this protein is intimately related to mitochondria. Indeed, it lives right *inside* the mitochondria membrane. Like some bacteria, our mitochondria have two membranes: a smooth outer one and a corrugated inner one. TOMM40 is embedded in the outer membrane.

The full name of TOMM40 explains what it does in that membrane: *40* indicates the protein's weight (40 kilodaltons), and TOMM stands for *translocase of (the) outer mitochondrial membrane*. *Translocase* refers to a protein that moves other molecules, usually across a membrane. TOMM40, then, is a protein that moves molecules across the mitochondrial outer membrane into mitochondria. It accomplishes this by opening a gate in the membrane for molecules to pass through.

As a gatekeeper, TOMM40 is essential. Although mitochondria have their own mitochondrial DNA, that DNA is so small that it begets only 1 percent of the proteins that mitochondria need for daily operation.[24] The other 99 percent is coordinated by nuclear DNA, manufactured outside mitochondria, and then shipped in through the outer membrane. If TOMM40 fails to open the gate, these essential proteins can't get in, which can disrupt mitochondria's powerhouse duties, starve neurons, and cause cognitive decline.

What separates a good gatekeeper TOMM40 from a bad one, according to Roses, is a region of its DNA called *poly-T*. Most of our DNA, in TOMM40 and elsewhere, looks like a random combination of the four DNA bases—*A* (adenine), *T* (thymine), *C* (cytosine), and *G* (guanine). But one region in TOMM40 stands out by having only *T*s, so literally, it is a poly-T (e.g., TTTTTTTTTTTTTTTTTTTTTTTT). The number of *T*s in this region differs from person to person. A short poly-T has fewer than twenty *T*s, while a long one goes over thirty. This varying length, Roses claimed, is the missing piece that, together with ApoE, determines when a given person will develop late-onset Alzheimer's.

The idea goes like this: TOMM40 and ApoE, the two neighbors on chromosome 19, mingle in interesting patterns. When there is a bad ApoE4, the next-door TOMM40 would have a long poly-T. When there is a neutral ApoE3, the next-door TOMM40 would have either a short or a long poly-T.[25]

This pattern doubles up in each individual person because each of us has two sets of chromosomes (two chromosome 19s, two ApoEs, and two TOMM40s). How these patterns supposedly speed up or delay Alzheimer's is depicted in figure 16.1. The main takeaway is that long poly-Ts are harmful and speed up Alzheimer's, with or without the presence of bad ApoE4; by contrast, short poly-Ts are protective and delay Alzheimer's even if ApoE4 is there.[26]

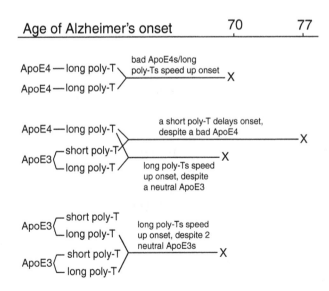

FIGURE 16.1 Combinations of ApoE and TOMM40 determine when a given person develops Alzheimer's.

Convinced that he'd found the missing gene, Roses founded a startup to further test and commercialize TOMM40. His goal was to create a genetic test formula that could predict a person's likelihood of developing Alzheimer's within five to seven years. A predictive test isn't a cure, but it still would be highly valuable. Researchers could use it to identify at-risk populations and test for treatments, and doctors could use it to improve patient diagnosis and care. Partial to red wine, Roses named his company Zinfandel Pharmaceuticals.

The rest of the world, however, was less impressed with Roses's idea. If you know the intricacies of how babies are made, you'll see why. I, for one, didn't. I thought that all eggs produced by a mother are identical and carry her DNA, but that's not true. During egg production, a mother's chromosomes randomly break, cross over, and recombine to produce an egg. This reshuffle produces a unique egg each time. The same is true with sperm. This reshuffling is nature's way to diversify and enrich our DNA. It also means that a chromosome is not passed down to a child intact and wholesale. Rather, it is inherited in segments. Segments with harmful genes can be selected against, while segments with useful genes may prevail.

When a chromosome randomly breaks into segments, genes on it that sit far from each other are more likely to be separated. Conversely, genes that sit near each other, like TOMM40 and ApoE, are more likely to end up in the same segment and be inherited together. Because the two genes often travel together, TOMM40's power to predict Alzheimer's may simply reflect the *same* power that ApoE has.[27] It is, in other words, the same finding dressed in different clothes.

Sure enough, in a larger study (with 20,000 participants, as opposed to the 300 in Roses's study), ApoE/TOMM40 was initially found to predicate a person's risk for Alzheimer's, but once the effect of ApoE was removed, TOMM40 no longer had predictive power.[28] This finding suggests that the detected risk for Alzheimer's comes solely from ApoE. Other studies, testing 1,000-1,500 participants each, had similar results.[29]

Roses was no stranger to such skepticism. Back in the 1990s, his ApoE finding was also challenged before it was widely proved. "It's like déjà vu all over again," Roses said, "several of the overt ApoE skeptics of 1993–1995 were also the same negative voices [now]."[30] To Roses's credit, some studies did connect long poly-Ts with a higher disease risk or an earlier age of onset.[31] But between these and Roses's studies, there is no agreement on what constitutes "long." In some studies, a long poly-T had more than twenty-one Ts; in others, it had more than twenty-seven or more than thirty Ts. There is also confusion on whether *very* long poly-Ts (presumably, thirty and above Ts) are protective, harmful, or both.[32]

In 2013, Roses's formula was used in a phase 3 trial aptly named TOMMORROW, a name that hints at TOMM40 and the hope for a better tomorrow for people with Alzheimer's. TOMMORROW is a unique trial. Coorganized by Roses's Zinfandel Pharmaceuticals and the Japanese company Takeda Pharmaceutical, the study tried to kill two birds with one stone: test both Roses's formula and Takeda's drug, pioglitazone. We have seen this drug before as an insulin sensitizer (see chapter 15), but there is also evidence that pioglitazone can improve mitochondria. In cell and animal studies, pioglitazone increased mitochondrial DNA, stimulated mitochondrial growth, reduced oxidative stress, and enhanced energy production.[33]

TOMMORROW enrolled 3,500 cognitively normal older adults and first ran Roses's test to determine their risk for Alzheimer's in the next five years. Participants determined to be high risk randomly received pioglitazone or placebo to determine if the drug can slow the disease onset. The low-risk participants received only placebo, in order to see if their TOMM40/ApoE combinations, as Roses's test predicts, are sufficient to stall Alzheimer's.

Unfortunately for Roses, this trial didn't come to fruition. In early 2018, Takeda called it quits. In interim analysis, their drug showed no hope for a treatment effect.[34] What this means for Roses's formula is unclear. No data have been published, and Roses is no longer with us to give his opinion. He succumbed to a heart attack in 2016, a tremendous loss to the field.

17

EAT YOUR VEGETABLES (AND BERRIES)

Growing up, I was constantly lectured by my mother that I didn't eat enough vegetables. Vegetables, I was told, contain vitamins essential for my health. In the name of health, I was proffered and cajoled (and sometimes coerced) to eat, among other green items, bell peppers, which I detested.

As an adult, I learned that people may not get enough vitamins from food if they don't stick to a healthy diet or if they don't absorb them well. Guilty of not eating particularly well (I still don't like bell peppers), I started taking daily multivitamins and have continued to do so for the last ten-plus years. It feels good to be doing *something* about my health, especially when it is something easy to do, unlike overhauling my diet.

Nevertheless, I wasn't clear on all the essential functions that vegetables and vitamins perform in my body. I had a vague sense that they are an important nutrient, aid my skin and bones in wonderful ways, and are often not to be found in the high-carb, high-fat modern diet. I was, therefore, pleasantly surprised that vegetables and vitamins, according to the mitochondrial hypothesis, can protect my brain and stall Alzheimer's.

Ironically, this is *not* a point that the hypothesis feels especially proud of. If the cause of Alzheimer's, as the hypothesis believes, is aged and defunct mitochondria, the real solution should be to replenish or repair those mitochondria. That would solve the root of the problem, increase energy production for the brain, and enhance neuron performance. And in theory, this can be done. It is well known that endurance exercise can generate new mitochondria in muscle and enhance athletic performance.[1] We can also slip healthy mitochondrial DNA into defunct mitochondria to repair their damage.

But in reality, neither option is practical. It turns out that brain cells are a lot harder to motivate than muscle cells. Exercising old mice on a treadmill may or may not regenerate brain mitochondria, depending on the study,[2] and no experiment has yet confirmed the effect in humans. As for DNA repair, the entrance into mitochondria is so tiny compared to the size of DNA that trying to slip in healthy DNA, as some researchers put it, is "akin to a camel passing through the eye of a needle."[3]

Given these reality checks, if the hypothesis wants something to work with *today*, it has to make do with an indirect solution: antioxidants. As mentioned in chapter 16, a suspected cause of aging and Alzheimer's is oxidative stress caused by reactive oxygen species, which react with DNA and protein and disrupt normal cell functions. Normally, reactive molecules are kept in check by antioxidants, which can donate or steal electrons to neutralize them. But when mitochondria become damaged, they produce excessive reactive molecules, tipping the balance. If we can't fix mitochondria to get at the root of the problem, maybe we can try increasing our body's antioxidant defenses so we can salvage at least some brain cells.

The most direct way to increase antioxidants is to boost them genetically so they overexpress in the body. So far, no such attempt has been made on humans—for good reason, judging by what happened to fruit flies that underwent the procedure. It did wonders for some flies; they increased metabolism and decreased oxidative damage, remained physically fit in old age, and enjoyed a one-third to one-half increase in life span.[4] But not all the flies were so lucky. In some, overexpressing the same antioxidants either failed to produce a benefit or actually *reduced* their life span.[5]

Before we can refine this procedure for humans, the hypothesis will once again have to settle. And that's where vegetables (and fruits) and vitamins come in. Vitamins E and C, which are plentiful in vegetables, are renowned antioxidants, and fruits, especially berries, are rich sources of several antioxidants, such as flavonols and anthocyanins. Lacking pharmaceutical treatments, the mitochondrial hypothesis diligently explored the cognitive benefits of these supplements and foods.

Because many people, like me, regularly take vitamin supplements, researchers had no difficulty organizing large population studies to assess vitamins' cognitive effects. In Cache County, Utah, 4,700 elderly residents were followed for three years for this purpose.[6] Among them, people who took both vitamins E and C saw a reduced risk for Alzheimer's. The separate use of vitamin E *or* vitamin C, however, had no effect. Researchers reasoned that the two vitamins needed to work as a team. Vitamin E donates an electron to a reactive oxygen molecule, neutralizing it. Having done that, vitamin E loses its antioxidant effect. Vitamin C then gives an electron back to vitamin E, effectively recycling it.

Unfortunately, this result didn't bear out elsewhere. In New York City[7] and Seattle, a combined 4,000 elderly residents were followed for an average of 4 and 5 years, respectively.[8] In both communities, the combined use of vitamins E and C did not reduce Alzheimer's risk—and neither did their separate use, in case you were wondering.

To settle the debate, we once again need clinical trials in which we can more precisely control what and how much of a supplement participants take and compare it with placebos. Multiple such trials have been completed. In one, some 6,000 nondemented elderly women were randomized to taking vitamin E or placebo for four years.[9] Despite the long treatment, the supplement failed to produce cognitive benefits. Men fared no better. Among some 7,000 nondemented elderly men, five years of taking vitamin E didn't reduce their risk for dementia.[10] When tried on patients with Alzheimer's and mild cognitive impairment, vitamins E and C delivered mixed results.[11] In some trials, they helped patients cope with daily activities, but in others, they made no difference or actually worsened cognitive test scores.

These results are personally disappointing, even though I never expected miracles from the vitamins that I take daily. If my mother finds out about these results, she will be very triumphant, for she always says that supplements are a copout. What I need, she insists, is eating more vegetables and fruits and getting my nutrients the natural way. However, if antioxidants are what we need to combat aging and cognitive decline, the amount that we can get from food is much lower than what's packed into concentrated pills, so why bother?

To my chagrin, I could be wrong about this—and my mother right. Apparently, among some 800 elderly residents of Chicago, those who consumed more vitamin E from *food* had a lower risk for Alzheimer's.[12] The same was found about higher *dietary* intake of vitamin C or vitamin E by 5,400 elderly Netherlanders.[13]

How can diets possibly help when larger amounts of vitamins from pills don't? Some researchers suggest that this is because *how long* is more important than *how much*: food intake reflects lifelong habits, while supplement use is of relatively shorter duration.[14] Other researchers say that this is because diets rich in vegetables and fruits give us not only vitamins, but other antioxidants—namely, polyphenols—that are the true savior of the brain.

Rich in plants, polyphenols are molecules with a ring-shaped chemical structure. Powerful antioxidants, they are used by plants to defend against ultraviolet radiation and disease-causing bacteria and viruses.[15] Based on more detailed chemical structures, polyphenols can be further divided into groups and subgroups, some of which exist in many plants, while others are specific to certain ones.[16] Everyday vegetables—spinach, onion, cabbage, and broccoli—are rich in polyphenols, as are berries of all kinds—blueberries, cherries, blackberries, and strawberries.

When old rats were fed these foods (specifically spinach and berries), they were rejuvenated in their cognitive and motor skills.[17] At twenty months of age, which is the human equivalent of sixty years, the rats escaped water mazes, balanced on rotating rods, and gracefully suspended from wires—quite the action-packed adventure. In transgenic Alzheimer's mice, eating blueberries likewise reversed cognitive impairment.[18]

In our efforts to verify these benefits in humans, we had an unexpected assistant: contraceptives. In 1960, the U.S. Food and Drug Administration (FDA) approved the first oral contraceptive. Five years later, the Supreme Court legalized its use among married couples; another seven years later, the ruling was extended to all women, married or otherwise. All of a sudden, millions of women across the country were on "the pill."

The safety of the pill, however, remained uncertain, and serious side effects, including stroke and heart disease, were suspected. To examine the long-term safety of oral contraceptives, in 1976, the National Institutes of Health sponsored the Nurses' Health Study. The study recruited 121,700 female nurses who were, at the time, thirty to fifty-five years old.[19] Every two years, the nurses completed questionnaires about their contraceptive use and health conditions.

In 1980, the study expanded to include diet and nutrition, so the nurses also completed questionnaires detailing what they regularly ate and how much.[20] The questionnaires were mailed to them every four years to capture their long-term dietary habits.

Four or five rounds of food questionnaires later, the original cohort of nurses had aged into their seventies. From 1995 to 2001,

16,000 of these elderly nurses underwent cognitive assessment. Among them, those who had eaten more blueberries and strawberries over the years had slower cognitive declines, passing for someone two years younger than their age.[21] In a separate assessment of 13,000 elderly nurses, those who ate more vegetables (though not fruits, in general) also had slower cognitive declines.[22] The most effective of these vegetables were leafy greens, such as spinach and romaine lettuce, and cruciferous vegetables, such as broccoli and cauliflower.

Similar results were found in mixed-gender populations. In a six-year study of 3,700 elderly residents from the south side of Chicago, eating more than two servings of vegetables a day delayed cognitive decline by a whopping five years.[23] Again, leafy greens had the strongest effect, and fruits were not helpful as a whole.

Outside the United States, researchers tracked the consumption of plant-based polyphenols through two national beverages: wine in France and green tea in Japan. Among the French, mild wine drinkers (one to two glasses a day) saw their Alzheimer's risk reduced by half, and moderate drinkers (three to four glasses a day) enjoyed a three-quarters risk reduction.[24] Among the Japanese, frequent green tea drinkers (five cups or more a day) reduced their risk for dementia by one-fourth.[25]

Together, these findings led some researchers to propose a "MIND diet," which emphasizes eating leafy greens and berries (but not fruits in general). It also encourages other polyphenol-rich, plant-based foods and beverages such as vegetables in general, nuts, beans, whole grains, and wine.[26] In a study of 900 elderly residents in the Chicago area, participants whose diet most resembled the MIND diet delayed their cognitive decline by up to 7.5 years and reduced their Alzheimer's risk by about 50 percent.[27]

Before you rush to the store and stock up on the items of the MIND diet, keep in mind that all these fantastic results come from "observational studies"—studies where we are merely observing people's diet and health outcomes, as opposed to providing them a dietary treatment and assessing health outcomes. Because there is no intervention, we can't say that polyphenol-rich diets actually *cause* cognitive benefits. For all we know, the two may simply coexist and are both caused by some other factors.

To prove cause and effect, we need studies that intentionally compare polyphenol-rich foods with placebos. So far, a handful of studies have been completed (all very small), but the results are more or less promising. Drinking blueberry juice every day for three months, compared with placebo, improved word learning and recall among nine older adults.[28] Drinking a mixed-berry beverage every day for five weeks improved forty older adults' ability to remember and process information.[29] Grape juice is iffier, though: in one study, drinking grape juice for three months improved memory in twelve older adults with mild cognitive impairment;[30] in another, four months of drinking the same juice had no such effect.[31]

Despite these studies—and even with more and larger studies—it's difficult to know precisely which vegetables or berries or other MIND foods can rescue our brain, or how much of them we need to eat. This is because plant foods do not come with neatly identified and measured polyphenols. Apples alone, for example, contain a dozen different polyphenols, and apples are one rare food item for which polyphenol compositions are known.[32] For most other foods, we don't know. The polyphenol contents of a given food also vary depending on sun exposure and rainfall, how ripe that food is, and how it is processed or cooked.[33] Further, once consumed, polyphenols are absorbed differently depending on a person's digestive system.[34] Interactions with other foods might make a difference too. Adding milk to tea, for example, has been reported to cancel out the effect of tea polyphenols.[35]

It is also possible that food, unlike purified drugs, works not through one but multiple mechanisms. In addition to being antioxidants, plant foods and their polyphenols are anti-inflammatory. And coincidentally, the Alzheimer's brain is known to harbor inflammation (see chapter 19 for details about the inflammation hypothesis).

Given the current state of knowledge, I am not particularly motivated to live by the MIND diet, but others may, especially if they already enjoy those foods. It's like what French researchers say: there's no reason to stop the elderly from enjoying wine, given its possible health benefits, but advising all elderly people to start drinking regularly would be premature.[36]

18

BLOOD, HEART, AND BRAIN

Ramesses II (1303–1213 BC) ruled Egypt for sixty-seven years. A fearless warrior, he led Egyptians into countless battles, conquering cities, repressing rebels, and expanding borders. A prolific builder, he erected more temples, took more wives, and fathered more children (allegedly more than 160) than any other pharaoh.[1] Many believe him to be *the* pharaoh who chased Moses and the Israelite slaves out of Egypt.

At a time when the average life span was no more than forty,[2] Ramesses II seemed an immortal god. He kept on living, and ruling, to the remarkable age of ninety. When he finally died, the pharaoh was meticulously mummified for a blissful afterlife. Priest embalmers washed his body with palm oil, made a cut in his torso to remove the inner organs, filled the body with incense and perfumes, and even stuffed the nose with peppercorns to preserve its angular shape.[3]

The removed liver, lungs, stomach, and intestines were preserved in jars and buried with the mummy. The pharaoh's brain, however, was nonchalantly discarded. In ancient Egypt, the brain was considered an inferior organ. It was cold and bloodless, and its function was supposedly to secrete water and mucus through the nose.[4] Surely the great pharaoh could do without it in his afterlife. To remove the brain, the embalmers inserted a long, metal hook up Ramesses II's nose into his head. They rotated the hook to blend and liquify the brain and then drained it through the nose.

Of all the organs, the heart was left inside the mummy. This was considered the one organ too precious to risk separation. To the ancient Egyptians, the heart was the source of consciousness, intelligence, and spirit—the passage to eternal life. Upon death, the heart would be weighed against the goddess Maat's feather, a symbol of truth and justice. If the heart were light, the deceased had lived virtuously and could enter eternal life. But if the heart were heavy with wrongdoing, the deceased would be devoured by the demoness Ammut, who is part lion, part crocodile, part hippo, and die the ultimate death.

Although the ancient Egyptians were mistaken in dismissing the brain, they were remarkably ahead of their time in understanding the heart. They knew that the heart was the pumping center of a system of vessels that distributes blood to the body.[5] Without a healthy heart, there could be no blood, no life, no mind. The heart and mind thus became synonymous. Diseases such as depression and dementia, which were already recorded in ancient Egyptian medical papyri, were entered accordingly as diseases of the heart.[6]

Several thousand years passed before modern medicine appreciated this connection between the heart and dementia. Although physically separate, the brain and the heart are connected by elaborate blood vessels, the so-called vascular system. An integral part of this system is arteries, blood vessels that carry blood from the heart to the rest of the body, including the brain, in order to deliver oxygen and nutrients. If arteries are clogged and hardened by cholesterol and other buildups, blood flow to the brain may be interrupted. Deprived of oxygen and nutrients, regions of the brain wither away and corresponding cognitive functions are lost. This condition is known as *vascular dementia*, the second-most-common dementia after Alzheimer's.[7]

Although they are both dementias, Alzheimer's disease and vascular dementia have notable differences.[8] Inside the brain, Alzheimer's features beta-amyloid plaques and tau tangles, while vascular dementia is recognized by lesions, or abnormal brain tissue. Further, Alzheimer's has a gradual onset, while vascular dementia can happen suddenly, when clogged arteries cause a stroke and damage brain functions. In Alzheimer's, memory is significantly affected, but in vascular dementia, memory may be mildly or not at all affected. Instead, patients suffer from early and severe impairment in executive function, characterized by an inability to pay attention, plan, organize, and follow through with tasks.

Despite these differences, separating Alzheimer's and vascular dementia is not always straightforward. Vascular dementia is a complex disease, with varied symptoms, and defies clear-cut diagnosis.[9] In addition, symptoms of the two diseases may coexist in, by some estimation, 40 percent of demented elderly people.[10] In fact, Dr. Alois Alzheimer's patient number one, Auguste Deter, and number two, Johann Feigl, both showed brain blood vessel damage, but the damage was minor compared to the numerous plaques found in their brains.[11] This, together with their gradual decline, led Alois Alzheimer to exclude vascular dementia and favor the diagnosis of a new disease that now bears his name.

Alois Alzheimer would be surprised that a century later, we are revisiting the idea that Alzheimer's disease may indeed find its origins in the heart and blood. Welcome to the vascular hypothesis.

We all know that in heart disease, cholesterol is a serious culprit. This waxy substance builds up inside blood vessels, blocks blood flow to the heart, and triggers heart attacks. In Alzheimer's, the vascular hypothesis also tags cholesterol as a suspect. In making this assertion, the hypothesis's number one evidence is ApoE. Recall that a bad ApoE, ApoE4, increases the risk for late-onset Alzheimer's, whereas a good ApoE, ApoE2, reduces that risk (see chapter 5 for more details). How ApoE does this is unknown, but the vascular hypothesis has put its money on cholesterol.

Contrary to the lore of healthy eating, cholesterol is not purely evil. It is, in particular, an essential nutrient for the brain, which stores 25 percent of the body's cholesterol despite its small weight.[12] In the brain, cholesterol is an important ingredient for building cell membranes and protecting axons, those long antennas that transmit brain signals.[13] Without cholesterol, there can be no information processing, memory, or learning.

Because cholesterol does not dissolve in the blood, in order to travel through the bloodstream, it has to ride with particles called *lipoproteins*, which have a soluble outer shell and can package cholesterol for transportation. An integral part of this outer shell is a protein called *apolipoprotein*. Different types of apolipoproteins exist, and one, commonly found in the brain, is *apolipoprotein E*. Where does this protein come from? It is produced per instructions of the ApoE gene! In other words, the late-onset Alzheimer's risk gene is intimately involved in cholesterol transportation within the brain. Coincidence? The vascular hypothesis doesn't think so.

Perhaps, the hypothesis suggests, Alzheimer's is simply the result of brain cholesterol deficiency. Perhaps different ApoEs are not equally good at transporting cholesterol. ApoE4, being an incompetent courier, does not allow enough cholesterol to be delivered to neurons, which weakens their function and leads to Alzheimer's.[14]

This is a beautifully straightforward story, but the reality is not that simple. When we examined the brains of deceased Alzheimer's patients, sometimes the cholesterol measurement came out lower than normal, as the story predicts, but at other times, it was higher than normal.[15]

With these contradictions, the hypothesis gets a little wishy-washy. In an alternative story of how cholesterol causes Alzheimer's, excessive rather than deficient cholesterol is blamed. Presumably, high cholesterol, by blocking blood vessels, impedes blood flow to the brain and causes damage and cognitive loss. Cholesterol may also beget beta-amyloid deposits. When mice were fed a high-cholesterol diet, beta-amyloid deposits in their brains increased; when they were given cholesterol-lowering drugs, the deposits decreased.[16] Population studies provide additional evidence. In Finland, among 1,500 middle-aged adults, high cholesterol increased their risk for Alzheimer's twenty years later.[17] Similar results were found among 9,800 northern Californians.[18]

Before you buy into this high-cholesterol story, know that it is not without its own contradictions. In Framingham, Massachusetts, researchers followed some 1,000 residents for forty years and found that high cholesterol had no effect on Alzheimer's.[19] In fact, among 1,100 elderly New Yorkers, high-density lipoprotein (HDL) cholesterol (aka the good cholesterol) was linked to a reduced risk for Alzheimer's.[20]

Vexed by these contradictions? Well, another caveat may render everything moot. In the population studies cited so far, what was measured was participants' blood cholesterol, not brain cholesterol—which we can't directly measure in living humans. As far as we know, blood and brain cholesterol are separate things. Blood cholesterol is made by the liver and influenced by diet, while brain cholesterol is made on-site by the brain itself. The two don't mingle because of the *blood-brain barrier*, a border of tightly packed cells separating circulating blood from the brain. Given this fact, how blood cholesterol, whether high or low, affects the brain is a mystery.

Perhaps, contrary to conventional wisdom, there *is* cross-talk between the two?[21] In one study, mice fed a high-cholesterol diet (and killed later for analysis) showed increased cholesterol not only in the blood, but also in their brains.[22] Does this happen in humans? We don't know.

Every time you go to the doctor's office these days, they take your blood pressure, even for routine cleaning appointments at the dentist. Every time, I am reminded that my blood pressure is a bit low. I duly nod each time and say, "It makes me dizzy when I stand up too fast." "Yes," a sympathetic doctor would say, "but that's better than *high* blood pressure."

High blood pressure, also known as *hypertension*, happens when blood traveling inside the arteries builds up excessive force and pressure. At first thought, a strong blood flow may not seem all that bad. But blood forcefully beating on the artery walls will damage and weaken their delicate tissues. Over time, arteries may rupture and blood clots may form, disrupting blood supply to the brain and causing stroke, brain damage, and a sudden onset of vascular dementia.

Given this series of events, taking antihypertensive drugs to lower blood pressure ought to help reduce vascular dementia. This theory was tested in a pan-European study in the early 1990s.[23] Over 2,400 cognitively normal older adults with high blood pressure randomly took antihypertensive drugs or placebo. After five years, those taking the drugs reduced their blood pressure and, as expected, had fewer cases of vascular dementia.

Unexpectedly, they also had fewer cases of Alzheimer's.

A world away, in Honolulu, Hawaii, more than 4,600 Japanese-Americans were recruited through their World War II military service registration for a decades-long aging study.[24] These veterans had their blood pressure measured back in the 1960s and 1970s, when they were middle-aged men. Then, in 1991 through 1993, the surviving participants were tested for dementia. Among them, those with high blood pressure in midlife were more likely to develop Alzheimer's in old age; taking antihypertensive drugs reduced this risk.

Together, these studies seem to say that Alzheimer's and vascular dementia have a shared origin in blood pressure. Perhaps high blood pressure not only causes stroke and sudden vascular dementia, it can gradually damage blood vessels, allow cholesterol and other substances to build up within them, diminish blood flow and oxygen supply to the brain, and beget a gradual dementia typical of Alzheimer's.[25] Damaged blood vessels also lose their ability to clear away beta-amyloid, allowing it to build up in the brain and cause further damage.[26]

This theory could explain why Alzheimer's risk increases with age. As we grow older, blood flow to the brain naturally declines. From the time we are twenty to when we are sixty-five, blood flow to the brain drops 15–20 percent.[27] In youth, the brain can handle some damaged vessels and reduced blood flow, but in old age, the already-struggling brain is pushed over the edge. Accordingly, compared with cognitively normal older adults, those with Alzheimer's show further decreased blood flow to parts of the brain that are important for memory and cognition, such as the hippocampus.[28]

Unfortunately, as you probably expected, contradictory evidence exists. If damaged blood vessels cause Alzheimer's over time, then vessel damage ought to coexist with hallmarks of the Alzheimer's brain—beta-amyloid plaques and tau tangles. But autopsy studies found no such relationship.[29] More ironically, some Alzheimer's patients show *increased* blood flow in their brains, sometimes in the same brain region where fellow patients have seen decreased flow.[30]

These contradictions, advocates of the vascular hypothesis argue, reflect the dynamic nature of the Alzheimer's brain. As the disease progresses, the brain is not simply dying, but also adjusting. Blood flow decrease is a sign of gradual decline; blood flow increase is the brain's effort to compensate, if only temporarily, for its destruction.[31] Ergo, decrease and increase coexist. While this explanation makes big-picture sense, it certainly is short on details: What precisely does this dynamic landscape look like? When and where does blood flow increase? When and where does it decrease?

If one looks closer at population studies that connect high blood pressure with Alzheimer's, one will uncover additional discrepancies. It is well known that a blood pressure reading comes in two parts (e.g., 120 over 80). 120 is the *systolic* pressure, which is the blood pressure on the artery wall when the heart beats, and 80 is the *diastolic* pressure, which is the blood pressure between heartbeats. In the Honolulu Japanese-American study, high diastolic pressure increased Alzheimer's risk. But in Finland, that pressure made no difference; rather, high systolic pressure increased Alzheimer's among 1,400 participants.[32]

All things considered, the current evidence for a vascular hypothesis, whether tying Alzheimer's to cholesterol or blood pressure, seems weak. But that didn't stop those who believed in it from joining in the search for a cure. Modern drug design is supposed to follow a rational process: identify disease-causing agents, clarify their working mechanisms, and then purposefully design drugs to target them. But in desperately trying to conquer Alzheimer's—and failing continuously—we seem to have entered a postrational era, in which expediency (and positive thinking) keep us going. Because drugs for high cholesterol and high blood pressure already exist, try them we must.

Multiple cholesterol-lowering drugs are on the market, but the most common are statins. To be precise, statins are not one drug, but rather a group of drugs with the same working mechanism. They reduce cholesterol by inhibiting an enzyme called HMGCR, which is required for cholesterol synthesis. Without HMGCR, cholesterol cannot form.

Statins rose to fame as an Alzheimer's treatment with the publication in 2000 of two studies claiming that the drugs could prevent Alzheimer's. One study was conducted in the United States,[33] and the other in the United Kingdom.[34] In both, researchers combed through large medical records to compare the diagnosis of Alzheimer's in statin users and nonusers. Three U.S. hospitals, 360 U.K. practitioners, and 120,000 records/patients later, the findings were fantastic: statin users reduced their risk for Alzheimer's by about 70 percent!

These results are widely cited and inspired much of the later work on statins. However, if one bothers to look closely, several of the details of these two studies don't quite add up. Between them, multiple types of statins prevented Alzheimer's except for one commonly prescribed statin called simvastatin. But simvastatin is similarly capable of reducing cholesterol. Moreover, people who took nonstatin cholesterol-lowing drugs, despite successfully controlling cholesterol, saw no cognitive benefits.

So, either statins worked wonders through ways unrelated to cholesterol, or, these famous studies were wrong and statins didn't work. The latter seems more likely, given what happened in Cache County, Utah.[35]

Starting in 1995, 5,000 elderly Cache County residents were enrolled in a population study and underwent a round of cognitive testing. A total of 200 Alzheimer's cases were diagnosed, and these cases were less common among residents who took statins. Apparently, then, statins reduced Alzheimer's. Three years later, the same group underwent another round of cognitive testing. By then, 104 new cases of Alzheimer's had developed, but these cases had no connection with statin use.

Why the difference within the same group? What happened to these residents three years later? Well, *that* happened: time. At the first cognitive testing, there was no passage of time. What we had was a fixed snapshot of people's states of mind and statin use. Because of this, the seeming connection between Alzheimer's and statins may be the exact opposite of what we thought. Rather than *statins preventing Alzheimer's*, it may be that *Alzheimer's patients are not taking statins*. Statins are primarily prescribed to lower cholesterol and prevent heart and vascular disease, so doctors are less likely to prescribe them to patients who show signs of cognitive impairment. The way doctors see it, these patients may not stick to the prescription, are more likely to have negative drug reactions, and may not get much benefit from the drug because of their limited life expectancy.[36]

More reliable than the snapshot is the second round of cognitive testing. This time, what were gathered were *newly developed* Alzheimer's cases in a three-year time window. If statins really did prevent Alzheimer's, people who were using statins during this time ought to have a lowered disease risk. But that didn't happen in Cache County, nor in other studies that considered the time factor.[37] As for the two famous, seminal studies claiming that statins prevented 70 percent of Alzheimer's cases, sure enough, they didn't fully account for the time factor.[38]

In more conclusive bad news, statins failed in multiple clinical trials. In an eighteen-month trial with 400 Alzheimer's patients, a statin (simvastatin) did no better than placebo at slowing the disease.[39] And in larger trials with tens of thousands of nondemented participants, years of statin use had no cognitive benefits over placebo.[40]

The ironic high point of the statin experiment came in 2012. That year, based on postmarketing reports, the U.S. Food and Drug Administration (FDA) announced that statins actually hamper the brain and cause "cognitive side effects (memory loss, confusion, etc.)."[41] The effects were supposedly "non-serious" and "reversible" once statins are stopped.[42] Still, those are pretty ridiculous side effects for a proposed dementia drug.

Antihypertensive drugs do not have these dramatic ups and downs in their quest to become an Alzheimer's drug. They are just all over the place. These drugs use different methods to lower blood pressure, such as relaxing blood vessels, slowing heart rate, and reducing blood volume. These differences make cross-study comparisons more difficult.

In the same Cache County study that cast doubt on statins, residents taking antihypertensive drugs actually saw reduced Alzheimer's during the three-year time window.[43] But this result did not hold up elsewhere. In Rotterdam, the Netherlands, antihypertensive drugs failed 7,000 elderly residents within a span of two years.[44]

Clinical trials did not clear things up. In a small trial, fourteen patients with high blood pressure and a family history of Alzheimer's randomly received an antihypertensive drug (ramipril, which works by relaxing blood vessels) or placebo for four months.[45] Compared with some other antihypertensive drugs, ramipril can cross from the blood to the brain, which makes it a more likely dementia drug. In the trial, ramipril reduced participants' blood pressure just fine, but it had no effect on cognitive performance.

In a larger trial with 160 mild-to-moderate Alzheimer's patients, six antihypertensive drugs were tested for one year.[46] This time, two of the drugs (perindopril and captopril), which work similarly as the previously failed ramipril, managed to slow cognitive decline. Curiously, the other four drugs lowered participants' blood pressure equally well but had no cognitive benefits, so whatever worked, it wasn't lowering the blood pressure.

With no consensus, but no catastrophic failure either, the pursuit of antihypertensive drugs continues. Several trials are in progress, and results are promised in the near future.[47] I, for one, do not feel especially hopeful.

19

A MISSED OPPORTUNITY

On June 4, 2019, the *Washington Post* ran a story[1] chastising Pfizer for hiding critical data about its blockbuster drug Enbrel, an anti-inflammatory drug with global sales of over $7 billion.[2] The fact that pharmaceutical companies would bury the side effects of their drug is old news. The *Post* wasn't falling for such clichés. It had a surprising, novel angle: Pfizer wasn't hiding *bad* effects of the drug; it was hiding *good* effects—very good effects.

Approved by the U.S. Food and Drug Adminstration (FDA) in 1998, Enbrel is a prescription drug for autoimmune diseases, notably rheumatoid arthritis. In these diseases, the immune system, for unknown reasons, mistakenly attacks healthy body tissue, especially joint tissue. Over time, this results in joint pain and swelling, cartilage and bone loss, and damage to organs such as the eyes and skin. A chief coordinator of this immune attack is a protein called *tumor necrosis factor (TNF)*. Enbrel works by blocking the activity of TNF, thus reducing inflammation and salvaging healthy tissue.

According to the *Post*, Enbrel is even more amazing than that. Evidently, it could "reduce the risk of Alzheimer's disease by 64 percent" and "could potentially safely prevent, treat and slow pro-gression of Alzheimer's disease."[3] This discovery apparently was made by a group of Pfizer researchers back in 2015, but Pfizer execu-tives decided to sit on it: they organized no clinical trials, no publica-tions, no announcements, no nothing. Supposedly, Pfizer buried the finding because its patent for Enbrel was expiring, and the company didn't want to tip off its competitors or pay for expensive trials, only for generics to swamp the market. An opportunity to advance sci-ence and benefit millions of patients, the *Post* lamented, was lost to corporate financial strategies and the bottom line.

The story drew considerable attention, as the *Post* anticipated. Reader responses ranged from the more tempered "This is so ethi-cally and morally wrong that it MUST stop. It is obvious that these business [sic] have lost their moral compass and must have oversight by professionals with no monetary benefit"[4] to the fully enraged "If there was potentially a treatment and NO ONE THOUGHT IT WAS IMPORTANT TO PURSUE IT THEY SHOULD BE PUNISHED, FIRED OR WHATEVER IS NECESSARY INCLUDING PROSECUTION!!! HOW DARE THEY PLAY WITH PEOPLE'S LIVES??? THIS IS ABSOLUTELY DISGUSTING."[5]

What these readers may not have realized, in the heat of the moment, is that the *Washington Post* is *also* a business, *also* out to make money—by selling stories and attracting readers and adver-tisements to its newspaper and website. Pfizer's story, rife with greed, deception, and human suffering, is precisely the kind of story that sells. It's a case of sensational reporting—with the trappings of investigative journalism.

Where did the number come from that Enbrel reduces Alzheim-er's risk by 64 percent? Medical insurance claims. In 2015, the Pfizer researchers in question analyzed a mountain of insurance claims from people with rheumatoid arthritis and other inflammatory diseases—and hence potential Enbrel users. The claims were divided into two groups of 127,000 each. The first group (let's call it the Alz group) included people who had Alzheimer's on top of their inflam-matory conditions; the other group (the non-Alz group) did not have Alzheimer's. Comparing the two groups, Pfizer researchers found 110 Enbrel users in the Alz group and 302—that is, 64 percent more—Enbred users in the non-Alz group. Taking its cue from this, the *Post* story concluded that Enbrel helped to land people in the non-Alz group by a margin of 64 percent.

Is this a sound conclusion? Not really. Examining insurance claims in this way constitutes a so-called case-control study: that is, we start with diseased people (the cases) and nondiseased people (the controls) and look back at their life experience to find factors (Enbrel use) that may have increased or decreased their disease risk. In case-control studies, data analysis takes the form of calculating the *odds ratio*. This scary-sounding concept is simply the ratio of *two* odds (odds 1 and odds 2), so it is calculated as odds 1 divided by odds 2. In this case, odds 1 is the likelihood of Enbrel users developing Alzheimer's, and odds 2 is the likelihood of non-Enbrel users developing Alzheimer's. Each of these is itself calculated by division. Odds 1, for example, is the number of Enbrel users with Alzheimer's divided by the number of Enbrel users *without* Alzheimer's. Crunching these numbers, we get an odds ratio of 0.36.

What does this number mean? It means that the odds of taking Enbrel and developing Alzheimer's is 0.36 times lower than *not* taking Enbrel and developing Alzheimer's. Wait—isn't that the same as "Taking Enbrel reduces Alzheimer's risk by 64 percent," which is precisely what the *Washington Post* claimed?

Well, no. Case-control studies start with *predetermined* diseased and nondiseased people, rather than an unknown population who may or may not develop a disease. As such, we cannot measure the risk of the disease happening to a random person. A 0.36 odds ratio simply means there is a reduced association between Enbrel and Alzheimer's. When a disease is rare, the number of nondiseased people approximates a whole population, so in those cases odds ratio may be considered an approximation of disease risk. However, Alzheimer's is *not* a rare disease among the elderly.

If this still sounds like statistical mumbo jumbo and hair-splitting, consider practical factors that will diminish the glow of the finding. To start, we do not know if the Alz group and the non-Alz group selected from the insurance claims were truly comparable: did they have the same age, the same genetic profiles, the same health histories? Without this "matching" technique, we don't know if it was Enbrel or any or all of these other factors that caused the difference in Alzheimer's onset. Think about it—given 127,000 Alzheimer's patients and 127,000 nonpatients, if we dig into every drug they ever took and every life experience they ever had, then we may, *by chance*, find something that seems to be associated with reduced Alzheimer's.

All this is not to say that Pfizer's internal finding is worthless. It *is* a potential lead, but a very initial one, and certainly not the automatic go-ahead for an $80 million clinical trial that the *Washington Post* made it out to be. It is also one among many, many potential leads that we can ponder to see if longer, bigger, and more expensive studies are warranted, which is exactly what Pfizer did: it pondered the question and decided that the warrant was not there—for reasons we now explore.

The Alzheimer's brain is an inflamed brain. This we suspected as early as 1975, when researchers found traces of immune reactions in beta-amyloid plaques.[6] By the mid-1990s, numerous markers of inflammation were confirmed in the diseased brain.[7] Chief among them is activated microglial cells.

Microglia are small, nonneuron supporting cells. They are the brain's resident immune cells, constantly scanning their surroundings for signs of injury and infection. At the first sign of danger, they are activated and mount aggressive attacks by swallowing and digesting pathogens, foreign intruders, dead neurons, and anything else that may harm the brain. In the Alzheimer's brain, activated microglia can be seen surrounding beta-amyloid deposits, presumably trying to get rid of them.[8] Activated microglia also release an army of pro-inflammatory proteins and molecules. These agents either join in the immune attack or cheerlead on the side, feeding microglia signals to keep on fighting.[9] It's a chaotic war zone.

In fact, it's more than chaotic; after some time, it gets downright confusing who is friend and who is foe.[10] As beta-amyloid keeps on coming, the microglia become overwhelmed and burned out. In their state of frustration and madness, they sweep through the brain, damaging healthy neurons in their path. Worse still, microglia themselves may start to release beta-amyloid![11] All these insults cause further brain inflammation, and thus more damage, and thus more inflammation, and so on and so forth until cognitive function is lost. This is the inflammation hypothesis for what causes Alzheimer's.

The hypothesis doesn't see inflammation only as a result of Alzheimer's, merely a reaction to beta-amyloid, tau, or other brain pathologies. Rather, it posits that inflammation is a necessary contributor—an actual cause of the disease. In other words, there is initial inflammation in the brain *before* the rise of Alzheimer's pathologies. Where does this inflammation come from? The hypothesis offers multiple possible sources: head trauma, aging, diabetes, or obesity.[12] These events trigger immune attacks, which worsen beta-amyloid and tau brain damage, which further triggers immune attacks. The hypothesis, then, turned the chicken-and-egg question on its head. Rather than inflammation causing Alzheimer's or Alzheimer's causing inflammation, it works both ways.

The hypothesis has *some* genetic evidence. As mentioned in chapter 16, other than APOE, some twenty other risk genes, with much smaller predictive powers, have been found to be connected to late-onset Alzheimer's. Several of these genes are involved in inflammatory responses, especially microglia activation and function.

One of them is TREM2. Multiple versions of this gene exist, and some are found to increase Alzheimer's risk.[13] How or why they do that remains a mystery, because (get this) TREM2 has both pro- and anti-inflammatory effects.[14] In transgenic Alzheimer's mice, TREM2 either activates microglia, removes beta-amyloid, and protects the brain, or, alternatively, it *over*activates the immune system, increases beta-amyloid, and damages the brain.[15] Another oft-cited gene, CR1, tells the same conflicting story. Certain versions of CR1 are supposed to increase Alzheimer's risk[16]—either because they increase immune attacks and brain damage or because they reduce immune responses and beta-amyloid removal.[17]

As genetic evidence remains confusing, the more compelling evidence for the inflammation hypothesis comes from population studies.

Most of us are no stranger to occasional (or regular) aches and pains: headaches, muscle aches, menstrual cramps, joint pain. When we are in pain, our go-to medicines are usually things like Advil and Aleve, which are nonsteroidal anti-inflammatory drugs (NSAIDs, pronounced "en-saids"). Among elder adults (65+), the use of NSAIDs is especially high: in general practice, up to 96 percent of older patients use NSAIDs.[18]

NSAIDs work by inhibiting two enzymes: COX-1 and COX-2. These enzymes produce compounds (called *prostaglandins*) that promote inflammation and pain. By inhibiting one or both of the COX enzymes, NSAIDs fight pain and inflammation in one strike. Some NSAIDs are available over the counter, such as aspirin, ibuprofen, and naproxen; others require prescriptions, such as celecoxib and meloxicam.

Multiple studies have found that people who take NSAIDs for unrelated conditions such as arthritis have a reduced risk for Alzheimer's.[19] In Rotterdam, the Netherlands, 7,000 older adults were followed from 1991 to 1998.[20] During this time, 293 of them developed Alzheimer's. Comparing NSAID use and Alzheimer's onset, researchers found a significant correlation: short-term NSAID use (i.e., less than one month) reduced the risk for Alzheimer's by 5 percent; midterm use (i.e., one to twenty-four months) reduced the risk by 17 percent; and long-term use (i.e., more than twenty-four months) reduced the risk by a whopping 80 percent.[21]

Likewise, in Cache County, Utah, 104 of 3,227 elderly residents developed Alzheimer's over a three-year period, and longer (i.e., more than two years) use of NSAIDs was associated with reduced risk.[22] The protective effect of NSAIDs was limited to participants who had been on the drugs *before* the study. Those who started taking NSAIDs only recently saw little to no protection.

Together, these findings motivated a series of clinical trials testing the cognitive effects of NSAIDs. In one trial, 700 mild-to-moderate Alzheimer's patients were recruited across the United States and randomly assigned to a prescription NSAID (rofecoxib) or placebo.[23] Unfortunately, after one year of dosing, rofecoxib did not arrest patients' decline. Another study in Italy tested common, over-the-counter ibuprofen among mild-to-moderate Alzheimer's patients.[24] Again, after one year of dosing, the drug performed no better than placebo.

Faced with these negative results, trial organizers suggested that maybe NSAIDs need to be taken longer and earlier—before irreversible brain damage has occurred—to be effective. After all, in population studies, NSAIDs only really protected people who took the drugs for years, and well before the onset of dementia symptoms.

Responding to this possibility, in 2001, the National Institute on Aging (NIA) sponsored a lengthy trial—up to seven years—and recruited elderly participants who were free of dementia symptoms.[25] Two NSAIDs were tested: naproxen and celecoxib. With over 2,500 participants, the trial was poised to offer some conclusive evidence on the effect of NSAIDs—or lack thereof. However, three years in, the trial was abruptly terminated when news broke from an unrelated study that celecoxib, at the same dosage tested, more than doubled the risk of death from heart attack, heart failure, and stroke.

In the aftermath of the trial, researchers examined what partial data they had managed to accumulate. Fortunately, participants taking NSAIDs didn't suffer the same alarming death rates, even though they did have increased risk for heart problems and stroke.[26] In terms of cognitive status, twenty-five people developed Alzheimer's. Unfortunately, not only did NSAIDs *not* reduce Alzheimer's risk, they seemed to *increase* it.[27]

Although these findings were as negative as they come, some had the lingering hope that had the trial been allowed to run its course, things might have looked different.[28] Perhaps these twenty-five diseased participants were on the cusp of showing cognitive symptoms. Perhaps their brains had already been laced with beta-amyloid. If so, by suppressing immune responses, NSAIDs would prevent microglia from clearing out beta-amyloid and would thus usher in the disease. Had there been more time, people who were truly years ahead of disease onset might have gotten a chance to show off the effect of NSAIDs.

Not to sound like a broken record, but this last hope was dashed in a 2011 Canadian trial dubbed INTREPAD.[29] INTREPAD tested naproxen in 195 participants with a family history of Alzheimer's. Not only were these participants free of cognitive complaints at the start of the trial, they were the "younger elderly" (i.e., on average 63 years old, a decade younger than their affected family members)[30] and unlikely to be on the verge of disease onset. After two years of dosing, the result was, as the trial organizers admitted, "extreme pessimism."[31] Naproxen offered no cognitive benefits. It did, however, have nasty side effects, including digestive tract problems and vascular-heart disorders.

When the INTREPAD results were announced, some saw it as "closing the book on NSAIDs for Alzheimer's prevention."[32] INTREPAD's leading researcher, John Breitner from McGill University, agreed: "I do not think there should be any more clinical trials of NSAIDs for prevention or treatment of AD [Alzheimer's disease]," he said. "There was a lot of hope, based on epidemiological data, that these drugs would prove to be beneficial, but we've learned over the last 10 to 15 years that when you test NSAIDs in trials, they are not beneficial and invariably, you make people sick."[33]

All these hopes and disillusions about NSAIDs, about the inflammation hypothesis, had played out well before the *Washington Post* story on Enbrel. Had the *Post* bothered to offer these details, its story would have been much less sensational and its readers more inclined to believe that science, rather than money alone, dissuaded Pfizer from pursuing Enbrel.

But wait—Enbrel is not an NSAID, is it? That's correct. As mentioned earlier, Enbrel is a TNF blocker: it reduces inflammation by blocking the pro-inflammatory signal TNF, not by inhibiting COX enzymes as NSAIDs do. So..., the failure of one doesn't predict that of the other, does it? That's also correct. But the thing is, Enbrel *had* been proved disappointing too. In 2011, it was tested against placebo among forty-one mild-to-moderate Alzheimer's patients.[34] After six months of dosing, it showed no benefit in either cognitive or overall function.

Still, if Pfizer has nothing to hide, it should at least publish its insurance claims data and let others judge the worth of Enbrel for themselves, shouldn't it? Well, similar findings *have* been published. In 2016, a group of non-Pfizer researchers published their examination of eight years of claims data from a large health database (Verisk Health), concluding that Enbrel users had a reduced risk for Alzheimer's.[35] Interestingly, two other TNF blockers examined in the study offered no such benefits. So, whatever worked (if anything did), it wasn't Enbrel's ability to block TNF and reduce inflammation.

So Pfizer was wrongfully accused this time, but where does this leave the inflammation hypothesis? Frankly, with lots of speculation—and confusion. We know for sure that inflammation exists in the Alzheimer's brain, but does it actually cause the disease, or is it merely a reaction to the disease? It is all very well for the hypothesis to say that it is both, but that answer offers no clear path toward drug development. In fact, it points in opposite directions of drug development. If inflammation is a *cause* of Alzheimer's and an overactivated immune system is what begets brain damage, then we need anti-inflammatory drugs to calm the system. If, instead, inflammation is a *result* of Alzheimer's and represents the immune system's effort to rid the brain of harmful buildup, then we need pro-inflammatory agents to help the system.

To this, the hypothesis would reply that it ain't so easy, so black and white, because inflammation plays different roles at different stages.[36] In a normal brain, reducing inflammation keeps things in tip-top condition and preempts Alzheimer's. If and when beta-amyloid starts to accumulate (because of brain trauma, general inflammation in the body, or whatever other reasons), brain inflammation is needed to clear out this anomaly. When beta-amyloid continues to pile up and microglia are turned against healthy tissue, brain inflammation once again becomes a liability.

In creating this elaborate scenario, the hypothesis has gotten a little paranoid about what is normal and what is diseased. Authors use the word *normal* in scare quotes to imply that some participants may *seem* "normal" but aren't *really* "normal" in the brain.[37] If so, can we ever figure out which precise moment to cast which (anti) inflammatory drug on which normal/diseased body? Seems tricky to me.

Then again, in surprising news, the Chinese thought they had it figured out!

On November 2, 2019, Shanghai Green Valley Pharmaceuticals announced that China's National Medical Products Administration granted preliminary approval of its drug, GV-971, for the treatment of mild-to-moderate Alzheimer's disease.[38] According to the news release, the drug would hit the Chinese market by the end of the year.

Scholarly publication on GV-971 is limited. According to the available data, GV-971, an ingredient derived from seaweed, treats Alzheimer's by treating the gut![39] The idea goes like this: When gut microorganisms are out of balance, they increase certain amino acids (phenylalanine and isoleucine); these amino acids encourage pro-inflammatory, T helper type 1 cells to move from the blood into the brain; once in the brain, these T-cells activate microglia, causing brain damage and cognitive decline. GV-971, by adjusting gut microorganisms, arrests this process.

In transgenic Alzheimer's mice, GV-971 reduced the two culprit amino acids and improved the animals' cognitive function.[40] In a six-month phase 2 trial in China with 255 mild-to-moderate Alzheimer's patients, the drug failed to raise cognitive test scores, but it did improve patients' overall function.[41] Yet this result was reversed in a nine-month phase 3 trial with 818 Chinese patients.[42] This time, GV-971 failed to improve patients' overall function but raised cognitive test scores. There is also something quite odd about the test scores: GV-971 mustered a significant effect only because the placebo group suffered a dramatic, uncharacteristic, last-minute decline.

Despite this inconclusive—and also very limited—data with a single, short, phase 3 trial, the Chinese authority had given GV-971 the green light. While Chinese social media sites overflowed with patriotic cheers and salutes, the rest of the world asks for a global trial for market entry. Green Valley plans to begin one in 2020, including participants from the United States, Europe, and Asia. No details are available as of this writing. According to the latest news, on April 8, 2020, Green Valley received clearance from the FDA to start the U.S. part of the global trial.[43]

20

PARADIGM SHIFT(?)

From Tauism to type 3 diabetes to mitochondrial dysfunction, multiple hypotheses predict the cause and cure of Alzheimer's. Yet the beta-amyloid hypothesis remains the presiding paradigm, the leading school of thought with the most researchers and research dollars. Continuously failing anti-amyloid drugs have not materially changed this reality.

Nonbelievers in beta-amyloid, or non-Baptists, see this as the failure of Alzheimer's research—of modern science in general—to correct its wrong. Once a large scientific enterprise has its wheels turning and momentum going, it can't easily stop itself and change direction. Hundreds of Baptists devoted their life's work to beta-amyloid, created a shared body of knowledge, secured grants and funding, and made a name for themselves—they will want to stick it out to the bitter end.

There is some truth behind this blaming. Despite the common belief that science is the epitome of innovation, day-to-day scientific work follows well-established protocols, and truly outside-the-box thinking is rare. Until we reveal the true face of Alzheimer's, we don't know what we don't know, and Baptists may well be the blind men feeling the elephant.

That said, in making Baptists the straw man, the insecure conservatives holding back "real" Alzheimer's research, we risk turning a scientific debate into a moral or political one. True, antiamyloid drugs have all failed, but so have other drugs. While the beta-amyloid hypothesis has gaps, so do alternative hypotheses.

As an outsider, I get why so many researchers identify with the beta-amyloid hypothesis. It has a lot going for it that other hypotheses don't. Besides the plain presence of beta-amyloid deposits in the Alzheimer's brain and the damage they cause to neurons, genetics constitutes a strong cornerstone of the hypothesis. Mutations in three genes (APP, PSEN1, and PSEN2) cause early-onset Alzheimer's (see chapter 4), and these mutations either increase beta-amyloid or the ratio of longer and stickier beta-amyloid 42, with the end result being increased deposits in the brain.

Non-Baptists don't object to these genetic findings, but they question their relevance. APP and PSEN mutations cause rare, early-onset Alzheimer's, which, non-Baptists say, may be inherently different from the common, late-onset Alzheimer's that is the public health concern. We can't completely deny this argument. For the most part, we have simply assumed that the two disease forms are the same because they *appear* the same, other than their age of onset. But you know what they say about appearances.

Then, in the summer of 2012, another mutation came about that finally made a concrete connection between beta-amyloid and *late-onset* Alzheimer's.[1] Combing through the genomes of 1,795 Icelanders, researchers discovered a rare mutation carried by 0.5 percent of them. Further testing in Scandinavian populations showed that the mutation exists in 0.2–0.5 percent of Finns, Swedes, and Norwegians. Carriers of this rare mutation are 5 times more likely to reach the age of eighty-five *without* developing late-onset Alzheimer's, and 7.5 times more likely to reach eighty-five *without* cognitive decline. In other words, this mutation protects against late-onset Alzheimer's.

Where does this mutation reside? Why, in the APP gene—the amyloid precursor protein gene responsible for making the precursor protein that is then cut by enzymes to shed beta-amyloid.

What does this mutation do? It begets a precursor protein that resists enzyme cutting, and thereby reduces beta-amyloid production by about 40 percent.

This discovery fits the beta-amyloid hypothesis and its pursuit of antiamyloid drugs more than perfectly. "It shows that if you reduce Aβ [beta-amyloid], you can prevent AD [Alzheimer's disease], and that is very much a proof of concept for the drug industry," says Gerald Schellenberg, a renowned Baptist and codiscoverer of the PSEN2 gene (see chapter 4).[2]

Against this backdrop, Baptists' defense for their repeatedly failed drugs sounded that much more convincing. In their view, the drug trials failed not because the beta-amyloid hypothesis was wrong, but because the trial drugs were faulty.

As discussed in chapter 6, when beta-amyloid monomers first build up, they form oligomers; as oligomers grow larger, they become plaques. In the latest version of the hypothesis, oligomers, rather than final plaques, are the toxic species that damages neurons and causes Alzheimer's.

But did antiamyloid drugs focus on oligomers? Not really. Vaccines that went through late-phase trials have tended to target multiple species of beta-amyloid. Eli Lilly's solanezumab, for example, targeted both monomers and oligomers, but preferred monomers; and Johnson & Johnson/Pfizer's bapineuzumab targeted all three: monomers, oligomers, and plaques.[3] Because monomers and plaques exist in much higher quantities in the Alzheimer's brain,[4] they compete for the vaccine's attention, with little left over to deal with oligomers. The situation is made worse when we have to limit drug doses for fear of side effects.

Nonselective drugs do not just neglect oligomers—by reducing monomers and plaques, they may actually *increase* toxic oligomers. This is because the three beta-amyloid species are supposed to exist in a dynamic balance.[5] When monomers are reduced, there are less of them to form oligomers. To maintain balance, plaques start to break down, replenishing monomers and oligomers. This could explain why beta-amyloid inhibitors, which prevent the forming of monomers, fail (see chapter 8). Supporting this theory, autopsies of beta-amyloid vaccine recipients found plaques with a "moth-eaten" appearance, as if the plaques were shedding beta-amyloid back into circulation in the brain.[6]

If this is true, in order for antiamyloid drugs to work, they need to target oligomers, ignore monomers, but also remove plaques the "beta-amyloid reserves". This is what Biogen, a biotech company based in Cambridge, Massachusetts, claimed to have achieved with its vaccine, aducanumab.[7] Aducanumab is collected from cognitively healthy, elderly human donors. The idea is that these donors, by preserving cognitive function in old age, must harbor protective antibodies that make for ideal drugs.

In an initial phase 1 trial, aducanumab showed good safety in one-time use.[8] Following this quick test, Biogen undertook a second phase 1 study in 2012, assessing the drug's safety in long-term (one year) use and its ability to reduce beta-amyloid. In an effort to hook investors, Biogen also tested the drug's cognitive effects, which is often reserved for late-phase trials.

This second phase 1 trial was a wild success.[9] Aducanumab drastically reduced participants' brain beta-amyloid to levels that other vaccines could only dream of. The higher the drug dose, the more reduction happened—a typical sign of a genuine drug effect. At the highest dose (10 mg), aducanumab almost wiped out beta-amyloid to levels that are considered normal. The cognitive effect was even more exciting. While placebo recipients worsened about 3 points on the Mini-Mental State Examination, a commonly used cognitive test, those on 1 mg and 6 mg of aducanumab worsened about

2 points, and those on 3 mg and 10 mg of aducanumab worsened less than 1 point.

As these results were announced, Biogen's stock soared, reaching $476 a share. The trial was hailed "a game-changer," the first confirmation that reducing enough beta-amyloid *will* bring about a clinical benefit.[10]

There is, however, a tiny problem. This supposed game-changer enrolled only 165 people, who were then put into different dosage and placebo groups of merely 30 or so per group. With these tiny sample sizes, the cognitive findings carry little value beyond marketing hype. As Biogen scientists admitted themselves, "This was a small study. . . . The trial was *not powered* for the exploratory clinical endpoints, thus the clinical cognitive results should be interpreted with caution."[11]

Also, in all the media buzz and talk of a game-changer, there was no mention of oligomers. As it turned out, the trial didn't even measure oligomers. It measured plaques, and that's what was reduced in the patients' brains. Ironically, this didn't prevent Biogen from speculating that the drug's success actually came from oligomer reduction: because plaques can break down into oligomers, plaque reduction might have slowed the release of oligomers.[12]

Riding on its success, Biogen skipped phase 2 and launched two phase 3 trials in 2015. Each trial recruited 1,600 patients, who would randomly receive either aducanumab or placebo for 1.5 years. Set to complete in 2022, these trials *are* powerful enough to assess the drug's cognitive effect—but we didn't have to wait that long. Following a familiar storyline, an interim analysis in March 2019 showed that the trials were not likely to succeed. Upon learning this, Biogen stopped both trials, as well as other plans to develop the drug.[13] Aducanumab was, for all intents and purposes, dead.

Then the plot really thickened: seven months after aducanumab was laid to rest, Biogen announced that the interim analysis was, well, wrong, and that the drug, in its highest dose of 10 mg, did slow cognitive decline in one of the phase 3 trials (but curiously, not in the other).[14] According to Biogen, *during* the trials, dosing changes were made that allowed more participants to receive the high dose. That data, however, weren't captured in the interim analysis. More remarkably, Biogen announced that based on its newly found success, it would be filing with the U.S. Food and Drug Administration (FDA) for approval for the drug—an announcement that stirred up tremendous speculation among researchers and investors.

As of October 2020, no approval has come. In the latest development, Biogen launched another phase 3 trial in March 2019, enrolling 2,400 participants who had participated in one of its previous aducanumab trials. The new trial is set to evaluate the drug's long-term safety and tolerability and is projected to complete in September 2023.[15]

Aside from aducanumab, other oligomer-selective vaccines are in early development. Some have showed promise in cell and animal studies, but it'll be a long time (if ever) before any of them go into late-phase human trials.[16] More discouragingly, we have yet to unveil the "real" oligomer. As mentioned in chapter 6, an oligomer is not one singular entity; it includes a range of beta-amyloid deposits of varying sizes and shapes. Exactly which oligomer is the cause of Alzheimer's and hence should be targeted is still unclear.

As the pursuit of oligomer drugs slowly continues, Baptists have a saving grace that's keeping companies and investors busy and hopeful—they argue that there is a second reason why anti-amyloid trials fail. This time, the problem is not the drugs; it is the participants.

Except for some early-phase trials that recruit healthy participants (who can better tolerate any unknown side effects of experimental drugs), Alzheimer's drug trials typically recruit mild or mild-to-moderate Alzheimer's patients to test the drugs' cognitive effects. Yet these typical patients, Baptists say, may be too far along to be saved. Their brains have accumulated so much beta-amyloid, which has caused too much damage that no amount of antiamyloiding would save them.

An analogy here is cholesterol-lowering drugs. These drugs reduce cholesterol and prevent heart disease. However, if cholesterol has already built up in the arteries, blocked blood flow, and caused a heart attack, cholesterol-lowering drugs—no matter how effective at lowering cholesterol—won't save us. By the same token, antiamyloid drugs—no matter how effective at lowering beta-amyloid—won't help us if we are already suffering from cognitive decline.

The analogy sounds reasonable—compelling even. Being compelling, it has ushered in a new way of thinking about beta-amyloid that resembles a paradigm shift.

In the old days, we thought that beta-amyloid was the driver of Alzheimer's.[17] In this hypothesis, beta-amyloid accumulates in the brain, driving cognitive function to go downhill gradually. By reducing beta-amyloid, we can hope to slow, or even reverse, cognitive symptoms. But according to autopsies and brain imaging studies, beta-amyloid is not exclusive to old age. It begins building up in the brain in our forties (or twenties even),[18] decades before possible dementia symptoms—just as cholesterol silently builds up in the arteries. It is only when buildup reaches a certain threshold that symptoms start to show.[19]

Given this fact, drugs fail if they don't reduce beta-amyloid to levels below the threshold. Even if they did, there's no guarantee of success because once beta-amyloid goes beyond the threshold, it might trigger events and damage that take on lives of their own: tau protein continues to build up, neurons continue to die, and brain function continues to fade.

So, in order for a drug to work, the safest bet is to take it *before* beta-amyloid hits that threshold. Baptists don't know where the threshold is, other than that it happens well before cognitive symptoms appear. Mild and mild-to-moderate patients have obvious symptoms ranging from memory lapses to confusion to difficulty with daily living activities. Given this fact, the amount of beta-amyloid deposits in their brains must have gone well beyond the threshold; ergo, it's already too late.

Future trials, Baptists urge, must recruit people who are in danger, but not yet showing symptoms. The goal should be to *prevent* Alzheimer's for them, not to *treat* Alzheimer's.

But how do we find people who need saving when they are still symptom free? One sure way is through the early-onset Alzheimer's communities: the Volga Germans, Colombian Paisas, and others. Using genetic testing, we can identify people from these communities who carry mutations that destine them for Alzheimer's. If a drug can delay or even prevent their fate, then we have a winner.

This is the method used by the Alzheimer's Prevention Initiative, an international research consortium, and its sponsor, Genentech, a San Francisco–based biotech subsidiary of the Swiss pharmaceutical company Roche. Launched in 2013, the trial recruited 242 Colombian Paisas. All participants were free of cognitive symptoms; some were mutation carriers, and others were not.

All the noncarriers received a placebo. Half the carriers would receive a trial drug, and the other half, placebo—the selection was random. Participants did not know what they would get, and neither did the researchers—a double-blind design that minimizes bias. In fact, the Paisas in the study didn't even know if they were mutation carriers. Genetic test results would not be disclosed because trial organizers worried about how people might react. As a twenty-four-year-old Paisa man once declared, "he would shoot himself in the head" if he tested positive for a mutation.[20]

Despite this trial design, which would keep their fate in the dark and possibly subject them to sugar pills, Paisas were eager to sign up. "They understand perfectly what they are getting—a placebo or a drug," said Lucia Madrigal, a local trial organizer. "It is hope for them. It is a way to do something and give back to their families. For them, this study is a source of hope. It's the only thing they've got."[21]

And what was the test drug? Crenezumab, a beta-amyloid vaccine developed by the trial sponsor, Genentech. Crenezumab boasts an excellent safety record, which is essential for prevention trials that dose cognitively normal people for a long period of time. In this case, dosing is planned to take five long years.

What about crenezumab's cognitive effects? In a phase 2 trial, crenezumab showed no benefit among 400 mild-to-moderate Alzheimer's patients. After some slicing and dicing, the drug did appear beneficial among the mild patients, which presumably supports its effects for early prevention.[22] But if history is any indication, this kind of subgroup analysis isn't particularly trustworthy. Like the Paisas, we can only *hope* that the drug will work. We will know for sure in early 2022, when the trial is set to conclude.

What we do know now is that a similar prevention trial is not faring too well so far. Organized by an international research consortium, the Dominantly Inherited Alzheimer's Network (DIAN), this trial is more ambitious. Instead of just Paisas, it recruited carriers of any early-onset mutations. Instead of one drug, it is testing three drugs separately.

The first of these drugs, we have met before: Eli Lilly's solanezumab, a beta-amyloid vaccine. This drug received lots of media attention during a roller coaster of three phase 3 trials, aptly named Expeditions 1–3 (see chapter 10). Expeditions 1 and 2, which recruited mild-to-moderate patients, both failed. When mild patients from the two trials were combined for analysis, the drug showed partial success, suggesting a prevention effect. But in Expedition 3, which recruited only mild patients, the drug failed yet again.

The second drug inspires even less confidence: Roche's gantenerumab, also a beta-amyloid vaccine. Publications on gantenerumab are scarce. The limited data available showed that it reduced brain beta-amyloid in a small, early-phase trial, but two of the six participants in the high-dose group suffered serious side effects, including brain inflammation.[23] More disconcertingly, Roche had terminated two phase 3 trials since 2014 because interim assessment showed that the drug had no cognitive effects among patients with mild or very mild symptoms—hardly promising for a prevention drug.[24]

The third drug, for a while, was Johnson & Johnson's JNJ-54861911. A beta-secretase inhibitor (see chapter 8), JNJ-54861911 prevents beta-amyloid from being created in the first place.

Refreshingly, unlike other companies that rush through early-phase trials, Johnson & Johnson undertook a dozen of them to test, in different fashions and with different doses and types of participants, the drug's safety, absorption, and basic function. According to these trials, the drug is safe and significantly reduces beta-amyloid.[25] Just when it seemed as though JNJ-54861911 would fit the adage "Slow and steady wins the race," news broke in summer 2018 of serious side effects: liver damage. Upon this discovery, JNJ-54861911 was taken out of the prevention trial.

Then, in February 2020, *more* bad news: in initial analysis, the two remaining drugs, solanezumab and gantenerumab, have failed to slow trial participants' cognitive decline—disappointing but not exactly surprising.[26] April brought more in-depth data, and things were looking up, somewhat.[27] Yes, the two drugs failed to improve all the participants' cognitive function. But, among participants who enrolled without symptoms (in this trial, not all participants were initially symptom free), cognitive functions were stabilizing. Moreover, gantenerumab (but not Lilly's solanezumab) also reduced participants' beta-amyloid and tau pathologies. On account of these, trial organizers are inviting all participants to a three-year extension study. The study will provide all participants with high-dose gantenerumab, in the hope that gantenerumab may eventually drive beta-amyloid below the assumed threshold for cognition to start benefiting. As the trial organizers put it, gantenerumab is still "very much alive" as a prevention treatment for Alzheimer's.[28] Lilly's solanezumab, though, is definitely out.

While early-onset Alzheimer's patients make ideal participants in prevention trials, these mutation carriers represent only 5 percent of Alzheimer's cases. To mobilize more participants and maximize the chances for success with late-onset cases, researchers are redrawing the boundaries of Alzheimer's—the line between "normal" and "diseased."

For a long time, Alzheimer's was considered a deviation from normal aging—something out there, all on its own. But in reality, the human brain naturally declines with age. It sounds ridiculous, but our ability to hold multiple pieces of information in the mind peaks at around ages eighteen to twenty and then starts going downhill; the ability to recall new information peaks at around forty and then weakens; after seventy, our ability to recall someone's name or a certain word diminishes.[29] At what point does "normal" stop and "diseased" begin?

Similarly, beta-amyloid starts building up in the brain during midlife and early adulthood. Other telltale signs—brain atrophy, synapse loss, neuron damage, and memory disruption—are experienced by presumably healthy older adults.[30] Again, at what point does "normal" stop and "diseased" begin?

To recognize the fuzzy edges between normal aging and Alzheimer's, researchers invented several new concepts in recent years.[31] We now have *asymptomatic Alzheimer's*, which refers to people who show no outward symptoms but harbor warning signs that can be picked up in brain scans: beta-amyloid buildup, tau abnormality, and brain atrophy. We also have *mild cognitive impairment*, which includes people who are overall independent in daily living but display subtle memory and cognitive impairment. *Prodromal Alzheimer's*, part of the mild cognitive impairment category, is the stage where one shows Alzheimer's-like symptoms, but the symptoms are not severe enough for an Alzheimer's diagnosis. All these groups make for potential prevention trial participants.

One of these trials is the Anti-Amyloid Treatment in Asymptomatic Alzheimer's trial, cofunded by the National Institute on Aging (NIA) and Eli Lilly. This trial uses brain scans to screen asymptomatic elderly people for elevated brain beta-amyloid. Qualified participants then randomly receive trial drug or placebo for 4.5 years. The drug is again Eli Lilly's controversial vaccine, solanezumab, although something else about the trial is refreshing.

The trial requires that one of every five people screened for participation comes from an under-represented minority group.[32]

In the United States, African Americans and Hispanics represent a large aging population that may face greater risk for Alzheimer's, but Alzheimer's trials by and large include whites: African Americans account for a mere 3–5 percent of participants, and Hispanics, 1.5 percent.[33] While clinical trials *are* arduous, they also offer a privilege, a hope, a chance to get better. Moreover, results from "white trials" may not apply, or apply equally well, to minority groups. For this reason alone, the results of this prevention trial, scheduled to arrive in early 2023, are eagerly anticipated.

The last prevention trial receiving media attention is the Generation Study, cosponsored by the Swiss pharma Novartis and American pharma Amgen. The Generation Study recruited cognitively healthy older adults who carried two copies of ApoE4 (which increases their risk for Alzheimer's by about twenty times) or carried one copy of ApoE4, plus elevated brain beta-amyloid.

Two drugs were being tested: CAD106 and CNP520. CAD106 is Novartis's beta-amyloid vaccine. In previous trials, it has been found to be generally safe and successfully raise antibodies; however, it didn't actually reduce brain beta-amyloid among mild-to-moderate patients, which seems worrisome.[34] As for CNP520, it is Novartis and Amgen's co-developed beta-secretase inhibitor. In early-phase trials, it was concluded safe and did reduce beta-amyloid.[35]

These two drugs could be dark horses come 2025, when the trial completes—except, in July 2019, a preplanned review showed CNP520 users experiencing cognitive worsening.[36] The drug was promptly taken out of the trial. No such announcements were made about the remaining drug CAD 106, yet, curiously, it was listed as "retired" in Novartis's 2019 third quarter financial report.

Changing their focus from treating to preventing Alzheimer's, pharmaceutical companies and research consortiums are not giving up on the beta-amyloid hypothesis. Millions of private and public dollars are invested, thousands of participants are involved, and hundreds of researchers are mobilized. If nothing else, the scale of this production alone mirrors a paradigm shift.

But, the major research efforts have not wavered from beta-amyloid. The drugs put to the test are still antiamyloid drugs. Is this enough of an innovation to get us on the right track, to uncover something that actually works, to finally conquer Alzheimer's?

I don't know. With all my heart, I sure hope so. In the depressing event that it isn't enough—and before we ever find out—we can at least do something today, the topic of my last chapter.

21

AN ENRICHED LIFE

I was born in Eau Claire, Wis, on May 24, 1913 and was baptized in St James Church . . . I prefer teaching music to any other profession.

The happiest day of my life so far was my First Communion Day which was in June nineteen hundred and twenty when I was but eight years of age, and four years later in the same month I was confirmed by Bishop D. D. McGavick. . . . Now I am wandering about in "Dove's Lane" waiting, yet only three more weeks, to follow in the footprints of my Spouse, bound to Him by the Holy Vows of Poverty, Chastity, and Obedience.[1]

The two passages here, each containing two sentences, were written by two women while in their early twenties. Knowing nothing else about them, if you had to take a wild guess at who is more likely to suffer from cognitive impairment and Alzheimer's in old age, which one would you pick?

If you picked the first woman, you would be correct. It wasn't too hard to guess, was it?

These writings are taken from the autobiographies of the School Sisters of Notre Dame, written by the sisters in the 1930s, a few weeks before they were to take their religious vows. Analyses of about 100 sisters' autobiographies, coupled with cognitive tests and autopsy studies, showed that idea density (the number of ideas expressed in every ten words) is strongly associated with the sisters' late-life cognitive health. A high idea density, as shown in the second passage, correlates with a low risk for cognitive decline and Alzheimer's. A low idea density, as shown in the first passage, correlates with a high risk for cognitive decline and Alzheimer's.

Why would this be? What does writing have to do with our cognitive fate?

According to the cognitive reserve hypothesis, how we write is a reflection of how much cognitive reserve our brain holds. Just as we can put money in the bank for a rainy day, we can collect cognitive power for old age, and idea density is one sign of how much of that power we have saved. High density equates to more savings to tap into to get by in old age. Low density suggests a scant supply that will soon run out.

This theory explains a long-standing mystery in Alzheimer's: cognitive symptoms do *not* always correlate with brain damage. Someone may show no cognitive impairment at life's end but reveal significant beta-amyloid and tau tangles in her brain upon autopsy.[2] Likewise, someone may tolerate more than the typical brain damage before he starts to show comparable cognitive decline.[3] This phenomenon is not unique to Alzheimer's either. Strokes of a certain magnitude are known to cause major impairment in one person, but very little in another.[4] In all these cases, cognitive reserve may be calling the shots: a large reserve allows people to withstand more damage, more neuron loss, and more synapse dysfunction before they succumb; a small reserve, not so much.

Where exactly does cognitive reserve come from? And how do we go about getting more of it?

Supposedly, part of this reserve is determined at birth. Studies show that people with larger heads are less likely to have Alzheimer's or have less severe Alzheimer's, presumably because a larger head holds more neurons, more synapses, and more innate capacities to spare.[5]

But if you aren't endowed with a large head, don't despair. According to the lowly mouse, how we live our lives makes a big difference too. Mice in the same research lab can wind up having very different fates. Some live solitary lives in a small cage, with access to food and water but nothing else. Some play all day with their fellow rodents in a large house equipped with running wheels, crawling tubes, colorful Legos and toys, which are changed periodically to maintain novelty.

This is the setup for the so-called enrichment study, designed to test if an enriched life, as opposed to a barren one, affects cognitive function. In mice, it sure does. When mice lived richly, their brains underwent physical changes: more neurons were generated in the brain's memory center, the hippocampus, and strong synapse activities emerged to support learning.[6] When transgenic Alzheimer's mice were allowed to have an enriched life, they also experienced enhanced brain activity and better performance in maze tests.[7] An enriched life, it seems, helps mice to develop and accrue cognitive power.

If the same applies to humans, by living richly, we have a fighting chance to ward off or delay Alzheimer's. What makes for an enriched life for humans? Ironically, it is not *that* different from what it takes for mice.

The first ingredient is cognitive stimulation, comparable to the Legos and toys in rodent houses.

In numerous population studies, a factor consistently linked to Alzheimer's risk is education.[8] Higher education reduces risk; lower education increases risk. Presumably, people with higher education are exposed to more mental challenges in early life, which helps to build cognitive reserve. Similarly, occupation matters. People who perform more complex, mentally challenging work (whether managerial, technical, or professional) have a reduced risk for Alzheimer's than people who perform manual labor, trade, or simple clerical work.[9]

While illuminating, these findings are also troubling. Although we are not lab mice, we don't exactly have free and equal choices in life either. Where we study and what we do for a living are closely tied to socioeconomic status—they *are* socioeconomic status. Depending on our early-life environment, some of us will have more opportunities, means, and motivation to pursue higher education and then find employment with an intellectual bent and upward mobility. Beginning-of-life factors such as maternal and infant nutrition could also influence brain development and affect our later attainment.[10] In a nutshell, we don't all have equal access to the cognitive benefits afforded by education and occupation.

As medical and natural scientists, Alzheimer's researchers are largely uninterested in such social-political concerns. Very few have explored the relationship between Alzheimer's and early-life factors such as our *parents'* occupations or the environment of our birth community.[11]

Fortunately, more egalitarian ways do exist to challenge the brain, such as reading books or newspapers, playing card or board games, and doing crosswords or other puzzles. Population studies show that people who frequently engage in these activities have a lower risk for Alzheimer's.[12] Indeed, in a study of older adults in Chicago, the benefits of these activities trumped the benefits of education and occupation, which suggests that schools and jobs per se may not be the real antidote; instead, people with higher education and more mentally challenging jobs may have a lower risk for Alzheimer's because they are also more cognitively active in everyday life.[13]

Physical changes in the brain support the effect of cognitive stimulation. In London, licensed taxi drivers must learn to navigate between thousands of locations in the city. Rigorous training and years on the job enlarged the region of their brains responsible for spatial navigation.[14] Apparently, the human brain, even in adulthood, remains plastic and can be pushed to increase its capacity.

And it doesn't take long for the brain to change either. In one interesting study,[15] twenty-four volunteers were given a brain scan and put into two groups: one group was to learn a three-ball cascade juggling routine in three months; the other was to do nothing different. When the three months were up, the learner group could skillfully perform the routine. Both groups then received a second brain scan, after which the learner group stopped practicing the routine. Another three months later, most in the learner group had lost their juggling skill, and a third scan was taken. What were the findings? There was no difference between the two groups at the initial brain scan. After three months of juggling, the learner group saw an increase in the brain region responsible for complex visual motion, but after three months of nonjuggling, the increased region had shrunk back.

What is less straightforward than these findings are the results of clinical studies.[16] These trials tested the effect of a variety of cognitive activities, such as engaging participants in conversation, memory drills, video games, and simulations of everyday activities.[17] Generally, exercises and drills didn't really work. Although people can be trained to do better on, say, math calculation, the effect doesn't translate into overall cognitive improvement or real-world benefits. On the other hand, activities that are more diffuse, such as group discussions of interesting topics or reminiscence activities, were more promising at improving participants' overall condition.

One caveat here is that current cognitive trials are of lesser quality. Their size is small: a few dozen participants, as opposed to the hundreds or thousands often recruited for drug trials. Money is probably a factor here because cognitive interventions are not easily patented for business profit and thus lack private investment.[18] There is also a lack of good placebos.[19] In drug trials, a look-alike sugar pill and a test drug are randomly assigned without people's knowledge to avoid human bias. But in cognitive trials, there is no ready-made sugar pill. The placebo group either receives no intervention—a fact that can't be hidden—or it receives some other intervention that may have effects of its own and muddy the trial results. To confirm the positive effect of cognitive stimulation as an Alzheimer's therapy, we need larger and more rigorous trials.

The second ingredient of an enriched human life, comparable to the running wheels in a rodent house, is physical activity.

Among the different forms of physical activity, most researchers favor *aerobic* (with oxygen) exercise, also known as *cardio* (related to the heart) exercise. This exercise gets the heart pumping to supply oxygenated blood to the muscles. Common examples include walking, running, cycling, swimming, and ball games. Aerobic exercise ranges in intensity but is not over-the-top strenuous (like weightlifting), so the body can sustain it for an extended time.

How might physical activity bank brain power? Multiple theories exist, drawing upon the various hypotheses introduced earlier in this book.[20] Physical activity can increase blood flow to the brain, boosting the oxygen/nutrient supply and thereby cognitive function. It can mobilize the antioxidant system and protect neurons from oxidative stress. It may also lower inflammation, promote neuron growth, and reduce abnormal beta-amyloid and tau in the brain.

Results from population studies support these benefits. In ten studies with a combined 23,000 participants who were followed anywhere between four and thirty-one years, physically active older adults were 40 percent less likely to develop Alzheimer's.[21] "Physically active" was defined based on international fitness guidelines, which require adults to do 150 minutes or more of moderate activity per week. For older adults, moderate activity includes anything ranging from walking (at a speed of more than three miles per hour), to leisure sports such as jogging and cycling, to intense housework such as shoveling snow.[22]

Clinical trial results, on the other hand (and as usual), are less consistent. In some trials, engaging older adults with or without Alzheimer's in physical activity improved their cognition and daily function,[23] but in other trials, it didn't.[24] When multiple trials are combined for comprehensive analyses—the so-called *meta-analyses*—the results didn't agree either; it depended on which trials were pooled. For instance, in one meta-analysis of thirty trials, physical activity improved cognitive function in people with dementia and related impairments.[25] In a second meta-analysis that looked at Alzheimer's patients, four trials also showed cognitive benefits.[26] But in a third that focused on aerobic exercise, twelve trials showed no cognitive effects among nondemented older adults.[27]

Usually, these clinical trials, with their controlled and randomized nature, would trump population study findings, but something about physical activity tips the scales. And once again, we have the lowly mouse to thank for this insight. When transgenic Alzheimer's mice were provided with a running wheel and exercised of their own free will, they experienced more cognitive benefits than if they were put on a motor-driven treadmill and made to run.[28] This happened despite the treadmill being set to match the mice's average, voluntary running speed. Presumably, forced running, even at a natural speed, induces stress, which offsets cognitive benefits. This may be why physical activity applied in clinical trials, which can be considered forced (or at least not completely voluntary), had inconsistent effects, whereas physical activity that people pursue and enjoy in everyday life showed more benefit in population studies.

Aside from potential cognitive effects, physical activity benefits Alzheimer's patients in other ways. It improves fitness, which is important to people with Alzheimer's, who have a higher tendency for falls, fractures, and immobility.[29] They also help reduce depression, which is common among demented patients.[30]

The third and last ingredient of an enriched human life is social engagement.

A fancy house, with all its toys and gadgets, apparently does the mouse brain little good if a mouse resides in it alone.[31] Conversely, a barren cage without amenities does the mouse brain little harm if a group of mice share it.[32] Housemates, it seems, are not a nice addition to, but the very fabric of, an enriched rodent life.

For humans, though, social engagement is more complex than just having housemates. Research generally breaks the concept down into two components: (1) maintaining a social network and (2) participating in social activities.

The size of our social network is typically determined based on the following factors: marital/relationship status (do we have and live with a partner?), parenthood (do we have children, how many, and how much do we interact with them?), extended family (how many relatives do we have, and how much do we interact with them?), and friendship (how many friends do we have, and how much do we interact with them?). Population studies show that older adults with larger social networks are less likely to experience cognitive decline.[33] In a study of 1,200 older adults in Stockholm, those who were married and living with someone were half as likely to have dementia than those who were single and living alone.[34] Even with more severe Alzheimer's brain damage, people with a larger social network tended to maintain higher levels of cognitive function.[35]

The second component, social participation, is measured by participation in group events and activities such as clubs, religious services, volunteer work, and group outings. Once again, population studies show that older adults who participate in more social activities have less cognitive decline and a lower risk for dementia.[36] In a study of 6,000 older residents in Chicago, frequent participation in social activities reduced cognitive decline by 90 percent.[37]

Some of these population studies adopted very long follow-up times in order to address the chicken-and-egg question: which comes first—low social engagement or the onset of dementia? When dementia sets in, people gradually lose their ability to handle complex social interactions, so they may withdraw from social circles and activities. In other words, low social engagement may be a result of dementia rather than the other way around. To address this possibility, studies used follow-up times over a decade to make sure that participants weren't becoming demented at the start of a study. In several such studies, the cognitive benefits of social engagement held true.[38]

Multiple explanations exist for why social engagement wields this power. One is that social engagement, in and of itself, is cognitively stimulating. Socializing requires us to effectively talk, listen, and recognize thoughts and feelings, mobilizing several regions of the brain. Many of these regions also support memory and other cognitive functions, so frequent social stimulation may build up cognitive reserve directly.[39] Indirectly, social support and a sense of belonging reduce stress, which in turn enhances cognitive function.[40] Social engagement also gives us a sense of purpose and fulfillment, which is thought to make older adults more cognitively resilient.[41]

At present, clinical trials comparing the effect of social engagement versus placebo are lacking.[42] In addition to the financial obstacle cited earlier, pure social intervention is difficult to manufacture in a trial setting. For example, one trial created social intervention in the form of group discussions. But by the researchers' own account, this was "lively intellectual discussion."[43] If so, the activity most likely involved cognitive stimulation in addition to group companionship.

To be fair, real-life social engagement is rarely pure—and neither are real-life cognitive stimulation and physical activity. Playing chess or attending a lecture is mentally stimulating, but it could also involve quite a bit of social interaction. Dancing and tai chi get the body moving, but they also require us to memorize choreography and be mindful of our movements. When practiced in a social or group setting, as they often are, they also build companionship and solidarity.

The complex nature of these events, then, runs counter to the philosophy of modern clinical trials, which is about isolating and purifying chemical treatments. This philosophy serves the development of drugs, but less so for nonpharmacological treatments. As we mentioned earlier, forced treadmill running is not quite the same as free-will running. This mismatch may be reason enough to ditch the gold standard of clinical trials and trust population studies that observe the real-world benefits of an enriched life.

Without clinical trials, we cannot know which cognitive, physical, or social activity, in what precise dosage, conserves the mind, but judging by existing evidence, it matters less *what* we do than *that* we do: read a book, travel with friends, learn ballroom dancing, go to the gym, join a choir, whatever—try to partake in multiple activities that combine multiple kinds of stimulation.[44] Live your life as though someone left the gate open. Isn't that what we are supposed to do anyway?

In 2018, Manhattan, Kansas, the "little apple" where I live and work, was ranked by Livability.com as the nation's second-best place to live. Home to Kansas State University, Manhattan is a bustling, yet also laid-back little town with affordable living and respectable entertainment and cultural venues. It is the birthplace of the made-up holiday "Fake Patty's Day," a boisterous drinking celebration held one week before the actual St. Patrick's Day because the real holiday falls during students' spring break.

Another Manhattan gem less well known to livability experts is a not-for-profit senior living community in the northeastern corner of the town. It has an idyllic name: Meadowlark.

In the summer of 2018, after having lived in Manhattan for ten years, I stepped inside Meadowlark for the first time. I must say that I was impressed by what I saw. Cozy, independent-living cottages flank a well-groomed campus, surrounding a spacious community center, assisted-living areas, and a large community garden. Inside the community center, a daily calendar in the lobby greets visitors and residents, displaying a busy day's schedule that includes activities such as book club meetings, discussion groups, Bible studies, and chair exercises.

Upbeat music drifted out of a room right next to the lobby. Through a half-opened door, I could see a dozen or so seniors enthusiastically lifting dumbbells and stretching their legs, while sitting in chairs. It must be the chair exercise class, I decided. Wandering further, I saw a restaurant, a billiard room, a theater, and small conference rooms lined with shelves full of books. Behind me, an elderly lady was wheelchaired into a salon, where another was getting her perm done. I wouldn't mind living here in old age, I thought.

But I wasn't here for a tour. I came for Meadowlark's Memory Program, a free program open to area seniors dealing with—or simply concerned about—memory and cognitive decline. My friend Debra, who experienced dementia care firsthand, told me about it. Intrigued, I reached out to Meadowlark, sharing my interest in Alzheimer's research and my hope for a visit. They kindly opened their doors, so there I was that morning, visiting the Memory Activities Class, an instructor-led weekly session that is part of the program.

Sitting around large folding tables arranged in a circle, I was joined by twenty-two senior citizens. As coffee and frozen yogurt were served, Don, the man sitting next to me, befriended me and filled me in on who was who. There was a retired school principal, a local carpenter, a husband-wife team. Many, including Don himself, come every week.

After the yogurt tray made its rounds, the class started with self-introduction. It was August 15, which apparently is National Relaxation Day, so folks were asked to also share what they like to do for relaxation.

One after another, people said their names. Some, usually the oldest of the old, also offered their age. They were proud, as they should be, that they were ninety-two years old and sitting in a classroom. Most of the people spoke fluently, but some struggled a bit to finish their sentences. One woman let on that she was recently diagnosed with Alzheimer's, to which several others said, "Me too."

Diagnosed or not, these elderly folks relaxed in style. By far, their favorite activity was reading: novels, newspapers, and magazines. A close second was playing puzzles: sudokus, crosswords, and jigsaws. Several also mentioned sightseeing, fishing, bird watching, and dog walking.

I too was invited to share my ways of relaxation. Under the curious looks of twenty-two seniors, I said—with self-consciousness and recognition of irony—"watching TV." Compared with the others' pastimes, mine was *not* particularly enriching. The kind of TV I—well, we—usually watch and the way we watch it is not cognitively, physically, or socially stimulating. This relaxation also takes time away from other, potentially more enriching activities. In fact, in one population study, watching television was linked with an increased risk for Alzheimer's.[45] That morning, I was the only person in the room who watched television for relaxation.

After self-introductions and relaxation sharing, the class entered a lively discussion of what daily strategies they use to cope with memory loss. And when I say lively, I mean *lively*—if only youthful students in my class could show this kind of enthusiasm. From left and right and every corner, people offered tales of their daily struggles and victories with sticky notes, a calendar, multiple calendars, voice recorders, and pill dispensers. It didn't matter if an answer had been given. No one seemed to notice (or if they did, they didn't care). When the hourlong class drew to an end, quite a few people still needed to speak. They compromised by talking loudly with their neighbors as the class gathered their belongings, sipped the last of their coffee, and filed into the corridor.

Another man (not my pal Don, who had dashed out to catch a doctor's appointment) was walking next to me as we retreated. He turned to look at me and, with a faint smile, asked what I thought about the class.

"I loved it," I said, genuinely. "It was fun . . . and stimulating!"

"Yes," he agreed, gesturing toward the woman walking on his other side. "I bring my wife Sherry every week."

Sherry was leaning on a walker, mustering all her attention and energy to move forward, with nothing left to give to us.

"I want to keep her involved for as long as possible, you know," the husband said. "I figured it can only help."

"Yes, it can only help," I said emphatically, for him and myself. "It can only help."

EPILOGUE

When I started this book five years ago, I had little idea what I was in for. I knew that the science of Alzheimer's was complicated, but I wasn't prepared for the amount of confusion and controversy that came up along the way. I knew media reports of "breakthroughs" should be approached with caution, but I didn't expect all the ambivalence, exaggeration, and even fraud in scholarly publications—publications that supposedly represent the integrity and rationality of modern science.

As someone who researches and teaches scientific communication, never was I more impressed with scientists' rhetorical skills. Say that a clinical trial uses four cognitive tests to judge the effect of a drug, and three tests fail while one succeeds. You can be sure that the three failed tests will be glossed over, while the successful one will receive tremendous attention. It will be highlighted in the title, emphasized in the abstract, elaborated in the discussion, and repeated in the conclusion. You absolutely could not miss it. But if you are not careful, you might just miss the fact that three of the tests did fail.

In the unfortunate event that all the tests look gloomy, we can still group, regroup, model, and remodel the data, and *something positive* will always (or at least usually) turn up. This something is then duly discussed, with warmth and enthusiasm. Connections are made to previously published studies to show that this something is actually quite significant and not at all unexpected. By golly, people should have been looking for it in the first place.

Suppose that the finding differs from (or indeed contradicts) previous findings and forbids drawing connections, explanations can always be found, usually without too much difficulty. The key is to hedge: differences *may* be caused by the use of different strains of mice; it is *possible* that the increased dosage had a negative effect. If the explanation throws doubt on opponents' work, all the better. *Perhaps* their sample was too small, their study too short, their method of detection insensitive. No respectable scientist would claim certainty without abundant facts, but everyone can earnestly speculate without breaking the so-called objectivity of science.

Not every publication tries this hard. Some come out loud and clear that a study didn't work, the expected result wasn't there, and all the findings were negative. But this usually happens when researchers are ready to bail out—to wash their hands of the matter and move on to something else. And *that* usually happens when grant funding or industry support runs out. At that point, ruthlessly denouncing a project shows that one is making sound, rational scientific decisions, not simply giving up when the going gets tough. However, should funding and support remain a possibility, seemingly obvious failures can still be ignored, and the clichéd conclusion "further studies are warranted" can be drawn.

Remember, too, that we are talking about the more influential and well-researched Alzheimer's hypotheses. Other hypotheses are out there that, for the time being, have more gaps than evidence.

For example, there's the cell cycle hypothesis, which likens Alzheimer's to cancer,[1] an intriguing idea just by the sound of it. *Cell cycle* refers to several phases that cells go through in order to divide and replicate. Most of the cells in the human body go through this cycle forty to sixty times in their lifetime.[2] But neurons, once mature, no longer divide, so they should have nothing to do with the cell cycle. If they enter the cycle by mistake, they may not exit properly and can wind up in a state of diseased immortality, like cancer cells. These cancerous neurons allow damage to accrue in the brain, causing Alzheimer's. Again, a very intriguing idea, but it needs a lot more evidence before it turns from intriguing to potentially promising.

Then there's the infection hypothesis linking Alzheimer's to viral or bacterial infection. The list of suspected viruses and bacteria is long, including herpes viruses, human immunodeficiency virus (HIV), and others whose names likely aren't remotely familiar to the average person.[3] Although infection-caused mental diseases are not unheard of (notably neurosyphilis, as mentioned elsewhere in this book), evidence for the infection hypothesis is presently rather weak. Some autopsy studies detected viruses and bacteria in the Alzheimer's brain, but others didn't. Even the studies that did couldn't agree on a common virus or bacterium.

Still, the hypothesis carries a certain allure, for it is truly outside the box and echoes some classical scientific breakthroughs. Back in the 1980s, it was widely believed that stomach ulcers were caused by stress. Then two Australian researchers, Barry Marshall and Robin Warren, proposed that a bacterium called *Helicobacter pylori* was the real culprit. Their theory was ridiculed until it was proved true, earning them the 2005 Nobel Prize in Physiology or Medicine.

In addition to viruses and bacteria, another proposed infectious agent is *prions*, a word derived from "protein" and "infectious." Prions are otherwise normal proteins with misfolded shapes that clump together. They ravage the brain, turning it into a sponge-like mess full of tiny holes. An arguably more terrifying fact about prions is that they are contagious. Acting like seeds, prions cause normal proteins that they come into contact with to misfold and thereby spread between cells, tissues, and individuals.[4]

The major type of prion disease in humans is Creutzfeldt-Jakob disease. A rare disease affecting one to two people per million each year, it causes rapid loss of cognitive and physical function that ends inevitably in death.[5] The disease can arise spontaneously, be inherited, or transmit through medical procedures—notably via the injection of cadaver-derived growth hormone, a now-obsolete practice. Prions also affect animals and are the cause of mad cow disease (in cattle) and scrapie (in sheep and goats).

Stanley Prusiner, a Nobel Prize winner in 1997 for his discovery of prions, is a chief advocate of the idea that prion diseases are not rare, that they include common neurological conditions such as Parkinson's and Alzheimer's (thus making prions a far more valuable—and fundable—area of study). He claims that beta-amyloid and tau proteins are essentially prions, and their spreading and aggregation in the brain is what causes Alzheimer's.[6]

What's the evidence for this claim? Beta-amyloid extracted from the Alzheimer's brain seeded the deposit of synthetic beta-amyloid.[7] Injecting Alzheimer's brain extract into mice caused the spread of beta-amyloid deposits.[8] More recently, Alzheimer's-like beta-amyloid deposits were found in four cadaver hormone recipients,[9] stoking fearful newspaper headlines that "Alzheimer's may be a transmissible infection" and "You can catch Alzheimer's."[10]

Now, if we really want, we can call beta-amyloid "prionlike," or even prions, because of their seeding behavior, but that doesn't make any material difference in drug development. Prion diseases don't have cures, so whether we call beta-amyloid prions or just abnormal proteins, we would still be trying to reduce beta-amyloid.[11] More important—and I can't stress this enough—calling beta-amyloid (or tau) prions does not make Alzheimer's a classic, infectious prion disease. So far as we know, blood transfusion does not increase one's risk for catching Alzheimer's,[12] and neither does cadaver hormone injection. Worldwide, 200 people have had the Creutzfeldt-Jakob prion disease transmitted to them due to receiving cadaver-derived growth hormone, but there was no report of Alzheimer's among 6,190 cadaver hormone recipients in the United States.[13] In fact, we can't seem to transmit Alzheimer's if we try. Injecting human Alzheimer's brain tissue into nonhuman primates failed to transmit the disease after dozens of attempts, whereas injecting Creutzfeldt-Jakob disease brain tissue into primates spread *that* disease in hundreds of cases.[14]

Still, the alarmist prion hypothesis gets a fair amount of media attention, as do other hypotheses that attribute Alzheimer's to pesticides, air pollution, sleep disorders, etc. To account for all these ideas seems premature at this time (and will require another book). But as scientists like to say, it *may* be warranted if the landscape of Alzheimer's research continues to shift, especially if the beta-amyloid paradigm actually crumbles.

Until then, I hope this book has been a useful and enjoyable read.

NOTES

PREFACE

1. The data given here come from Alzheimer's Association, "2018 Alzheimer's Disease Facts and Figures," *Alzheimers Dement* 14, no. 3 (2018): 367–429; Alzheimer's Association, "2019 Alzheimer's Disease Facts and Figures," *Alzheimers Dement* 15, no. 3 (2019): 321–87; Martin Prince et al., *World Alzheimer Report 2015: The Global Impact of Dementia: An Analysis of Prevalence, Incidence, Cost and Trends* (London: Alzheimer's Disease International, 2015); World Health Organization, "The Top 10 Causes of Death," World Health Organization, May 24, 2018, https://www.who.int/news-room/fact-sheets/detail/the-top-10-causes-of-death.

1. LOST WIVES

1. Sean Page and Tracey Fletcher, "Auguste D: One Hundred Years On: 'The Person' Not 'the Case,'" *Dementia* 5, no. 4 (2006): 571–83; Konrad Maurer and Ulrike Maurer, *Alzheimer: The Life of a Physician and the Career of a Disease*, trans. Neil Levi and Alistair Burns (New York: Columbia University Press, 2003).
2. Page and Fletcher, "Auguste D."
3. Maurer and Maurer, *Alzheimer.*

4. Maurer and Maurer, *Alzheimer*.
5. Maurer and Maurer, *Alzheimer*.
6. Maurer and Maurer, *Alzheimer*, 2–3.
7. Claire O'Brien, "Auguste D. and Alzheimer's Disease," *Science* 273, no. 5271 (1996): 28.
8. Rainulf A. Stelzmann, H. Norman Schnitzlein, and F. Reed Murtagh, "An English Translation of Alzheimer's 1907 Paper, 'Über eine eigenartige Erkankung der Hirnrinde,' " *Clinical Anatomy* 8, no. 6 (1995): 429–31.
9. Stelzmann et al., "An English Translation," 430.
10. Konrad Maurer, Stephan Volk, and Hector Gerbaldo, "Auguste D and Alzheimer's Disease," *The Lancet* 349 (1997): 1546–49.
11. Maurer and Maurer, *Alzheimer*.
12. Maurer and Maurer, *Alzheimer*, 109, 133, respectively.
13. Maurer and Maurer, *Alzheimer*, 109.
14. Maurer and Maurer, *Alzheimer*.
15. Stelzmann et al., "An English Translation."
16. Stelzmann et al., "An English Translation."
17. Germán Elías Berrios, "Alzheimer's Disease: A Conceptual History," *International Journal of Geriatric Psychiatry* 5, no. 6 (1990): 355–65.
18. Berrios, "Alzheimer's Disease"; Michel Goedert, "Oskar Fischer and the Study of Dementia," *Brain* 132, no. 4 (2009): 1102–11.
19. Maurer and Maurer, *Alzheimer*, 177–79.
20. Maurer and Maurer, *Alzheimer*, 179 (emphasis mine).

2. CURSED INHERITANCE

1. Solomon C. Fuller, "Alzheimer's Disease (Senium Praecox) I: The Report of a Case and Review of Published Cases," *Journal of Nervous & Mental Disease* 39, no. 7 (1912): 440–55; Solomon C. Fuller, "Alzheimer's Disease (Senium Praecox) II: The Report of a Case and Review of Published Cases," *Journal of Nervous & Mental Disease* 39, no. 8 (1912): 536–57.
2. For the details of Johann's case, see Fuller, "Alzheimer's Disease (Senium Praecox) II," 541–43.
3. Hans-Jürgen Möller and Manuel B. Graeber, "The Case Described by Alois Alzheimer in 1911," *European Archives of Psychiatry and Clinical Neuroscience* 248, no. 3 (1998): 111–22.
4. Lawrence A. Hansen, Eliezer Masliah, Douglas Galasko, and Robert D. Terry, "Plaque-Only Alzheimer Disease Is Usually the Lewy Body Variant, and Vice Versa," *Journal of Neuropathology and Experimental Neurology* 52, no. 6 (1993): 648–54.

5. Hansen et al., "Plaque-Only."

6. Quoted in Möller and Graeber, "Case Described," 117, italics in original.

7. Quoted in Möller and Graeber, "Case Described," 117.

8. Germán Elías Berrios, "Alzheimer's Disease: A Conceptual History," *International Journal of Geriatric Psychiatry* 5, no. 6 (1990): 355–65; Thomas Beach, "The History of Alzheimer's Disease: Three Debates," *Journal of the History of Medicine and Allied Sciences* 42, no. 3 (1987): 327–49.

9. Berrios, "Alzheimer's Disease"; Beach, "History of Alzheimer's Disease."

10. Manuel B. Graeber, "Alois Alzheimer (1864–1915)," International Brain Research Organization, 2003, http://ibro.org/wp-content/uploads/2018/07/Alzheimer-Alois-2003.pdf.

11. Graeber, "Alois Alzheimer," 7.

12. All accounts of the German settlers' experiences in this chapter, except for specific details related to the Reiswigs, were drawn from Fred C. Koch, *The Volga Germans: In Russia and the Americas, from 1763 to the Present* (University Park: Pennsylvania State University Press, 1977).

13. Koch, *Volga Germans*, 13.

14. All accounts of the Reiswigs' experiences in this chapter were drawn from Gary Reiswig, *The Thousand Mile Stare: One Family's Journey Through the Struggle and Science of Alzheimer's* (Boston: Nicholas Brealey, 2010).

15. Of the three spared, one developed late-onset Alzheimer's.

16. Thomas D. Bird et al., "Familial Alzheimer's Disease in Germans from Russia: A Model of Genetic Heterogeneity in Alzheimer's Disease," in *Heterogeneity of Alzheimer's Disease*, ed. Francois Boller, F. Forette, Z. S. Khachaturian, Michel Poncet, and Yves Christen (Berlin and Heidelberg: Springer Berlin Heidelberg, 1992), 118–29.

17. Hans H. Klünemann et al.. "Alzheimer's Second Patient: Johann F. and His Family," *Annals of Neurology* 52, no. 4 (2002): 520–23.

18. Chang-En Yu et al., "The N141I Mutation in PSEN2: Implications for the Quintessential Case of Alzheimer Disease," *Archives of Neurology* 67, no. 5 (2010): 631–33.

19. Yu et al., "N141I Mutation."

3. LEARNING TO WALK

1. Robert D. Terry, "Alzheimer's Disease at Mid-Century (1927–1977) and a Little More," in *Alzheimer: 100 Years and Beyond*, ed. Mathias Jucker et al. (Heidelberg, Germany: Springer, 2006), 59–61; Zaven S. Khachaturian, "A Chapter in the Development on Alzheimer's Disease Research," in

Alzheimer: 100 Years and Beyond, ed. Mathias Jucker et al. (Heidelberg, Germany: Springer, 2006), 63–86.

2. Petra Kaufmann, Anne R. Pariser, and Christopher Austin, "From Scientific Discovery to Treatments for Rare Diseases—the View from the National Center for Advancing Translational Sciences—Office of Rare Diseases Research," *Orphanet Journal of Rare Diseases* 13, no. 196 (2018): 1–8.

3. Germán Elías Berrios, "Alzheimer's Disease: A Conceptual History," *International Journal of Geriatric Psychiatry* 5, no. 6 (1990): 355–65; Michel Goedert, "Oskar Fischer and the Study of Dementia," *Brain* 132, no. 4 (2009): 1102–11.

4. Robert D. Terry, "Dementia: A Brief and Selective Review," *Archives of Neurology* 33, no. 1 (1976): 3.

5. Robert Katzman and Katherine Bick, *Alzheimer Disease: The Changing View* (San Diego: Academic Press, 2000), 39.

6. Khachaturian, "Chapter in the Development."

7. Garry Blessed, Bernard E. Tomlinson, and Martin Roth, "The Association Between Quantitative Measures of Dementia and of Senile Change in the Cerebral Grey Matter of Elderly Subjects," *British Journal of Psychiatry: The Journal of Mental Science* 114, no. 512 (1968): 797–811.

8. Terry, "Dementia," 2.

9. Robert Katzman, "Editorial: The Prevalence and Malignancy of Alzheimer Disease. A Major Killer," *Archives of Neurology* 33, no. 4 (1976), 217.

10. Once again, we have to applaud the wisdom of Alois Alzheimer and Emil Kraepelin. Half a century ago, they had suspected that the so-called presenile and senile dementias may be related, dementia can occur independent of age, and varying ages of onset exist on a continuum.

11. Terry, "Dementia."

12. Terry, "Dementia."

13. Katzman, "Prevalence and Malignancy."

14. Barron H. Lerner, "Rita Hayworth's Misdiagnosed Struggle," *Los Angeles Times*, November 20, 2006, https://www.latimes.com/archives/la-xpm-2006-nov -20-he-myturn20-story.html.

15. Albin Krebs, "Rita Hayworth, Movie Legend, Dies," *New York Times*, May 16, 1987, https://www.nytimes.com/1987/05/16/obituaries/rita-hayworth-movie -legend-dies.html.

16. Khachaturian, "Chapter in the Development," 78.

17. Katzman and Bick, *Alzheimer Disease*, 305–6.

18. Katzman and Bick, *Alzheimer Disease*, 285.

19. Khachaturian, "Chapter in the Development."

20. Patrick Fox, "From Senility to Alzheimer's Disease: The Rise of the Alzheimer's Disease Movement," *The Milbank Quarterly* 67, no. 1 (1989): 58–102.

21. Fox, "Senility to Alzheimer's Disease."
22. Alzheimer's Association, "IRS Information Returns: Form 990. Year Ended June 30, 2019," December 18, 2019, https://www.alz.org/media/Documents /form-990-fy-2019.pdf.
23. Katzman and Bick, *Alzheimer Disease*, xiv.
24. Katzman and Bick, *Alzheimer Disease*, xiv.
25. Katzman and Bick, *Alzheimer Disease*. According to Fox, "Senility to Alzheimer's Disease," the number was more like between 30,000 and 40,000.
26. Katzman and Bick, *Alzheimer Disease*, 344–45.
27. Katzman and Bick, *Alzheimer Disease*, 345.
28. Peter Whitehouse, *The Myth of Alzheimer's: What You Aren't Being Told About Today's Most Dreaded Diagnosis* (New York: St. Martin's Press, 2008), 114.
29. Whitehouse, *Myth of Alzheimer's*, 104.
30. Margaret Lock, *The Alzheimer Conundrum: Entanglements of Dementia and Aging* (Princeton, NJ: Princeton University Press, 2013), 63.
31. Fox, "Senility to Alzheimer's Disease."
32. Peter Davies and A. J. Maloney, "Selective Loss of Central Cholinergic Neurons in Alzheimer's Disease," *The Lancet* 308, no. 8000 (1976): 1403; P. White et al., "Neocortical Cholinergic Neurons in Elderly People," *The Lancet* 309, no. 8013 (1977): 668–71; Elaine K. Perry, Robert H. Perry, Garry Blessed, and Bernard E. Tomlinson, "Necropsy Evidence of Central Cholinergic Deficits in Senile Dementia," *The Lancet* 1, no. 8004 (1977): 189.
33. Peter J. Whitehouse et al., "Alzheimer's Disease and Senile Dementia: Loss of Neurons in the Basal Forebrain," *Science* 215, no. 4537 (1982): 1237–39.
34. Edith L. Cohen and Richard J. Wurtman, "Brain Acetylcholine: Control by Dietary Choline," *Science* 191, no. 4227 (1976): 561–62.
35. Edith L. Cohen and Richard J. Wurtman, "Brain Acetylcholine: Increase After Systemic Choline Administration," *Life Sciences* 16, no. 7 (1975): 1095–102.
36. Cohen and Wurtman, "Brain Acetylcholine: Control."
37. W. D. Boyd et al., "Clinical Effects of Choline in Alzheimer Senile Dementia," *The Lancet* 310, no. 8040 (1977): 711; C. M. Smith et al., "Choline Therapy in Alzheimer's Disease," *The Lancet* 312, no. 8084 (1978): 318; L. J. Thal, W. Rosen, N. S. Sharpless, and H. Crystal, "Choline Chloride Fails to Improve Cognition in Alzheimer's Disease," *Neurobiology of Aging* 2, no. 3 (1981): 205–8.
38. M. W. Dysken et al., "Lecithin Administration in Alzheimer Dementia," *Neurology* 32, no. 10 (1982): 1203–4; S. Weintraub et al., "Lethicin in the Treatment of Alzheimer's Disease," *Archives of Neurology* 40, no. 8 (1983): 527–28.
39. Nawab Qizilbash et al., "Cholinesterase Inhibition for Alzheimer Disease: A Meta-Analysis of the Tacrine Trials," *JAMA* 280, no. 20 (1998): 1777–82.

40. Qizilbash et al., "Cholinesterase Inhibition."
41. M. Lynn Crismon, "Tacrine: First Drug Approved for Alzheimer's Disease," *Annals of Pharmacotherapy* 28, no. 6 (1994): 744–51.
42. Sharon L. Rogers, Rachelle S. Doody, Richard C. Mohs, and Lawrence T. Friedhoff, "Donepezil Improves Cognition and Global Function in Alzheimer Disease: A 15-Week, Double-Blind, Placebo-Controlled Study," *Archives of Internal Medicine* 158, no. 9 (1998): 1021–31; Michael Rösler et al., "Efficacy and Safety of Rivastigmine in Patients with Alzheimer's Disease: International Randomised Controlled Trial," *BMJ* 318, no. 7184 (1999): 633–38; Gordon K. Wilcock, Sean Lilienfeld, and Els Gaens, "Efficacy and Safety of Galantamine in Patients with Mild to Moderate Alzheimer's Disease: Multicentre Randomised Controlled Trial," *BMJ* 321, no. 7274 (2000): 1445–49.
43. Raymond Dingledine and Chris McBain, "Glutamate and Aspartate," in *Basic Neurochemistry: Molecular, Cellular and Medical Aspects*, ed. George Siegle et al. (Philadelphia: Lippincott-Raven, 1999).
44. J. Timothy Greenamyre et al., "Glutamate Transmission and Toxicity in Alzheimer's Disease," *Progress in Neuropsychopharmacology & Biological Psychiatry* 12, no. 4 (1988): 421–30; N. B. Farber, J. W. Newcomer, and J. W. Olney, "The Glutamate Synapse in Neuropsychiatric Disorders. Focus on Schizophrenia and Alzheimer's Disease," *Progress in Brain Research* 116 (1998): 421–37.
45. Bengt Winblad and N. Poritis, "Memantine in Severe Dementia: Results of the 9M-Best Study (Benefit and Efficacy in Severely Demented Patients During Treatment with Memantine)," *International Journal of Geriatric Psychiatry* 14, no. 2 (1999): 135–46; Barry Reisberg et al., "Memantine in Moderate-to-Severe Alzheimer's Disease," *New England Journal of Medicine* 348, no. 14 (2003): 1333–41.
46. A combination of memantine and donepezil (trade name Namzaric) was also approved by FDA.
47. John H. Growdon, "Acetylcholine in AD: Expectations Meet Reality," in *Alzheimer: 100 Years and Beyond*, ed. Mathias Jucker et al. (Heidelberg: Springer, 2006), 127–32.

4. SEARCHING FOR THE ALZHEIMER'S GENE

1. Sex determination can get more complicated than XY versus XX; see Claire Ainsworth, "Sex Redefined: The Idea of 2 Sexes Is Overly Simplistic," *Scientific American*, October 22, 2018, https://www.scientificamerican.com/article/sex-redefined-the-idea-of-2-sexes-is-overly-simplistic1/.

2. Chromosome 22 was once thought to be the smallest chromosome, but it was later found to be slightly bigger than chromosome 21. The numbering, however, remained the same.

3. Rare exceptions have been reported. See Thomas D. Bird, "Alzheimer Disease Overview," in *GeneReviews*®, ed. Margaret P. Adam, Holly H. Ardinger, Roberta A. Pagon, and Stephanie E. Wallace (Seattle: University of Washington, 2018).

4. Pam Belluck, "Alzheimer's Stalks a Colombian Family," *New York Times*, June 1, 2010, https://www.nytimes.com/2010/06/02/health/02alzheimers.html.

5. Belluck, "Alzheimer's Stalks."

6. Belluck, "Alzheimer's Stalks."

7. Rudolph E. Tanzi and Ann B. Parson, *Decoding Darkness: The Search for the Genetic Causes of Alzheimer's Disease* (Cambridge, MA: Perseus Publishing, 2000).

8. Rainulf A. Stelzmann, H. Norman Schnitzlein, and F. Reed Murtagh, "An English Translation of Alzheimer's 1907 Paper, 'Über eine eigenartige Erkankung der Hirnrinde," *Clinical Anatomy* 8, no. 6 (1995): 429–31.

9. Paul Divry, "Etude histochimique des plaques séniles," *Journal Belge de Neurologie et de Psychiatrie* 27 (1927): 643–57.

10. Maarit Tanskanen, " 'Amyloid'—Historical Aspects," in *Amyloidosis*, ed. Dali Feng (London: IntechOpen, 2013), 3–24.

11. Tanzi and Parson, *Decoding Darkness*.

12. George G. Glenner and Caine W. Wong, "Alzheimer's Disease: Initial Report of the Purification and Characterization of a Novel Cerebrovascular Amyloid Protein," *Biochemical and Biophysical Research Communications* 120, no. 3 (1984): 885–90.

13. Jie Kang et al., "The Precursor of Alzheimer's Disease Amyloid A4 Protein Resembles a Cell-Surface Receptor," *Nature* 325, no. 6106 (1987): 733–36.

14. Kang et al., "Precursor of Alzheimer's."

15. Alison Goate et al., "Segregation of a Missense Mutation in the Amyloid Precursor Protein Gene with Familial Alzheimer's Disease," *Nature* 349, no. 6311 (1991): 704–6.

16. Tanzi and Parson, *Decoding Darkness*, 128.

17. Gerard D. Schellenberg et al., "Genetic Linkage Evidence for a Familial Alzheimer's Disease Locus on Chromosome 14," *Science* 258, no. 5082 (1992): 668–71.

18. This account of the rumors was drawn upon Tanzi and Parson, *Decoding Darkness*.

19. Peter St George-Hyslop et al., "Genetic Evidence for a Novel Familial Alzheimer's Disease Locus on Chromosome 14," *Nature Genetics* 2, no. 4 (1992): 330–34.

20. Peter H. St George-Hyslop et al., "The Genetic Defect Causing Familial Alzheimer's Disease Maps on Chromosome 21," *Science* 235, no. 4791 (1987): 885–90.

21. Tanzi and Parson, *Decoding Darkness*.

22. R. Sherrington et al., "Cloning of a Gene Bearing Missense Mutations in Early-Onset Familial Alzheimer's Disease," *Nature* 375, no. 6534 (1995): 754–60.

23. Bird, "Alzheimer Disease Overview."

24. The account of the discovery of PSEN2 was drawn upon Tanzi and Parson, *Decoding Darkness*.

25. Ephrat Levy-Lahad et al., "Candidate Gene for the Chromosome 1 Familial Alzheimer's Disease Locus," *Science* 269, no. 5226 (1995): 973–77.

26. Bird, "Alzheimer Disease Overview."

27. Gary Reiswig, *The Thousand Mile Stare: One Family's Journey Through the Struggle and Science of Alzheimer's* (Boston: Nicholas Brealey, 2010), 128.

5. LATE-ONSET ALZHEIMER'S

1. National Institutes of Health, "Estimates of Funding for Various Research, Condition, and Disease Categories (RCDC)," February 24, 2020, from https://report.nih.gov/categorical_spending.aspx.

2. Kit Yee Chan et al., "Epidemiology of Alzheimer's Disease and Other Forms of Dementia in China, 1990–2010: A Systematic Review and Analysis," *The Lancet* 381, no. 9882 (2013): 2016–23.

3. Alzheimer's Association, "2018 Alzheimer's Disease Facts and Figures," *Alzheimers Dement* 14, no. 3 (2018): 367–429.

4. Elizabeth Arias, "United States Life Tables, 2011," *National Vital Statistics Reports* 64, no. 11 (2015): 1–62.

5. M. Pericak-Vance et al., "Linkage Studies in Familial Alzheimer Disease: Evidence for Chromosome 19 Linkage," *American Journal of Human Genetics* 48, no. 6 (1991): 1034–50.

6. Warren Strittmatter et al., "Apolipoprotein E: High-Avidity Binding to β-Amyloid and Increased Frequency of Type 4 Allele in Late-Onset Familial Alzheimer Disease," *Proceedings of the National Academy of Sciences of the United States of America* 90, no. 5 (1993): 1977–81.

7. P. P. Singh, M. Singh, and S. S. Mastana, "ApoE Distribution in World Populations with New Data from India and the UK," *Annals of Human Biology* 33, no. 3 (2006): 279–308.

8. Strittmatter et al., "Apolipoprotein E"; A. Saunders et al., "Association of Apolipoprotein E Allele ε-4 with Late-Onset Familial and Sporadic Alzheimer's Disease," *Neurology* 43, no. 8 (1993): 1467–72.

9. E. H. Corder et al., "Gene Dose of Apolipoprotein E Type 4 Allele and the Risk of Alzheimer's Disease in Late Onset Families," *Science* 261, no. 5123 (1993): 921–23.

10. Corder et al., "Gene Dose."

11. Corder et al., "Gene Dose."

12. This statistic does not count ApoE2/ApoE4 combinations. E. H. Corder et al., "Protective Effect of Apolipoprotein E Type 2 Allele for Late Onset Alzheimer Disease," *Nature Genetics* 7, no. 2 (1994): 180–4.

13. Jason J. Corneveaux et al., "Association of CR1, CLU and PICALM with Alzheimer's Disease in a Cohort of Clinically Characterized and Neuropathologically Verified Individuals," *Human Molecular Genetics* 19, no. 16 (2010): 3295–301. Numerical risk ratio estimates differ across studies, but the same trend holds true.

14. Lindsay A. Farrer et al., "Effects of Age, Sex, and Ethnicity on the Association Between Apolipoprotein E Genotype and Alzheimer Disease: A Meta-Analysis," *JAMA* 278, no. 16 (1997): 1349–56; Ming-Xin Tang et al., "The APOE- ε4 Allele and the Risk of Alzheimer Disease Among African Americans, Whites, and Hispanics," *JAMA* 279, no. 10 (1998): 751–55; Mengying Liu, Chen Bian, Jiqiang Zhang, and Feng Wen, "Apolipoprotein E Gene Polymorphism and Alzheimer's Disease in Chinese Population: A Meta-Analysis," *Scientific Reports* 4, no. 4383 (2014): 1–7.

15. Andre Altmann, Lu Tian, Victor W. Henderson, and Michael D. Greicius, "Sex Modifies the APOE-Related Risk of Developing Alzheimer Disease," *Annals of Neurology* 75, no. 4 (2014): 563–73.

16. Farrer et al., "Effects of Age."

17. Strittmatter et al., "Apolipoprotein E"; Takahisa Kanekiyo, Huaxi Xu, and Guojun Bu, "ApoE and Aβ in Alzheimer's Disease: Accidental Encounters or Partners?" *Neuron* 81, no. 4 (2014): 740–54.

18. Kanekiyo et al., "ApoE and Aβ."

19. Jungsu Kim, Jacob M. Basak, and David M. Holtzman, "The Role of Apolipoprotein E in Alzheimer's Disease," *Neuron* 63, no. 3 (2009): 287–303; Kanekiyo et al., "ApoE and Aβ."

20. Sam Roberts, "Allen Roses, Who Upset Common Wisdom on Cause of Alzheimer's, Dies at 73," *New York Times*, October 5, 2016, https://www.nytimes.com/2016/10/06/science/allen-roses-who-upset-common-wisdom-on-cause-of-alzheimers-dies-at-73.html.

21. Roberts, "Allen Roses."

6. THE PARADIGM

1. Rebecca Hiscott, "At the Bench: John Hardy, PhD, on Unraveling the Genetics of Alzheimer's Disease and Attending the 'Oscars of Science,'" *Neurology Today* 16, no. 1 (2016): 21–22.
2. James F. Gusella et al., "A Polymorphic DNA Marker Genetically Linked to Huntington's Disease," *Nature* 306, no. 5940 (1983): 234–38.
3. "Nicholas Wood Interviews John Hardy," YouTube, February 19, 2013, https://www.youtube.com/watch?v=YZThB_M8DXw.
4. "Nicholas Wood Interviews John Hardy."
5. "Nicholas Wood Interviews John Hardy."
6. Alison Goate et al., "Segregation of a Missense Mutation in the Amyloid Precursor Protein Gene with Familial Alzheimer's Disease," *Nature* 349, no. 6311 (1991): 704–6.
7. Shigeki Kawabata, Gerald Higgins, and Jon Gordon, "Amyloid Plaques, Neurofibrillary Tangles and Neuronal Loss in Brains of Transgenic Mice Overexpressing a C-Terminal Fragment of Human Amyloid Precursor Protein," *Nature* 354, no. 6353 (1991): 476–78.
8. John Hardy, "Alzheimer's Disease: The Amyloid Cascade Hypothesis: An Update and Reappraisal," *Journal of Alzheimer's Disease* 9, no. 3 (2006): 151–53.
9. Hardy, "Alzheimer's Disease."
10. These citation statistics are according to Scopus.
11. John A. Hardy and Gerald A. Higgins, "Alzheimer's Disease: The Amyloid Cascade Hypothesis," *Science* 256, no. 5054 (1992): 184–85.
12. John Hardy and David Allsop, "Amyloid Deposition as the Central Event in the Aetiology of Alzheimer's Disease," *Trends in Pharmacological Sciences* 12 (1991): 383–88; Dennis J. Selkoe, "The Molecular Pathology of Alzheimer's Disease," *Neuron* 6, no. 4 (1991): 487–98.
13. Hardy, "Alzheimer's Disease," 152.
14. Martin Citron et al., "Mutation of the β-Amyloid Precursor Protein in Familial Alzheimer's Disease Increases β-Protein Production," *Nature* 360, no. 6405 (1992): 672–74.
15. Hardy, "Alzheimer's Disease," 151.
16. Thomas Kuhn, *The Structure of Scientific Revolutions*, 4th ed. (Chicago and London: University of Chicago Press, 2012).
17. Kuhn, *Structure of Scientific Revolutions*.
18. Kuhn, *Structure of Scientific Revolutions*.
19. Sascha Weggen and Dirk Beher, "Molecular Consequences of Amyloid Precursor Protein and Presenilin Mutations Causing Autosomal-Dominant Alzheimer's Disease," *Alzheimer's Research & Therapy* 4, no. 9 (2012): 1–14.

20. M. Paul Murphy and Harry LeVine III, "Alzheimer's Disease and the β-Amyloid Peptide," *Journal of Alzheimer's Disease* 19, no. 1 (2010): 311–23.

21. Murphy and LeVine, "Alzheimer's Disease."

22. Weggen and Beher, "Molecular Consequences."

23. Bart De Strooper and Wim Annaert, "Novel Research Horizons for Presenilins and γ-Secretases in Cell Biology and Disease," *Annual Review of Cell and Developmental Biology* 26 (2010): 235–60; Weggen and Beher, "Molecular Consequences."

24. Peter T. Nelson, Heiko Braak, and William R. Markesbery, "Neuropathology and Cognitive Impairment in Alzheimer Disease: A Complex but Coherent Relationship," *Journal of Neuropathology and Experimental Neurology* 68, no. 1 (2009): 1–14.

25. Karen Rodrigue, Kristen Kennedy, and Denise Park, "Beta-Amyloid Deposition and the Aging Brain," *Neuropsychology Review* 19, no. 4 (2009): 436–50.

26. The description of Sister Mary was drawn from David A. Snowdon, "Aging and Alzheimer's Disease: Lessons from the Nun Study," *The Gerontologist* 37, no. 2 (1997): 150–56.

27. Snowdon, "Aging and Alzheimer's," 150.

28. Snowdon, "Aging and Alzheimer's," 151.

29. Bruce A. Yankner et al., "Neurotoxicity of a Fragment of the Amyloid Precursor Associated with Alzheimer's Disease," *Science* 245, no. 4916 (1989): 417–20; Neil W. Kowall et al., "An in Vivo Model for the Neurodegenerative Effects of Beta Amyloid and Protection by Substance P," *Proceedings of the National Academy of Sciences of the United States of America* 88, no. 16 (1991): 7247–51.

30. Masafumi Sakono and Tamotsu Zako, "Amyloid Oligomers: Formation and Toxicity of Aβ Oligomers," *FEBS Journal* 277 (2010): 1348–58; William L. Klein, "Synaptotoxic Amyloid-β Oligomers: A Molecular Basis for the Cause, Diagnosis, and Treatment of Alzheimer's Disease?" *Journal of Alzheimer's Disease* 33 Suppl 1 (2013): S49–S65.

31. Elizabeth Agnvall, "New Science Sheds Light on the Cause of Alzheimer's Disease," AARP, January 24, 2012, http://www.aarp.org/health/conditions-treatments/info-05-2010/alzheimers_disease.html.

32. Maria Laura Giuffrida et al., "Beta-Amyloid Monomers Are Neuroprotective," *Journal of Neuroscience* 29, no. 34 (2009): 10582–87; Bruce A. Yankner, Lawrence K. Duffy, and Daniel A. Kirschner, "Neurotrophic and Neurotoxic Effects of Amyloid β Protein: Reversal by Tachykinin Neuropeptides," *Science* 250, no. 4978 (1990): 279–82.

33. Inna Kuperstein et al., "Neurotoxicity of Alzheimer's Disease Aβ Peptides Is Induced by Small Changes in the $A\beta_{42}$ to $A\beta_{40}$ Ratio," *EMBO Journal* 29, no. 19 (2010): 3408–420.

34. Emilie Cerf et al., "High Ability of Apolipoprotein E4 to Stabilize Amyloid-β Peptide Oligomers, the Pathological Entities Responsible for Alzheimer's Disease," *FASEB Journal: Official Publication of the Federation of American Societies for Experimental Biology* 25, no. 5 (2011): 1585–95.

35. Tadafumi Hashimoto et al., "Apolipoprotein E, Especially Apolipoprotein E4, Increases the Oligomerization of Amyloid β Peptide," *Journal of Neuroscience: The Official Journal of the Society for Neuroscience* 32, no. 43 (2012): 15181–92. The amount of oligomers in ApoE4/ApoE4 patient brains is 2.7 times higher than that in ApoE3/ApoE3 patient brains and 6.9 times higher than that in patients with a protective ApoE2.

36. K. Rajasekhar, Malabika Chakrabarti, and T. Govindaraju, "Function and Toxicity of Amyloid Beta and Recent Therapeutic Interventions Targeting Amyloid Beta in Alzheimer's Disease," *Chemical Communications* 51, no. 70 (2015): 13434–50; Klein, "Synaptotoxic Amyloid-β Oligomers"; Sakono and Zako, "Amyloid Oligomers"; Dominic M. Walsh et al., "Naturally Secreted Oligomers of Amyloid β Protein Potently Inhibit Hippocampal Long-Term Potentiation in Vivo." *Nature* 416, no. 6880 (2002): 535–39.

37. James Cleary et al., "Natural Oligomers of the Amyloid-β Protein Specifically Disrupt Cognitive Function," *Nature Neuroscience* 8, no. 1 (2004): 79–84.

38. Sakono and Zako, "Amyloid Oligomers"; Iryna Benilova, Eric Karran, and Bart De Strooper, "The Toxic Aβ Oligomer and Alzheimer's Disease: An Emperor in Need of Clothes," *Nature Neuroscience* 15, no. 3 (2012): 349–57.

39. Hashimoto et al., "Apolipoprotein E"; Mikko Hölttä et al., "Evaluating Amyloid-β Oligomers in Cerebrospinal Fluid as a Biomarker for Alzheimer's Disease," *PLoS One* 8, no. 6 (2013): E66381.

40. Benilova et al., "The Toxic Aβ Oligomer and Alzheimer's Disease"; Kirsten Viola and William Klein, "Amyloid β Oligomers in Alzheimer's Disease Pathogenesis, Treatment, and Diagnosis," *Acta Neuropathologica; Pathology and Mechanisms of Neurological Disease* 129, no. 2 (2015): 183–206.

41. Benilova et al., "The Toxic Aβ Oligomer and Alzheimer's Disease"; Franz Hefti et al., "The Case for Soluble Aβ Oligomers as a Drug Target in Alzheimer's Disease," *Trends in Pharmacological Sciences* 34, no. 5 (2013): 261–66; Sylvain E. Lesné et al., "Brain Amyloid-β Oligomers in Ageing and Alzheimer's Disease," *Brain* 136, no. 5 (2013): 1383–98.

7. ELI LILLY AND MICE

1. John Simons, "Lilly Goes off Prozac: The Drugmaker Bounced Back from the Loss of Its Blockbuster, but the Recovery Had Costs," *Fortune*, June 28, 2004, https://archive.fortune.com/magazines/fortune/fortune_archive/2004/06/28/374398/index.htm.

2. Eli Lilly, "Lilly Reports Fourth-Quarter and Full-Year 2016 Results," January 31, 2017, https://investor.lilly.com/news-releases/news-release-details /lilly-reports-fourth-quarter-and-full-year-2016-results?releaseID=1009682.

3. Tracy Staton, "Eli Lilly—10 Largest U.S. Patent Losses," FiercePharma, October 24, 2011, https://www.fiercepharma.com/special-report/eli-lilly -10-largest-u-s-patent-losses; Eli Lilly, "Lilly Reports."

4. United Nations, Department of Economic and Social Affairs Population Division, *World Population Ageing* (New York: United Nations, 2015).

5. Alzheimer's Association, "2018 Alzheimer's Disease Facts and Figures," *Alzheimers Dement* 14, no. 3 (2018): 367–429.

6. Lars M. Ittner and Jürgen Götz, "Amyloid-β and Tau—A Toxic Pas de Deux in Alzheimer's Disease," *Nature Reviews Neuroscience* 12, no. 2 (2011): 67–72; David M. Holtzman, John C. Morris, and Alison M. Goate, "Alzheimer's Disease: The Challenge of the Second Century," *Science Translational Medicine* 3, no. 77 (2011): 77sr1.

7. U.S. Food and Drug Administration, "The Drug Development Process Step 3: Clinical Research," FDA, January 4, 2018, https://www.fda.gov/ForPatients /Approvals/Drugs/ucm405622.htm.

8. U.S. Food and Drug Administration, "Drug Development Process."

9. D. O. Wirak et al., "Deposits of Amyloid β Protein in the Central Nervous System of Transgenic Mice," *Science* 253, no. 5017 (1991): 323–25.

10. Mathias Jucker et al., "Age-Associated Inclusions in Normal and Transgenic Mouse Brain," *Science* 255, no. 5050 (1992): 1443–45.

11. Shigeki Kawabata, Gerald Higgins, and Jon Gordon, "Amyloid Plaques, Neurofibrillary Tangles and Neuronal Loss in Brains of Transgenic Mice Overexpressing a C-Terminal Fragment of Human Amyloid Precursor Protein," *Nature* 354, no. 6353 (1991): 476–78.

12. John Rennie, "The Mice that Missed," *Scientific American* 266, no. 6 (1992), 20, 26.

13. John Hardy, "Alzheimer's Disease: The Amyloid Cascade Hypothesis: An Update and Reappraisal," *Journal of Alzheimer's Disease* 9, no. 3 (2006): 151–53.

14. Lawrence Fisher, "Athena Neurosciences Makes Itself Heard in Fight Against Alzheimer's," *New York Times*, February 15, 1995, https://www .nytimes.com/1995/02/15/business/business-technology-athena-neuro -sciences-makes-itself-heard-fight-against.html; Gina Kolata, "Landmark in Alzheimer Research: Breeding Mice with the Disease," *New York Times*, February 9, 1995, https://www.nytimes.com/1995/02/09/us/landmark-in -alzheimer-research-breeding-mice-with-the-disease.html.

15. Dora Games et al., "Alzheimer-Type Neuropathology in Transgenic Mice Overexpressing V717F β-Amyloid Precursor Protein," *Nature* 373, no. 6514 (1995): 523–27.

8. INHIBITORS THAT CAN'T INHIBIT

1. See Clinicaltrials.gov, identifiers NCT00762411 and NCT00594568.
2. Rachelle S. Doody et al., "A Phase 3 Trial of Semagacestat for Treatment of Alzheimer's Disease," *New England Journal of Medicine* 369, no. 4 (2013): 341–50. Cerebrospinal fluid beta-amyloid measurements were used.
3. Doody et al., "Phase 3 Trial of Semagacestat."
4. Eric Karran and John Hardy, "A Critique of the Drug Discovery and Phase 3 Clinical Programs Targeting the Amyloid Hypothesis for Alzheimer Disease," *Annuals of Neurology* 76, no. 2 (2014): 185–205.
5. Karran and Hardy, "Critique of the Drug Discovery."
6. Randall J. Bateman et al., "A γ-Secretase Inhibitor Decreases Amyloid-β Production in the Central Nervous System," *Annals of Neurology* 66, no. 1 (2009): 48–54.
7. Eric R. Siemers et al., "Effects of a Gamma-Secretase Inhibitor in a Randomized Study of Patients with Alzheimer Disease," *Neurology* 66, no. 4 (2006): 602–4.
8. Adam S. Fleisher et al., "Phase 2 Safety Trial Targeting Amyloid β Production with a γ-Secretase Inhibitor in Alzheimer Disease," *Archives of Neurology* 65, no. 8 (2008): 1031–38.
9. Fleisher et al., "Phase 2 Safety Trial," 1037.
10. Bart De Strooper, "Lessons from a Failed γ-Secretase Alzheimer Trial," *Cell* 159, no. 4 (2014): 721–26.
11. Justin D. Lathia, Mark P. Mattson, and Aiwu Cheng, "Notch: From Neural Development to Neurological Disorders," *Journal of Neurochemistry* 107, no. 6 (2008): 1471–81.
12. De Strooper, "Lessons."
13. Vladimir Coric et al., "Safety and Tolerability of the γ-Secretase Inhibitor Avagacestat in a Phase 2 Study of Mild to Moderate Alzheimer Disease," *Archives of Neurology* 69, no. 11 (2012): 1430–40.
14. Coric et al., "Safety and Tolerability."
15. Patrick C. May et al., "Robust Central Reduction of Amyloid-β in Humans with an Orally Available, Non-Peptidic β-Secretase Inhibitor," *Journal of Neuroscience: The Official Journal of the Society for Neuroscience* 31, no. 46 (2011): 16507–16.
16. May et al., "Robust Central Reduction."
17. May et al., "Robust Central Reduction"; Patrick C. May et al., "The Potent BACE1 Inhibitor LY2886721 Elicits Robust Central Aβ Pharmacodynamic Responses in Mice, Dogs, and Humans," *Journal of Neuroscience: The Official Journal of the Society for Neuroscience* 35, no. 3 (2015): 1199–210.

18. May et al., "Potent BACE1 Inhibitor LY2886721."
19. Alzforum, "LY2886721," accessed March 26, 2020, http://www.alzforum .org/therapeutics/ly2886721.
20. Devendra Kumar et al., "Secretase Inhibitors for the Treatment of Alzheimer's Disease: Long Road Ahead," *European Journal of Medicinal Chemistry* 148 (2018): 436–52; Tom Fagan, "Liver Tox Ends Janssen BACE Program," Alzforum, May 18, 2018, https://www.alzforum.org/news/research-news/liver-tox-ends -janssen-bace-program.
21. Kumar et al., "Secretase Inhibitors."
22. Eli Lilly, "Update on Phase 3 Clinical Trials of Lanabecestat for Alzheimer's Disease," June 12, 2018, https://investor.lilly.com/news-releases/news -release-details/update-phase-3-clinical-trials-lanabecestat-alzheimers -disease

9. POISON OR CURE: AN ALZHEIMER'S VACCINE

1. Lauren R. Platt, Concepción F. Estívariz, and Roland W. Sutter, "Vaccine-Associated Paralytic Poliomyelitis: A Review of the Epidemiology and Estimation of the Global Burden," *Journal of Infectious Diseases* 210 Suppl 1 (2014): S380–S389.
2. A. Wakefield et al., "Ileal-Lymphoid-Nodular Hyperplasia, Non-Specific Colitis, and Pervasive Developmental Disorder in Children," *The Lancet* 351, no. 9103 (1998): 637–41.
3. R. Gasparini, D. Panatto, P. L. Lai, and D. Amicizia, "The 'Urban Myth' of the Association Between Neurological Disorders and Vaccinations," *Journal of Preventive Medicine and Hygiene* 56, no. 1 (2015), E3.
4. Gasparini et al., "Urban Myth."
5. Porter Anderson, "Dale Schenk: Alzheimer's Researcher," CNN, December 11, 2001, http://www.cnn.com/2001/CAREER/jobenvy/07/23/dale.schenk/.
6. Anderson, "Dale Schenk."
7. Tom Fagan and Gabrielle Strobel, "Dale Schenk, 59, Pioneer of Alzheimer's Immunotherapy," Alzforum, October 3, 2016, https://www.alzforum.org /news/community-news/dale-schenk-59-pioneer-alzheimers-immunotherapy.
8. Rudolph E. Tanzi and Ann B. Parson, *Decoding Darkness: The Search for the Genetic Causes of Alzheimer's Disease* (Cambridge, MA: Perseus Publishing, 2000).
9. Dale Schenk et al., "Immunization with Amyloid-β Attenuates Alzheimer-Disease-Like Pathology in the PDAPP Mouse," *Nature* 400, no. 6740 (1999): 173–77.

10. Schenk et al., "Immunization with Amyloid-β."

11. Christopher Janus et al., "Aβ Peptide Immunization Reduces Behavioural Impairment and Plaques in a Model of Alzheimer's Disease," *Nature* 408, no. 6815 (2000): 979–82.

12. Antony J. Bayer et al., "Evaluation of the Safety and Immunogenicity of Synthetic Abeta42 (AN1792) in Patients with AD," *Neurology* 64, no. 1 (2005): 94–101.

13. J. M. Orgogozo et al., "Subacute Meningoencephalitis in a Subset of Patients with AD After Aβ42 Immunization," *Neurology* 61, no. 1 (2003): 46–54; S. Gilman et al., "Clinical Effects of Aβ Immunization (AN1792) in Patients with AD in an Interrupted Trial," *Neurology* 64, no. 9 (2005): 1553–62.

14. Orgogozo et al., "Subacute Meningoencephalitis."

15. Gilman et al., "Clinical Effects of Aβ Immunization."

16. Isidre Ferrer et al., "Neuropathology and Pathogenesis of Encephalitis Following Amyloid β Immunization in Alzheimer's Disease," *Brain Pathology* 14, no. 1 (2004): 11–20.

17. Bengt Winblad et al., "Active Immunotherapy Options for Alzheimer's Disease," *Alzheimer's Research & Therapy* 6, no. 7 (2014):1–12.

18. Winblad et al., "Active Immunotherapy Options."

19. Bengt Winblad et al., "Safety, Tolerability, and Antibody Response of Active Aβ Immunotherapy with CAD106 in Patients with Alzheimer's Disease: Randomised, Double-Blind, Placebo-Controlled, First-in-Human Study," *The Lancet Neurology* 11, no. 7 (2012): 597–604; Martin R. Farlow et al., "Long-Term Treatment with Active Aβ Immunotherapy with CAD106 in Mild Alzheimer's Disease," *Alzheimer's Research & Therapy* 7, no. 1 (2015):1–13.

20. Madolyn Bowman Rogers, "Immunotherapy II: Active Approaches Down, New Passive Crops Up," Alzforum, December 17, 2014, http://www.alzforum .org/news/conference-coverage/immunotherapy-ii-active-approaches -down-new-passive-crops.

21. Markus Mandler, Walter Schmidt, and Frank Mattner, "Development of AFFITOPE Alzheimer Vaccines: Results of Phase I Studies with AD01 and AD02," *Alzheimer's & Dementia* 7, no. 4 (2011): S793.

22. Gwyneth Dickey Zakaib, "In Surprise, Placebo, Not Aβ Vaccine, Said to Slow Alzheimer's," Alzforum, June 6, 2014, https://www.alzforum.org/news /research-news/surprise-placebo-not-av-vaccine-said-slow-alzheimers.

23. Zakaib, "In Surprise."

24. Rogers, "Immunotherapy II."

25. Igor Klatzo, Henryk Wisniewski, and Eugene Streicher, "Experimental Production of Neurofibrillary Degeneration. I. Light Microscopic Observations," *Journal of Neuropathology and Experimental Neurology* 24 (1965): 187–99.

26. D. R. Crapper, S. S. Krishnan, and A. J. Dalton, "Brain Aluminum Distribution in Alzheimer's Disease and Experimental Neurofibrillary Degeneration," *Science* 180, no. 4085 (1973): 511–13.
27. Theodore Lidsky, "Is the Aluminum Hypothesis Dead?" *Journal of Occupational and Environmental Medicine* 56 (2014): S73–S79.
28. Judith Landsberg, Brendan McDonald, Geoff Grime, and Frank Watt, "Microanalysis of Senile Plaques Using Nuclear Microscopy," *Journal of Geriatric Psychiatry and Neurology* 6, no. 2 (1993): 97–104; Virginie Rondeau, "A Review of Epidemiologic Studies on Aluminum and Silica in Relation to Alzheimer's Disease and Associated Disorders," *Reviews on Environmental Health* 17, no. 2 (2002): 107–22.
29. Stephen C. Bondy, "Prolonged Exposure to Low Levels of Aluminum Leads to Changes Associated with Brain Aging and Neurodegeneration," *Toxicology* 315 (2014): 1–7; Rondeau, "Review of Epidemiologic Studies"; Lidsky, "Aluminum Hypothesis Dead?"
30. Lidsky, "Aluminum Hypothesis Dead?"

10. ELI LILLY'S THREE EXPEDITIONS

1. Tom Fagan and Gabrielle Strobel, "Dale Schenk, 59, Pioneer of Alzheimer's Immunotherapy," Alzforum, October 3, 2016, https://www.alzforum.org/news/community-news/dale-schenk-59-pioneer-alzheimers-immunotherapy.
2. Eli Lilly, "Lilly Halts Development of Semagacestat for Alzheimer's Disease Based on Preliminary Results of Phase III Clinical Trials," Eli Lilly, August 17, 2010, https://investor.lilly.com/news-releases/news-release-details/lilly-halts-development-semagacestat-alzheimers-disease-based?releaseid=499794.
3. Jeff Swiatek, "Lean Years Behind It, Eli Lilly Sees Growth in New Drugs," Indy-Star, May 31, 2015, https://www.indystar.com/story/money/2015/06/01/lean-years-behind-eli-lilly-sees-growth-new-drugs/28172457/.
4. Swiatek, "Lean Years Behind."
5. Swiatek, "Lean Years Behind."
6. Dale Schenk, Michael Hagen, and Peter Seubert, "Current Progress in Beta-Amyloid Immunotherapy," *Current Opinion in Immunology* 16, no. 5 (2004): 599–606.
7. Ronald B. Demattos et al., "Peripheral Anti-Aβ Antibody Alters CNS and Plasma Aβ Clearance and Decreases Brain Aβ Burden in a Mouse Model of Alzheimer's Disease." *Proceedings of the National Academy of Sciences of the United States of America* 98, no. 15 (2001): 8850–55.

8. Jean-Cosme Dodart et al., "Immunization Reverses Memory Deficits Without Reducing Brain Aβ Burden in Alzheimer's Disease Model," *Nature Neuroscience* 5, no. 5 (2002): 452–57.

9. Dodart et al., "Immunization Reverses Memory Deficits."

10. Eric R. Siemers et al., "Safety and Changes in Plasma and Cerebrospinal Fluid Amyloid Beta After a Single Administration of an Amyloid Beta Monoclonal Antibody in Subjects with Alzheimer Disease," *Clinical Neuropharmacology* 33, no. 2 (2010): 67–73.

11. Martin Farlow et al., "Safety and Biomarker Effects of Solanezumab in Patients with Alzheimer's Disease," *Alzheimer's & Dementia: The Journal of the Alzheimer's Association* 8, no. 4 (2012): 261–71.

12. Farlow et al., "Safety and Biomarker Effects," 267.

13. Rachelle S. Doody et al., "Phase 3 Trials of Solanezumab for Mild-to-Moderate Alzheimer's Disease," *New England Journal of Medicine* 370, no. 4 (2014): 311–21.

14. The Alzheimer's Disease Cooperative Study–Activities of Daily Living.

15. The Alzheimer's Disease Assessment Scale–Cognitive Subscale, the Mini-Mental State Examination, and the Clinical Dementia Rating Sum of Boxes.

16. Eric R. Siemers et al., "Phase 3 Solanezumab Trials: Secondary Outcomes in Mild Alzheimer's Disease Patients," *Alzheimer's & Dementia: The Journal of the Alzheimer's Association* 12, no. 2 (2016): 110–20.

17. The Alzheimer's Disease Assessment Scale–Cognitive Subscale 11 and Cognitive Subscale 14, the Mini–Mental State Examination, and the Alzheimer's Disease Cooperative Study–Instrumental Activities of Daily Living.

18. Siemers et al., "Phase 3 Solanezumab Trials."

19. The Alzheimer's Disease Cooperative Study–Activities of Daily Living, the Alzheimer's Disease Cooperative Study–Basic Activities of Daily Living, and the Clinical Dementia Rating Sum of Boxes.

20. Fergus Walsh, "Alzheimer's Drug Solanezumab Could Slow Patients' Decline," BBC News, July 23, 2015, https://www.bbc.com/news/av/health-33618682/alzheimer-s-drug-solanezumab-could-slow-patients-decline.

21. Sarah Knapton, "First Drug to Slow Alzheimer's Disease Unveiled in Landmark Breakthrough," *The Telegraph*, July 22, 2015, https://www.telegraph.co.uk/news/health/news/11755380/First-drug-to-slow-Alzheimers-Disease-unveiled-in-landmark-breakthrough.html.

22. Lawrence S. Honig et al., "Trial of Solanezumab for Mild Dementia Due to Alzheimer's Disease," *New England Journal of Medicine* 378, no. 4 (2018): 321–30.

23. Eli Lilly, "Lilly Announces Top-Line Results of Solanezumab Phase 3 Clinical Trial," November 23, 2016, https://investor.lilly.com/news-releases

/news-release-details/lilly-announces-top-line-results-solanezumab-phase-3-clinical?ReleaseID=1000871.

24. Ransdell Pierson, "Lilly's Drug for Alzheimer's Fails Big Trial; Shares Drop," *Reuters*, November 23, 2016, https://www.reuters.com/article/us-health-alzheimer-s-lilly/lillys-drug-for-alzheimers-fails-big-trial-shares-drop-idUSKBN13I146.

25. Pam Belluck, "Eli Lilly's Experimental Alzheimer's Drug Fails in Large Trial," *New York Times*, November 23, 2016, https://www.nytimes.com/2016/11/23/health/eli-lillys-experimental-alzheimers-drug-failed-in-large-trial.html.

26. Matthew J. Belvedere, "Eli Lilly Shares Tank After Alzheimer's Drug Fails in Late-Stage Trial," *CNBC*, November 23, 2016, https://www.cnbc.com/2016/11/23/eli-lilly-shares-tank-after-alzheimers-drug-fails-in-late-stage-trial.html.

27. Frédérique Bard et al., "Peripherally Administered Antibodies Against Amyloid β-Peptide Enter the Central Nervous System and Reduce Pathology in a Mouse Model of Alzheimer Disease," *Nature Medicine* 6, no. 8 (2000): 916–19.

28. Ronald S. Black et al., "A Single Ascending Dose Study of Bapineuzumab in Patients with Alzheimer Disease," *Alzheimer Disease and Associated Disorders* 24, no. 2 (2010): 198–203.

29. Stephan Salloway et al., "A Phase 2 Multiple Ascending Dose Trial of Bapineuzumab in Mild to Moderate Alzheimer Disease," *Neurology* 73, no. 24 (2009): 2061–70.

30. Stephen Salloway et al., "Two Phase 3 Trials of Bapineuzumab in Mild-to-Moderate Alzheimer's Disease," *New England Journal of Medicine* 370, no. 4 (2014): 322–33.

31. Francine Gervais et al., "Targeting Soluble Aβ Peptide with Tramiprosate for the Treatment of Brain Amyloidosis," *Neurobiology of Aging* 28, no. 4 (2007): 537–47.

32. Paul S. Aisen et al., "Tramiprosate in Mild-to-Moderate Alzheimer's Disease—a Randomized, Double-Blind, Placebo-Controlled, Multi-Centre Study (the Alphase Study)," *Archives of Medical Science* 7, no. 1 (2011): 102–11; John Hey et al., "Phase 1 Development of ALZ-801, a Novel Beta Amyloid Anti-Aggregation Prodrug of Tramiprosate with Improved Drug Properties, Supporting Bridging to the Phase 3 Program," *Alzheimer's & Dementia: The Journal of the Alzheimer's Association* 12, no. 7 (2016): P613; Martin Tolar et al. "Efficacy of Tramiprosate in APOE4 Heterozygous Patients with Mild to Moderate AD: Combined Sub-Group Analyses from Two Phase 3 Trials," *Neurobiology of Aging* 39 (2016): S22.

33. Bertrand Marotte, "Neurochem Plummets as Clinical Trial Flops," *The Globe and Mail*, August 28, 2007, https://www.theglobeandmail.com/report -on-business/neurochem-plummets-as-clinical-trial-flops/article4098414/.

11. TAOISM AND TAU MICE

1. Jolene Brackey, *Creating Moments of Joy for the Person with Alzheimer's or Dementia* (West Lafayette, IN: Purdue University Press, 2007).

2. Inge Grundke-Iqbal et al., "Microtubule-Associated Protein Tau—A Component of Alzheimer Paired Helical Filaments," *Journal of Biological Chemistry* 261, no. 13 (1986): 6084–89; Inge Grundke-Iqbal et al., "Abnormal Phosphorylation of the Microtubule-Associated Protein τ (Tau) in Alzheimer Cytoskeletal Pathology," *Proceedings of the National Academy of Sciences of the United States of America* 83, no. 13 (1986): 4913–17.

3. Murray D. Weingarten, Arthur H. Lockwood, Shu-Ying Hwo, and Marc W. Kirschner, "A Protein Factor Essential for Microtubule Assembly," *Proceedings of the National Academy of Sciences of the United States of America* 72, no. 5 (1975): 1858–62.

4. Carlo Ballatore, Virginia M. Y. Lee, and John Q. Trojanowski, "Tau-Mediated Neurodegeneration in Alzheimer's Disease and Related Disorders," *Nature Reviews Neuroscience* 8, no. 9 (2007): 663–72; Khalid Iqbal, Fei Liu, Cheng-Xin Gong, and Inge Grundke-Iqbal, "Tau in Alzheimer Disease and Related Tauopathies," *Current Alzheimer Research* 7, no. 8 (2010): 656–64; Meaghan Morris, Sumihiro Maeda, Keith Vossel, and Lennart Mucke, "The Many Faces of Tau," *Neuron* 70, no. 3 (2011): 410–26.

5. Ballatore et al., "Tau-Mediated Neurodegeneration"; Iqbal et al., "Tau in Alzheimer Disease"; Morris et al., "Many Faces of Tau."

6. Iqbal et al., "Tau in Alzheimer Disease."

7. Alberto Serrano-Pozo, Matthew P. Frosch, Eliezer Masliah, and Bradley T. Hyman, "Neuropathological Alterations in Alzheimer Disease," *Cold Spring Harbor Perspectives in Medicine* 1, no. 1 (2011): a006189.

8. Rudolph E. Tanzi and Ann B. Parson, *Decoding Darkness: The Search for the Genetic Causes of Alzheimer's Disease* (Cambridge, MA: Perseus Publishing, 2000).

9. Cécile Dumanchin et al., "Segregation of a Missense Mutation in the Microtubule-Associated Protein Tau Gene with Familial Frontotemporal Dementia and Parkinsonism," *Human Molecular Genetics* 7, no. 11 (1998): 1825–29; Mike Hutton et al., "Association of Missense and 5'-Splice-Site Mutations in Tau with the Inherited Dementia FTDP-17," *Nature* 393, no. 6686 (1998): 702–5.

10. Ballatore et al., "Tau-Mediated Neurodegeneration."

11. Jean-Louis Guénet, Annie Orth, and François Bonhomme, "Origins and Phylogenetic Relationships of the Laboratory Mouse," in *The Laboratory Mouse*, ed. Hans Hedrich (Cambridge, MA: Academic Press, 2012), 3–20.

12. Guénet et al., "Origins and Phylogenetic Relationships."

13. Leila McNeill, "The History of Breeding Mice for Science Begins with a Woman in a Barn," *Smithsonian Magazine*, March 20, 2018, https://www.smithsonianmag.com/science-nature/history-breeding-mice-science-leads-back-woman-barn-180968441/.

14. McNeill, "History of Breeding Mice."

15. Carol C. Linder and Muriel T. Davisson, "Historical Foundations," in *The Laboratory Mouse*, ed. Hans Hedrich (Cambridge, MA: Academic Press, 2012), 21–35.

16. The biographical details about Lathrop given here were drawn from David P. Steensma, Robert A. Kyle, and Marc A. Shampo, "Abbie Lathrop, the 'Mouse Woman of Granby': Rodent Fancier and Accidental Genetics Pioneer," *Mayo Foundation for Medical Education and Research* 85, no. 11 (2010): e83.

17. McNeill, "History of Breeding Mice."

18. National Human Genome Research Institute, "Why Mouse Matters," National Institutes of Health. Last modified July 23, 2010. https://www.genome.gov/10001345/importance-of-mouse-genome.

19. Morris et al., "Many Faces of Tau."

20. Cheng-Xin Gong and Khalid Iqbal, "Hyperphosphorylation of Microtubule-Associated Protein Tau: A Promising Therapeutic Target for Alzheimer Disease," *Current Medicinal Chemistry* 15, no. 23 (2008): 2321–28.

21. Michael K. Ahlijanian et al., "Hyperphosphorylated Tau and Neurofilament and Cytoskeletal Disruptions in Mice Overexpressing Human P25, an Activator of cdk5," *Proceedings of the National Academy of Sciences of the United States of America* 97, no. 6 (2000): 2910–15.

22. Astrid Sydow et al., "Tau-Induced Defects in Synaptic Plasticity, Learning, and Memory Are Reversible in Transgenic Mice After Switching off the Toxic Tau Mutant," *Journal of Neuroscience* 31, no. 7 (2011): 2511–25.

23. K. Santacruz et al., "Tau Suppression in a Neurodegenerative Mouse Model Improves Memory Function," *Science* 309, no. 5733 (2005): 476–81.

24. Morris et al., "Many Faces of Tau."

25. Gregory A. Elder, Miguel A. Gama Sosa, and Rita De Gasperi, "Transgenic Mouse Models of Alzheimer's Disease," *Mount Sinai Journal of Medicine: A Journal of Translational and Personalized Medicine* 77, no. 1 (2010): 69–81.

26. Alzforum, "Tau P301S (Line PS19)." Last updated April 13, 2018, http://www.alzforum.org/research-models/tau-p301s-line-ps19.

12. APPLES, OYSTERS, AND UNDERDOGS

1. Chris Smyth, "Scientists Create the First Drug to Halt Alzheimer's," *Times* (London), July 28, 2016, https://www.thetimes.co.uk/article/scientists -create-the-first-drug-to-halt-alzheimers-xzlkvrkvp.
2. Chronis Fatouros et al., "Inhibition of Tau Aggregation in a Novel *Caenorhabditis elegans* Model of Tauopathy Mitigates Proteotoxicity," *Human Molecular Genetics* 21, no. 16 (2012): 3587–603.
3. Elias Akoury et al., "Mechanistic Basis of Phenothiazine-Driven Inhibition of Tau Aggregation," *Angewandte Chemie International Edition* 52, no. 12 (2013): 3511–15.
4. Erin E. Congdon et al., "Methylthioninium Chloride (Methylene Blue) Induces Autophagy and Attenuates Tauopathy in Vitro and in Vivo," *Autophagy* 8, no. 4 (2012): 609–22.
5. ClinicalTrials.gov, identifier: NCT01626391.
6. Claude M. Wischik et al., "Tau Aggregation Inhibitor Therapy: An Exploratory Phase 2 Study in Mild or Moderate Alzheimer's Disease," *Journal of Alzheimer's Disease* 44, no. 2 (2015): 705–20.
7. Serge Gauthier et al., "Efficacy and Safety of Tau-Aggregation Inhibitor Therapy in Patients with Mild or Moderate Alzheimer's Disease: A Randomised, Controlled, Double-Blind, Parallel-Arm, Phase 3 Trial," *The Lancet* 388, no. 10062 (2016): 2873–84.
8. Tom Fagan, "In First Phase 3 Trial, the Tau Drug LMTM Did Not Work. Period," Alzforum, July 29, 2016, https://www.alzforum.org/news/conference -coverage/first-phase-3-trial-tau-drug-lmtm-did-not-work-period.
9. Fagan, "First Phase 3 Trial."
10. Fagan, "First Phase 3 Trial."
11. Fagan, "First Phase 3 Trial."
12. Gauthier et al., "Efficacy and Safety."
13. Fagan, "First Phase 3 Trial."
14. Gordon K. Wilcock et al., "Potential of Low Dose Leuco-Methylthioninium Bis(Hydromethanesulphonate) (LMTM) Monotherapy for Treatment of Mild Alzheimer's Disease: Cohort Analysis as Modified Primary Outcome in a Phase III Clinical Trial," *Journal of Alzheimer's Disease* 61, no. 1 (2017): 435–57.
15. Tom Fagan, "Tau Inhibitor Fails Again—Subgroup Analysis Irks Clinicians at CTAD," Alzforum, December 16, 2016, https://www.alzforum.org /news/conference-coverage/tau-inhibitor-fails-again-subgroup-analysis -irks-clinicians-ctad.
16. Fagan, "Tau Inhibitor Fails Again."
17. Fagan, "Tau Inhibitor Fails Again."

18. Gauthier et al., "Efficacy and Safety."
19. Gauthier et al., "Efficacy and Safety"; Wilcock et al., "Potential of Low Dose Leuco-Methylthioninium Bis(Hydromethanesulphonate)."
20. Cheng-Xin Gong and Khalid Iqbal, "Hyperphosphorylation of Microtubule-Associated Protein Tau: A Promising Therapeutic Target for Alzheimer Disease," *Current Medicinal Chemistry* 15, no. 23 (2008): 2321–28.
21. Gong and Iqbal, "Hyperphosphorylation."
22. Teodoro Del Ser et al., "Treatment of Alzheimer's Disease with the GSK-3 Inhibitor Tideglusib: A Pilot Study," *Journal of Alzheimer's Disease* 33, no. 1 (2013): 205–15.
23. Simon Lovestone et al., "A Phase II Trial of Tideglusib in Alzheimer's Disease," *Journal of Alzheimer's Disease* 45, no. 1 (2015): 75–88.
24. S. Quraishe, C. M. Cowan, and A. Mudher, "NAP (Davunetide) Rescues Neuronal Dysfunction in a Drosophila Model of Tauopathy," *Molecular Psychiatry* 18, no. 7 (2013): 834–42.
25. Pat McCaffrey, "Boston: Neuroprotective Peptide Inches Forward in Clinic," Alzforum, May 6, 2008, https://www.alzforum.org/news/conference-coverage /boston-neuroprotective-peptide-inches-forward-clinic; Allon Therapeutics, "Allon's Phase II Clinical Trial Shows Statistically Significant Efficacy on Human Cognition and Memory," BioSpace, February 27, 2008, https://www .biospace.com/article/releases/allon-therapeutics-inc-s-phase-ii-clinical-trial -shows-statistically-significant-efficacy-on-human-cognition-and-memory-/.
26. Adam L. Boxer et al., "Davunetide in Patients with Progressive Supranuclear Palsy: A Randomised, Double-Blind, Placebo-Controlled Phase 2/3 Trial," *Lancet Neurology* 13, no. 7 (2014): 676–85.
27. Allon Therapeutics, "Allon Announces PSP Clinical Trial Results," CISION, December 18, 2012, http://www.prnewswire.com/news-releases/allon -announces-psp-clinical-trial-results-183980141.html.
28. Michala Kolarova et al., "Structure and Pathology of Tau Protein in Alzheimer Disease," *International Journal of Alzheimer's Disease* (2012): article 731526; Peter Filipcik et al., "Cortical and Hippocampal Neurons from Truncated Tau Transgenic Rat Express Multiple Markers of Neurodegeneration," *Cellular and Molecular Neurobiology* 29, no. 6 (2009): 895–900.
29. Norbert Zilka et al., "Truncated Tau from Sporadic Alzheimer's Disease Suffices to Drive Neurofibrillary Degeneration in Vivo," *FEBS Letters* 580, no. 15 (2006): 3582–88; Kristina Paholikova et al., "N-Terminal Truncation of Microtubule Associated Protein Tau Dysregulates Its Cellular Localization," *Journal of Alzheimer's Disease* 43, no. 3 (2015): 915–26.
30. Eva Kontsekova et al., "First-in-Man Tau Vaccine Targeting Structural Determinants Essential for Pathological Tau-Tau Interaction Reduces Tau

Oligomerisation and Neurofibrillary Degeneration in an Alzheimer's Disease Model," *Alzheimer's Research & Therapy* 6, no. 44 (2014): 1–12.

31. Petr Novak et al., "Safety and Immunogenicity of the Tau Vaccine AADvac1 in Patients with Alzheimer's Disease: A Randomised, Double-Blind, Placebo-Controlled, Phase 1 Trial," *The Lancet Neurology* 16, no. 2 (2017): 123–34.

32. AXON Neuroscience, "AXON Neuroscience's Vaccine to Halt Alzheimer's Finishes Phase 1 Clinical Trial," *Businesswire*, July 8, 2015, http://www.businesswire.com/news/home/20150708005060/en/AXON-Neuroscience%E2%80%99s-Vaccine-Halt-Alzheimer%E2%80%99s-Finishes-Phase.

33. AXON Neuroscience, "Axon Announces Positive Results from Phase II ADAMANT Trial for AADvac1 in Alzheimer's Disease," CISION, September 9, 2019, http://www.axon-neuroscience.eu/docs/press_release_Axon_announces_positive_result_9-9-2019.pdf.

34. AXON Neuroscience, "Axon Announces Positive Results," emphasis added.

35. Alectos, "Alectos Therapeutics Announces Achievement of Phase I Clinical Milestone in Merck Alzheimer's collaboration," December 12, 2014, http://alectos.com/content/phase1-milestone-merck-alzheimers/.

36. Bin Zhang et al., "The Microtubule-Stabilizing Agent, Epothilone D, Reduces Axonal Dysfunction, Neurotoxicity, Cognitive Deficits, and Alzheimer-Like Pathology in an Interventional Study with Aged Tau Transgenic Mice," *Journal of Neuroscience* 32, no. 11 (2012): 3601–11.

37. ClinicalTrials.gov, identifier: NCT01492374.

38. Ludovic Collin et al., "Neuronal Uptake of Tau/PS422 Antibody and Reduced Progression of Tau Pathology in a Mouse Model of Alzheimer's Disease," *Brain* 137, no. 10 (2014): 2834–46.

39. Einar M. Sigurdsson, "Tau Immunotherapies for Alzheimer's Disease and Related Tauopathies: Progress and Potential Pitfalls," *Journal of Alzheimer's Disease* 64, no. S1 (2018): S555–S565.

13. TYPE 3 DIABETES

1. This analogy of diabetes is developed based on Kim Chilman-Blair and John Taddeo, *What's Up with Ella? Medikidz Explain Type 1 Diabetes* (London: Medikidz Ltd., 2009).

2. A temporary form of diabetes, called *gestational diabetes*, affects pregnant women and increases the risk of type 2 diabetes later in life.

3. Sónia Correia et al., "Insulin Signaling, Glucose Metabolism and Mitochondria: Major Players in Alzheimer's Disease and Diabetes Interrelation," *Brain Research* 1441 (2012): 64–78.

4. Omar Ali, "Genetics of Type 2 Diabetes," *World Journal of Diabetes* 4, no. 4 (2013): 114–23.

5. Suzanne Craft, "The Role of Metabolic Disorders in Alzheimer Disease and Vascular Dementia: Two Roads Converged," *Archives of Neurology* 66, no. 3 (2009): 300–5.

6. Enrique J. Rivera et al., "Insulin and Insulin-Like Growth Factor Expression and Function Deteriorate with Progression of Alzheimer's Disease: Link to Brain Reductions in Acetylcholine," *Journal of Alzheimer's Disease* 8, no. 3 (2005): 247–68.

7. Andisheh Abedini et al., "Time-Resolved Studies Define the Nature of Toxic IAPP Intermediates, Providing Insight for Anti-Amyloidosis Therapeutics," *eLife* 5, (2016): 1–28.

8. Juliette Janson et al., "Increased Risk of Type 2 Diabetes in Alzheimer Disease," *Diabetes* 53, no. 2 (2004): 474–81.

9. Correia et al., "Insulin Signaling"; José Alejandro Luchsinger, "Adiposity, Hyperinsulinemia, Diabetes and Alzheimer's Disease: An Epidemiological Perspective," *European Journal of Pharmacology* 585, no. 1 (2008): 119–29.

10. Dongfeng Cao, Hailin Lu, Terry L. Lewis, and Ling Li, "Intake of Sucrose-Sweetened Water Induces Insulin Resistance and Exacerbates Memory Deficits and Amyloidosis in a Transgenic Mouse Model of Alzheimer Disease," *Journal of Biological Chemistry* 282, no. 50 (2007): 36275–82.

11. Z. Arvanitakis et al., "Diabetes Is Related to Cerebral Infarction But Not to AD Pathology in Older Persons," *Neurology* 67, no. 11 (2006): 1960–65.

12. City of Rochester, "History of Rochester," City of Rochester, accessed March 27, 2020, http://www.rochestermn.gov/about-the-city/history-of -rochester.

13. City of Rochester, "History of Rochester."

14. National Weather Service, "Rochester Tornado August 21 1883," accessed March 31, 2020, https://www.weather.gov/arx/aug211883.

15. National Weather Service, "Rochester Tornado."

16. Matthew Dacy, "What's in a Name? The Story of 'Mayo Clinic,'" Mayo Clinic, February 9, 2009, https://sharing.mayoclinic.org/2009/02/09/whats-in-a -name-the-story-of-mayo-clinic/.

17. C. L. Leibson et al., "Risk of Dementia Among Persons with Diabetes Mellitus: A Population-Based Cohort Study," *American Journal of Epidemiology* 145, no. 4 (1997): 301–8.

18. Leibson et al., "Risk of Dementia."

19. Erasmus University Medical Center, "The Rotterdam Study," Erasmus University Rotterdam, accessed March 27, 2020, http://www.epib.nl /research/ergo.htm.

20. A. Ott et al., "Diabetes Mellitus and the Risk of Dementia: The Rotterdam Study," *Neurology* 53, no. 9 (1999): 1937–42.
21. Toshiharu Ninomiya, "Japanese Legacy Cohort Studies: The Hisayama Study," *Journal of Epidemiology* 28, no. 11 (2018): 444–51.
22. Hirotsugu Ueshima, "Hisayama Study," University of Minnesota, accessed April 3, 2020, http://www.epi.umn.edu/cvdepi/study-synopsis/hisayama -study/.
23. Ueshima, "Hisayama Study."
24. T. Yoshitake et al., "Incidence and Risk Factors of Vascular Dementia and Alzheimer's Disease in a Defined Elderly Japanese Population: The Hisayama Study," *Neurology* 45, no. 6 (1995): 1161–68.
25. Miia Kivipelto et al., "Obesity and Vascular Risk Factors at Midlife and the Risk of Dementia and Alzheimer Disease," *Archives of Neurology* 62, no. 10 (2005): 1556–60; Luchsinger, "Adiposity, Hyperinsulinemia, Diabetes and Alzheimer's Disease."
26. Gisele Wolf-Klein et al., "Are Alzheimer Patients Healthier?" *Journal of the American Geriatrics Society* 36, no. 3 (1988): 219–24.

14. KETONES: THE BRAIN FUEL

1. Eric Kossoff et al., *The Ketogenic and Modified Atkins Diets: Treatments for Epilepsy and Other Disorders* (New York: Demos Medical Publishing, 2016), 19.
2. Hippocrates, "On the Sacred Disease," Classics Archive, accessed March 30, 2020, http://classics.mit.edu/Hippocrates/sacred.html.
3. John Freeman et al., "The Ketogenic Diet: From Molecular Mechanisms to Clinical Effects," *Epilepsy Research* 68 (2006): 145–80; Mark Greener, "Food for Thought: The Ketogenic Diet for Epilepsy," *Progress in Neurology and Psychiatry* 18, no. 3 (2014): 6–9.
4. Alton Goldbloom, "Some Observations on the Starvation Treatment of Epilepsy," *Canadian Medical Association Journal* 12, no. 8 (1922): 539–40.
5. James W. Wheless, "History of the Ketogenic Diet," *Epilepsia* 49 (2008): 3–5.
6. Wheless, "History of the Ketogenic Diet."
7. Stephen C. Cunnane et al., "Can Ketones Help Rescue Brain Fuel Supply in Later Life? Implications for Cognitive Health During Aging and the Treatment of Alzheimer's Disease," *Frontiers in Molecular Neuroscience* 9 (2016): 1–21.
8. S. Hoyer, R. Nitsch, and K. Oesterreich, "Predominant Abnormality in Cerebral Glucose Utilization in Late-Onset Dementia of the Alzheimer Type: A Cross-Sectional Comparison Against Advanced Late-Onset and Incipient Early-Onset Cases," *Journal of Neural Transmission—Parkinson's Disease and*

Dementia Section 3, no. 1 (1991): 1–14; Stephen Cunnane et al., "Brain Fuel Metabolism, Aging, and Alzheimer's Disease," Nutrition 27, no. 1 (2011): 3–20.

9. Samuel T. Henderson et al., "Study of the Ketogenic Agent AC-1202 in Mild to Moderate Alzheimer's Disease: A Randomized, Double-Blind, Placebo-Controlled, Multicenter Trial," Nutrition & Metabolism 6, no. 31 (2009): 1–25.

10. Cunnane et al., "Brain Fuel Metabolism"; Kaushik Shah, Shanal DeSilva, and Thomas Abbruscato, "The Role of Glucose Transporters in Brain Disease: Diabetes and Alzheimer's Disease," International Journal of Molecular Sciences 13 (2012): 12629–55.

11. Michael Schöll et al., "Glucose Metabolism and PIB Binding in Carriers of a His163Tyr Presenilin 1 Mutation," Neurobiology of Aging 32, no. 8 (2011): 1388–99.

12. Eric M. Reiman et al., "Functional Brain Abnormalities in Young Adults at Genetic Risk for Late-Onset Alzheimer's Dementia," Proceedings of the National Academy of Sciences of the United States of America 101, no. 1 (2004): 284–89.

13. G. Stennis Watson and Suzanne Craft, "Modulation of Memory by Insulin and Glucose: Neuropsychological Observations in Alzheimer's Disease," European Journal of Pharmacology 490, no. 1 (2004): 97–113; Sandra I. Sünram-Lea, Jonathan K. Foster, Paula Durlach, and Catalina Perez, "The Effect of Retrograde and Anterograde Glucose Administration on Memory Performance in Healthy Young Adults," Behavioural Brain Research 134, no. 1 (2002): 505–16; Carol A. Manning, Michael E. Ragozzino, and Paul E. Gold, "Glucose Enhancement of Memory in Patients with Probable Senile Dementia of the Alzheimer's Type," Neurobiology of Aging 14, no. 6 (1993): 523–28.

14. Michael Vitek et al., "Advanced Glycation End Products Contribute to Amyloidosis in Alzheimer Disease," Proceedings of the National Academy of Sciences of the United States of America 91, no. 11 (1994): 4766–70; M. Dolores Ledesma, Pedro Bonay, Camilo Colaço, and Jesús Avila, "Analysis of Microtubule-Associated Protein Tau Glycation in Paired Helical Filaments," Journal of Biological Chemistry 269, no. 34 (1994): 21614–19.

15. Freeman et al., "Ketogenic Diet."

16. William R. Leonard, J. Josh Snodgrass, and Marcia L. Robertson, "Evolutionary Perspectives on Fat Ingestion and Metabolism in Humans," in Fat Detection: Taste, Texture, and Post Ingestive Effects, ed. Jean-Pierre Montmayeur and Johannes le Coutre (Boca Raton, FL: CRC Press, 2010), 3–18.

17. Loren Cordain, Michael R. Eades, and Mary D. Eades, "Hyperinsulinemic Diseases of Civilization: More Than Just Syndrome X," Comparative Biochemistry and Physiology, Part A 136, no. 1 (2003): 95–112.

18. Cunnane et al., "Can Ketones Help?"
19. Mary Newport, *Alzheimer's Disease: What If There Was a Cure? The Story of Ketones* (Laguna Beach, CA: Basic Health Publications, Inc., 2013).
20. ClinicalTrials.gov, identifier: NCT01883648.
21. José Enrique de la Rubia Ortí et al., "Improvement of Main Cognitive Functions in Patients with Alzheimer's Disease After Treatment with Coconut Oil Enriched Mediterranean Diet: A Pilot Study," *Journal of Alzheimer's Disease* 65, no. 2 (2018): 577–87.
22. Andreas Eenfeldt, "20 and 50 Grams of Carbs—How Much Food Is That?" *Diet Doctor*, February 14, 2020, https://www.dietdoctor.com/low-carb/20-50 -how-much.
23. Robert Krikorian et al., "Dietary Ketosis Enhances Memory in Mild Cognitive Impairment," *Neurobiology of Aging* 33, no. 2 (2012): 425.e19–25.e27.
24. Samuel T. Henderson et al., "Study of the Ketogenic Agent AC-1202 in Mild to Moderate Alzheimer's Disease: A Randomized, Double-Blind, Placebo-Controlled, Multicenter Trial," *Nutrition & Metabolism* 6, no. 31 (2009): 1–25.
25. Mark A. Reger et al., "Effects of β-Hydroxybutyrate on Cognition in Memory-Impaired Adults," *Neurobiology of Aging* 25, no. 3 (2004): 311–14.
26. Clinicaltrials.gov, identifier: NCT00660088.
27. Henderson et al., "Ketogenic Agent AC-1202."
28. Henderson et al., "Ketogenic Agent AC-1202," 22.
29. U.S. Food and Drug Administration, "Frequently Asked Questions About Medical Foods," May 2016, https://www.fda.gov/downloads/food/guidance -regulation/guidancedocumentsregulatoryinformation/ucm500094.pdf.
30. U.S. Food and Drug Administration, "FDA Warning Letter to Accera, Inc," *Casewatch*, December 26, 2013, https://quackwatch.org/cases/fdawarning /prod/fda-warning-letters-about-products-2013/accera/.
31. U.S. Food and Drug Administration, "FDA Warning Letter."
32. Esther Landhuis, "Medical Foods—Fallback Option for Elusive AD Drug Status?" Alzforum, October 14, 2009, https://www.alzforum.org/news/research -news/medical-foods-fallback-option-elusive-ad-drug-status.
33. Landhuis, "Medical Foods—Fallback Option?"
34. Steven Douglas Maynard and Jeff Gelblum, "Retrospective Case Studies of the Efficacy of Caprylic Triglyceride in Mild-to-Moderate Alzheimer's Disease," *Neuropsychiatric Disease and Treatment* 9 (2013): 1629–35.
35. U.S. Food and Drug Administration, "FDA Warning Letter."
36. Clinicaltrials.gov, identifier: NCT01741194.
37. Accera, "Accera Announces Results of Its First Phase 3 Study in Mild-to-Moder -ate Alzheimer's Disease," Cerecin, February 28, 2017, http://www.cerecin.com /newsroom/accera-announces-results-of-its-first-phase-3-study.html.

38. Cerecin, "Accera Closes New Investment Led by Asia's Leading Agribusiness Group, Wilmar, and Rebrands as Cerecin," October 4, 2018, http://www .cerecin.com/newsroom/accera-closes-new-investment-led-by-asia-leading -agribusiness-group.html.
39. Cerecin, "Accera Closes New Investment."
40. Cerecin, "Accera Closes New Investment."

15. INSULIN FIXES

1. This case is embellished based on the details given in Katherine Tuttle, "A 60-Year-Old Man with Type 2 Diabetes, Hypertension, Dyslipidemia, and Albuminuria," *Advanced Studies in Medicine* 5, no. 1A (2005): S34–S35.
2. Kristina Schoonjans and Johan Auwerx, "Thiazolidinediones: An Update," *The Lancet* 355, no. 9208 (2000): 1008–10.
3. Neil J. Elgee, Robert H. Williams, and Norman D. Lee, "Distribution and Degradation Studies with Insulin-I131," *Journal of Clinical Investigation* 33, no. 9 (1954): 1252–60.
4. Richard U. Margolis and Norman Altszuler, "Insulin in the Cerebrospinal Fluid," *Nature* 215, no. 5108 (1967): 1375–76.
5. Sarah M. Gray, Rick I. Meijer, and Eugene J. Barrett, "Insulin Regulates Brain Function, But How Does It Get There?" *Diabetes* 63, no. 12 (2014): 3992–97.
6. Susana Cardoso et al., "Insulin Is a Two-Edged Knife on the Brain," *Journal of Alzheimer's Disease* 18, no. 3 (2009): 483–507.
7. M. Salkovic-Petrisic and S. Hoyer, "Central Insulin Resistance as a Trigger for Sporadic Alzheimer-Like Pathology: An Experimental Approach," *Journal of Neural Transmission* Suppl 72 (2007): 217–33; Suzanne M. de la Monte et al., "Therapeutic Rescue of Neurodegeneration in Experimental Type 3 Diabetes: Relevance to Alzheimer's Disease," *Journal of Alzheimer's Disease* 10, no. 1 (2006): 89–109; Nataniel Lester-Coll et al., "Intracerebral Streptozotocin Model of Type 3 Diabetes: Relevance to Sporadic Alzheimer's Disease," *Journal of Alzheimer's Disease* 9, no. 1 (2006): 13–33.
8. Eric Steen et al., "Impaired Insulin and Insulin-Like Growth Factor Expression and Signaling Mechanisms in Alzheimer's Disease—Is This Type 3 Diabetes?" *Journal of Alzheimer's Disease* 7, no. 1 (2005): 63–80.
9. Enrique J. Rivera et al., "Insulin and Insulin-Like Growth Factor Expression and Function Deteriorate with Progression of Alzheimer's Disease: Link to Brain Reductions in Acetylcholine," *Journal of Alzheimer's Disease* 8, no. 3 (2005): 247–68.

10. Suzanne Craft et al., "Insulin Dose–Response Effects on Memory and Plasma Amyloid Precursor Protein in Alzheimer's Disease: Interactions with Apolipoprotein E Genotype," *Psychoneuroendocrinology* 28, no. 6 (2003): 809–22.

11. Ana I. Duarte, Paula I. Moreira, and Catarina R. Oliveira, "Insulin in Central Nervous System: More Than Just a Peripheral Hormone," *Journal of Aging Research* (2012): 1–21; Karl Kaiyla et al., "Obesity Induced by a High-Fat Diet Is Associated with Reduced Brain Insulin Transport in Dogs," *Diabetes* 49, no. 9 (2000): 1525–33; William M. Pardridge, "Receptor-Mediated Peptide Transport Through the Blood-Brain Barrier," *Endocrine Reviews* 7, no. 3 (1986): 314–30.

12. Duarte et al., "Insulin in Central Nervous System," 1.

13. Fernanda G. De Felice et al., "Protection of Synapses Against Alzheimer's-Linked Toxins: Insulin Signaling Prevents the Pathogenic Binding of Abeta Oligomers," *Proceedings of the National Academy of Sciences of the United States of America* 106, no. 6 (2009): 1971–76.

14. Ward A. Pedersen et al., "Rosiglitazone Attenuates Learning and Memory Deficits in Tg2576 Alzheimer Mice," *Experimental Neurology* 199, no. 2 (2006): 265–73.

15. G. Stennis Watson et al., "Preserved Cognition in Patients with Early Alzheimer Disease and Amnestic Mild Cognitive Impairment During Treatment with Rosiglitazone: A Preliminary Study," *American Journal of Geriatric Psychiatry* 13, no. 11 (2005): 950–58.

16. Steven E. Nissen and Kathy Wolski, "Effect of Rosiglitazone on the Risk of Myocardial Infarction and Death from Cardiovascular Causes," *New England Journal of Medicine* 356, no. 24 (2007): 2457–71.

17. Sonal Singh, Yoon K. Loke, and Curt D. Furberg, "Long-Term Risk of Cardiovascular Events with Rosiglitazone: A Meta-Analysis," *JAMA* 298, no. 10 (2007): 1189–95; Steven E. Nissen and Kathy Wolski, "Rosiglitazone Revisited: An Updated Meta-Analysis of Risk for Myocardial Infarction and Cardiovascular Mortality," *Archives of Internal Medicine* 170, no. 14 (2010): 1191–201; Edoardo Mannucci et al., "Cardiac Safety Profile of Rosiglitazone: A Comprehensive Meta-Analysis of Randomized Clinical Trials," *International Journal of Cardiology* 143, no. 2 (2010): 135–40.

18. Tracy Staton, "GSK Settles Bulk of Avandia Suits for $460m," FiercePharma, July 14, 2010, https://www.fiercepharma.com/pharma/gsk-settles-bulk-of-avandia-suits-for-460m.

19. ClinicalTrials.gov, identifier: NCT00884533.

20. Sofia Tzimopoulou et al., "A Multi-Center Randomized Proof-of-Concept Clinical Trial Applying [18F]FDG-PET for Evaluation of Metabolic Therapy

with Rosiglitazone XR in Mild to Moderate Alzheimer's Disease," *Journal of Alzheimer's Disease* 22, no. 4 (2010): 1241–56.

21. M. E. Risner et al., "Efficacy of Rosiglitazone in a Genetically Defined Population with Mild-to-Moderate Alzheimer's Disease," *Pharmacogenomics Journal* 6, no. 4 (2006): 246–54.

22. Risner et al., "Efficacy of Rosiglitazone."

23. C. Harrington et al., "Rosiglitazone Does Not Improve Cognition or Global Function When Used as Adjunctive Therapy to AChE Inhibitors in Mild-to-Moderate Alzheimer's Disease: Two Phase 3 Studies," *Current Alzheimer Research* 8, no. 5 (2011): 592–606; Michael Gold et al., "Rosiglitazone Monotherapy in Mild-to-Moderate Alzheimer's Disease: Results from a Randomized, Double-Blind, Placebo-Controlled Phase III Study," *Dementia and Geriatric Cognitive Disorders* 30, no. 2 (2010): 131–46.

24. Nissen and Wolski, "Effect of Rosiglitazone"; Nissen and Wolski, "Rosiglitazone Revisited."

25. David S. Geldmacher et al., "A Randomized Pilot Clinical Trial of the Safety of Pioglitazone in Treatment of Patients with Alzheimer Disease," *Archives of Neurology* 68, no. 1 (2011): 45–50.

26. Nektaria Nicolakakis et al., "Complete Rescue of Cerebrovascular Function in Aged Alzheimer's Disease Transgenic Mice by Antioxidants and Pioglitazone, a Peroxisome Proliferator-Activated Receptor Gamma Agonist," *Journal of Neuroscience: The Official Journal of the Society for Neuroscience* 28, no. 37 (2008): 9287–96.

27. Tomohiko Sato et al., "Efficacy of PPAR-γ Agonist Pioglitazone in Mild Alzheimer Disease," *Neurobiology of Aging* 32, no. 9 (2011): 1626–33.

28. Haruo Hanyu et al., "Pioglitazone Improved Cognition in a Pilot Study on Patients with Alzheimer's Disease and Mild Cognitive Impairment with Diabetes Mellitus," *Journal of the American Geriatrics Society* 57, no. 1 (2009): 177–79.

29. Geldmacher et al., "A Randomized Pilot Clinical Trial."

30. ClinicalTrials.gov, identifier: NCT01931566.

31. Suzanne Craft et al., "Memory Improvement Following Induced Hyperinsulinemia in Alzheimer's Disease," *Neurobiology of Aging* 17, no. 1 (1996): 123–30; Suzanne Craft et al., "Insulin Dose–Response Effects."

32. Mark A. Reger et al., "Intranasal Insulin Administration Dose-Dependently Modulates Verbal Memory and Plasma Amyloid-β in Memory-Impaired Older Adults," *Journal of Alzheimer's Disease* 13, no. 3 (2008): 323–31.

33. Jan Born et al., "Sniffing Neuropeptides: A Transnasal Approach to the Human Brain," *Nature Neuroscience* 5, no. 6 (2002): 514–16.

34. Mark A. Reger et al., "Intranasal Insulin Improves Cognition and Modulates β-Amyloid in Early AD," *Neurology* 70, no. 6 (2008): 440–48.

35. Christian Benedict et al., "Intranasal Insulin Improves Memory in Humans," *Psychoneuroendocrinology* 29, no. 10 (2004): 1326–34.
36. Suzanne Craft et al., "Intranasal Insulin Therapy for Alzheimer Disease and Amnestic Mild Cognitive Impairment: A Pilot Clinical Trial," *Archives of Neurology* 69, no. 1 (2012): 29–38.
37. Amy Claxton et al., "Long-Acting Intranasal Insulin Detemir Improves Working Memory for Adults with Mild Cognitive Impairment or Early-Stage Alzheimer's Dementia," *Journal of Alzheimer's Disease* 44, no. 3 (2015): 897–906.
38. Suzanne Craft et al., "Effects of Regular and Long-Acting Insulin on Cognition and Alzheimer's Disease Biomarkers: A Pilot Clinical Trial," *Journal of Alzheimer's Disease* 57, no. 4 (2017): 1325–34.
39. Claxton et al., "Long-Acting Intranasal Insulin," 904 (emphasis mine).
40. Suzanne Craft et al., "Safety, Efficacy, and Feasibility of Intranasal Insulin for the Treatment of Mild Cognitive Impairment and Alzheimer Disease Dementia: A Randomized Clinical Trial," *JAMA Neurology*, online first, https://jamanetwork.com/journals/jamaneurology/article-abstract/2767376.
41. GLP-1 also slows stomach and intestine movement, which reduces nutrient absorption, lowers glucose, and decreases the need for insulin. See Patrick E. MacDonald et al., "The Multiple Actions of GLP-1 on the Process of Glucose-Stimulated Insulin Secretion," *Diabetes* 51 Suppl 3 (2002): S434–S442.
42. Paula L. McClean, Vadivel Parthsarathy, Emilie Faivre, and Christian Hölscher, "The Diabetes Drug Liraglutide Prevents Degenerative Processes in a Mouse Model of Alzheimer's Disease," *Journal of Neuroscience: The Official Journal of the Society for Neuroscience* 31, no. 17 (2011): 6587–94.
43. Henrik H. Hansen et al., "Long-Term Treatment with Liraglutide, a Glucagon-Like Peptide-1 (GLP-1) Receptor Agonist, Has No Effect on β-Amyloid Plaque Load in Two Transgenic APP/PS1 Mouse Models of Alzheimer's Disease," *Plos One* 11, no. 7 (2016): E0158205.
44. Michael Gejl et al., "In Alzheimer's Disease, Six-Month Treatment with GLP-1 Analogue Prevents Decline of Brain Glucose Metabolism: Randomized, Placebo-Controlled, Double-Blind Clinical Trial," *Frontiers in Aging Neuroscience* 8 (2016): 1–10.
45. ClinicalTrials.gov, identifier: NCT01843075.
46. Edy Kornelius et al., "DPP-4 Inhibitor Linagliptin Attenuates Aβ-Induced Cytotoxicity Through Activation of AMPK in Neuronal Cells," *CNS Neuroscience & Therapeutics* 21, no. 7 (2015): 549–57; Jayasankar Kosaraju et al., "Saxagliptin: A Dipeptidyl Peptidase-4 Inhibitor Ameliorates Streptozotocin-Induced

Alzheimer's Disease," *Neuropharmacology* 72 (2013): 291–300; Jayasankar Kosaraju et al., "Vildagliptin: An Anti-Diabetes Agent Ameliorates Cognitive Deficits and Pathology Observed in Streptozotocin-Induced Alzheimer's Disease," *Journal of Pharmacy and Pharmacology* 65, no. 12 (2013): 1773–84; Jayasankar Kosaraju, R. M. Damian Holsinger, Lixia Guo, and Kin Tam, "Linagliptin, a Dipeptidyl Peptidase-4 Inhibitor, Mitigates Cognitive Deficits and Pathology in the 3xTg-AD Mouse Model of Alzheimer's Disease," *Molecular Neurobiology* 54, no. 8 (2017): 6074–84.

47. Maria Rosaria Rizzo et al., "Dipeptidyl Peptidase-4 Inhibitors Have Protective Effect on Cognitive Impairment in Aged Diabetic Patients with Mild Cognitive Impairment," *Journals of Gerontology Series A: Biomedical Sciences and Medical Sciences* 69, no. 9 (2014): 1122–31; Ahmet Turan Isik, Pinar Soysal, Adnan Yay, and Cansu Usarel, "The Effects of Sitagliptin, a DPP-4 Inhibitor, on Cognitive Functions in Elderly Diabetic Patients With or Without Alzheimer's Disease," *Diabetes Research and Clinical Practice* 123 (2017): 192–98.

48. Elizabeth Mietlicki-Baase, "Amylin-Mediated Control of Glycemia, Energy Balance, and Cognition," *Physiology & Behavior* 162 (2016): 130–40; Laura Hieronymus and Stacy Griffin, "Role of Amylin in Type 1 and Type 2 Diabetes," *The Diabetes Educator* 41, no. 1 (2015): 47S–56S.

49. Brittany L. Adler et al., "Neuroprotective Effects of the Amylin Analogue Pramlintide on Alzheimer's Disease Pathogenesis and Cognition," *Neurobiology of Aging* 35, no. 4 (2014): 793–801.

50. Kaleena Jackson et al., "Amylin Deposition in the Brain: A Second Amyloid in Alzheimer Disease?" *Annals of Neurology* 74, no. 4 (2013): 517–26.

51. Adler et al., "Neuroprotective Effects of the Amylin Analogue"; H. Zhu et al., "Intraperitoneal Injection of the Pancreatic Peptide Amylin Potently Reduces Behavioral Impairment and Brain Amyloid Pathology in Murine Models of Alzheimer's Disease," *Molecular Psychiatry* 20, no. 2 (2014): 232–39.

52. Ripudaman S. Hundal et al., "Mechanism by Which Metformin Reduces Glucose Production in Type 2 Diabetes," *Diabetes* 49, no. 12 (2000): 2063–69.

53. Amit Gupta, Bharti Bisht, and Chinmoy Sankar Dey, "Peripheral Insulin-Sensitizer Drug Metformin Ameliorates Neuronal Insulin Resistance and Alzheimer's-Like Changes," *Neuropharmacology* 60, no. 6 (2011): 910–20; Eva Kickstein et al., "Biguanide Metformin Acts on Tau Phosphorylation Via mTOR/Protein Phosphatase 2A (PP2A) Signaling," *Proceedings of the National Academy of Sciences of the United States of America* 107, no. 50 (2010): 21830–35; Jing Wang et al., "Metformin Activates an Atypical PKC-CBP Pathway to Promote Neurogenesis and Enhance Spatial Memory Formation," *Cell Stem Cell* 11, no. 1 (2012): 23–35.

54. Yaomin Chen et al., "Antidiabetic Drug Metformin (Glucophage®) Increases Biogenesis of Alzheimer's Amyloid Peptides Via Up-Regulating *BACE1* Transcription," *Proceedings of the National Academy of Sciences of the United States of America* 106, no. 10 (2009): 3907–12; Erica Barini et al., "Metformin Promotes Tau Aggregation and Exacerbates Abnormal Behavior in a Mouse Model of Tauopathy," *Molecular Neurodegeneration* 11, no. 16 (2016):1–20; Nopporn Thangthaeng et al., "Metformin Impairs Spatial Memory and Visual Acuity in Old Male Mice," *Aging and Disease* 8, no. 1 (2017): 17–30.

55. Chih-Cheng Hsu, Mark L Wahlqvist, Meei-Shyuan Lee, and Hsin-Ni Tsai, "Incidence of Dementia Is Increased in Type 2 Diabetes and Reduced by the Use of Sulfonylureas and Metformin," *Journal of Alzheimer's Disease* 24, no. 3 (2011): 485–93.

56. Tze Pin Ng et al., "Long-Term Metformin Usage and Cognitive Function Among Older Adults with Diabetes," *Journal of Alzheimer's Disease* 41, no. 1 (2014): 61–68.

57. Patrick Imfeld, Michael Bodmer, Susan S. Jick, and Christoph R. Meier, "Metformin, Other Antidiabetic Drugs, and Risk of Alzheimer's Disease: A Population-Based Case-Control Study," *Journal of the American Geriatrics Society* 60, no. 5 (2012): 916–21.

58. Eileen M. Moore et al., "Increased Risk of Cognitive Impairment in Patients with Diabetes Is Associated with Metformin," *Diabetes Care* 36, no. 10 (2013): 2981–87.

59. Aaron M. Koenig et al., "Effects of the Insulin Sensitizer Metformin in Alzheimer Disease: Pilot Data from a Randomized Placebo-Controlled Crossover Study," *Alzheimer Disease and Associated Disorders* 31, no. 2 (2017): 107–13.

60. José A. Luchsinger et al., "Metformin in Amnestic Mild Cognitive Impairment: Results of a Pilot Randomized Placebo Controlled Clinical Trial," *Journal of Alzheimer's Disease* 51, no. 2 (2016): 501–14.

61. Luchsinger et al., "Metformin in Amnestic Mild Cognitive Impairment," 501 (emphasis mine).

16. BACTERIA IN THE BRAIN

1. Geoffrey Cooper, *The Cell: A Molecular Approach* (Sunderland, MA: Sinauer Associates, 2000).

2. Cooper, *The Cell*.

3. Michael W. Gray, "Mitochondrial Evolution," *Cold Spring Harbor Perspectives in Biology* 4, no. 9 (2012): a011403.

4. Ana Navarro and Alberto Boveris, "The Mitochondrial Energy Transduction System and the Aging Process," *American Journal of Physiology: Cell Physiology* 292, no. 2 (2007): C670–C686.

5. Navarro and Boveris, "Mitochondrial Energy Transduction System."

6. Navarro and Boveris, "Mitochondrial Energy Transduction System."

7. Tamás Szabados et al., "A Chronic Alzheimer's Model Evoked by Mitochondrial Poison Sodium Azide for Pharmacological Investigations," *Behavioural Brain Research* 154, no. 1 (2004): 31–40.

8. S. Hoyer, "Brain Glucose and Energy Metabolism Abnormalities in Sporadic Alzheimer Disease. Causes and Consequences: An Update," *Experimental Gerontology* 35, no. 9 (2000): 1363–72.

9. Xinglong Wang et al., "Oxidative Stress and Mitochondrial Dysfunction in Alzheimer's Disease," *Biochimica et Biophysica Acta* 1842, no. 8 (2014): 1240–47; Carmelina Gemma, Jennifer Vila, Adam Bachstetter, and Paula Bickford, "Oxidative Stress and the Aging Brain: From Theory to Prevention," in *Brain Aging: Models, Methods, and Mechanisms*, ed. David Riddle (Boca Raton, FL: CRC Press/Taylor & Francis, 2007), 353–74.

10. Paula I. Moreira et al., "Mitochondrial Dysfunction Is a Trigger of Alzheimer's Disease Pathophysiology," *Biochimica et Biophysica Acta* 1802, no. 1 (2010): 2–10; Wang et al., "Oxidative Stress and Mitochondrial Dysfunction."

11. Moreira et al., "Mitochondrial Dysfunction"; Wang et al., "Oxidative Stress and Mitochondrial Dysfunction."

12. J. Wang et al., "Increased Oxidative Damage in Nuclear and Mitochondrial DNA in Alzheimer's Disease," *Journal of Neurochemistry* 93, no. 4 (2005): 953–62; David L. Marcus et al., "Increased Peroxidation and Reduced Antioxidant Enzyme Activity in Alzheimer's Disease," *Experimental Neurology* 150, no. 1 (1998): 40–44.

13. Russell H. Swerdlow, Jeffrey M. Burns, and Shaharyar M. Khan, "The Alzheimer's Disease Mitochondrial Cascade Hypothesis: Progress and Perspectives," *Biochimica et Biophysica Acta* 1842, no. 8 (2014): 1219–31.

14. Russell H. Swerdlow, Jeffrey M. Burns, and Shaharyar M. Khan, "The Alzheimer's Disease Mitochondrial Cascade Hypothesis," *Journal of Alzheimer's Disease* 20 Suppl 2 (2010): S265–S279.

15. Swerdlow et al., "Alzheimer's Disease Mitochondrial Cascade Hypothesis."

16. Friderun Ankel-Simons and Jim Cummins, "Misconceptions About Mitochondria and Mammalian Fertilization: Implications for Theories on Human Evolution," *Proceedings of the National Academy of Sciences of the United States of America* 93, no. 24 (1996): 13859–63.

17. David C. Chan and Eric A. Schon, "Eliminating Mitochondrial DNA from Sperm," *Developmental Cell* 22, no. 3 (2012): 469–70.

18. Wang et al., "Oxidative Stress and Mitochondrial Dysfunction"; Pinar E. Coskun et al., "Systemic Mitochondrial Dysfunction and the Etiology of Alzheimer's Disease and Down Syndrome Dementia," *Journal of Alzheimer's Disease* 20 Suppl 2 (2010): S293–S310; Michelangelo Mancuso, Daniele Orsucci, Gabiele Siciliano, and Luigi Murri, "Mitochondria, Mitochondrial DNA, and Alzheimer's Disease. What Comes First?" *Current Alzheimer Research* 5, no. 5 (2008): 457–68.

19. Coskun et al., "Systemic Mitochondrial Dysfunction"; Mancuso et al., "Mitochondria, Mitochondrial DNA, and Alzheimer's Disease."

20. Sebastián Cervantes et al., "Genetic Variation in APOE Cluster Region and Alzheimer's Disease Risk," *Neurobiology of Aging* 32, no. 11 (2011): 2107.e7–2107e17.

21. Caroline Van Cauwenberghe, Christine Van Broeckhoven, and Kristel Sleegers, "The Genetic Landscape of Alzheimer Disease: Clinical Implications and Perspectives," *Genetics in Medicine* 18, no. 5 (2016): 421–30.

22. Mark T. W. Ebbert et al., "Population-Based Analysis of Alzheimer's Disease Risk Alleles Implicates Genetic Interactions," *Biological Psychiatry* 75, no. 9 (2014): 732–37.

23. Allen D. Roses et al., "A TOMM40 Variable-Length Polymorphism Predicts the Age of Late-Onset Alzheimer's Disease," *Pharmacogenomics Journal* 10, no. 5 (2010): 375–84.

24. Nils Wiedemann, Ann E. Frazier, and Nikolaus Pfanner, "The Protein Import Machinery of Mitochondria," *Journal of Biological Chemistry* 279, no. 15 (2004): 14473–76.

25. All the subjects in Roses's study were of European descent.

26. Roses et al., "A TOMM40 Variable-Length Polymorphism"; Tom Fagan, "Las Vegas: AD, Risk, ApoE—Tomm40 No Tomfoolery," Alzforum, November 15, 2009,https://www.alzforum.org/news/conference-coverage/las-vegas-ad-risk-apoe-tomm40-no-tomfoolery.

27. Rita J. Guerreiro and John Hardy, "TOMM40 Association with Alzheimer Disease: Tales of APOE and Linkage Disequilibrium," *Archives of Neurology* 69, no. 10 (2012): 1243–44.

28. Gyungah Jun et al., "Comprehensive Search for Alzheimer Disease Susceptibility Loci in the APOE Region," *Archives of Neurology* 69, no. 10 (2012): 1270–79.

29. Su Hee Chu et al., "TOMM40 Poly-T Repeat Lengths, Age of Onset and Psychosis Risk in Alzheimer Disease," *Neurobiology of Aging* 32, no. 12 (2011): 2328.e1–2328.e9; Cervantes et al., "Genetic Variation in APOE."

30. Madolyn Bowman Rogers, "Large Study Questions Tomm40's Effect on AD Age of Onset," Alzforum, August 15, 2011, https://www.alzforum.org/news/research-news/large-study-questions-tomm40s-effect-ad-age-onset.

31. Sterling C. Johnson et al., "The Effect of *TOMM40* Poly-T Length on Gray Matter Volume and Cognition in Middle-Aged Persons with *APOE* ε3/ε3 Genotype," *Alzheimer's & Dementia: The Journal of the Alzheimer's Association* 7, no. 4 (2011): 456–65; Richard J. Caselli et al., "TOMM40, ApoE, and Age of Onset of Alzheimer's Disease," *Alzheimer's and Dementia* 6, no. 4 (2010): S202.

32. Carlos Cruchaga et al., "Association and Expression Analyses with Single-Nucleotide Polymorphisms in *TOMM40* in Alzheimer Disease," *Archives of Neurology* 68, no. 8 (2011): 1013–19; Rogers, "Large Study Questions Tomm40's Effect."

33. Sangeeta Ghosh et al., "The Thiazolidinedione Pioglitazone Alters Mitochondrial Function in Human Neuron-Like Cells," *Molecular Pharmacology* 71, no. 6 (2007): 1695–702; Jay C. Strum et al., "Rosiglitazone Induces Mitochondrial Biogenesis in Mouse Brain," *Journal of Alzheimer's Disease* 11, no. 1 (2007): 45–51; Leanne Wilson-Fritch et al., "Mitochondrial Remodeling in Adipose Tissue Associated with Obesity and Treatment with Rosiglitazone," *Journal of Clinical Investigation* 114, no. 9 (2004): 1281–89.

34. Takeda Pharmaceutical Company, "Takeda and Zinfandel Pharmaceuticals Discontinue TOMMROW Trial Following Planned Futility Analysis," January 25, 2018, https://www.takeda.com/newsroom/newsreleases/2018/takeda-tommorrow-trial/.

17. EAT YOUR VEGETABLES (AND BERRIES)

1. Isabella Irrcher et al., "Regulation of Mitochondrial Biogenesis in Muscle by Endurance Exercise," *Sports Medicine* 33, no. 11 (2003): 783–93.

2. E. Lezi, Jeffrey M. Burns, and Russell H. Swerdlow, "Effect of High-Intensity Exercise on Aged Mouse Brain Mitochondria, Neurogenesis, and Inflammation," *Neurobiology of Aging* 35, no. 11 (2014): 2574–83; Aaron M. Gusdon et al., "Exercise Increases Mitochondrial Complex I Activity and DRP1 Expression in the Brains of Aged Mice," *Experimental Gerontology* 90 (2017): 1–13.

3. Shaharyar Khan, Rafal Smigrodzki, and Russell Swerdlow, "Cell and Animal Models of mtDNA Biology: Progress and Prospects," *American Journal of Physiology* 292, no. 2 (2007): C664.

4. William C. Orr and Rajindar S. Sohal, "Extension of Life-Span by Overexpression of Superoxide Dismutase and Catalase in *Drosophila melanogaster*," *Science* 263, no. 5150 (1994): 1128–30; Jingtao Sun and John Tower, "FLP Recombinase-Mediated Induction of Cu/Zn-Superoxide Dismutase Transgene Expression Can Extend the Life Span of Adult *Drosophila melanogaster* Flies," *Molecular and Cellular Biology* 19, no. 1 (1999): 216–28; Jingtao Sun

et al., "Induced Overexpression of Mitochondrial Mn-Superoxide Dismutase Extends the Life Span of Adult *Drosophila melanogaster*," *Genetics* 161, no. 2 (2002): 661–72.

5. William C. Orr and Rajindar S. Sohal, "The Effects of Catalase Gene Overexpression on Life Span and Resistance to Oxidative Stress in Transgenic *Drosophila melanogaster*," *Archives of Biochemistry and Biophysics* 297, no. 1 (1992): 35–41; William C. Orr and Rajindar S. Sohal, "Effects of Cu-Zn Superoxide Dismutase Overexpression on Life Span and Resistance to Oxidative Stress in Transgenic *Drosophila melanogaster*," *Archives of Biochemistry and Biophysics* 301, no. 1 (1993): 34–40; William C. Orr, Robin J. Mockett, Judith J. Benes, and Rajindar S. Sohal, "Effects of Overexpression of Copper-Zinc and Manganese Superoxide Dismutases, Catalase, and Thioredoxin Reductase Genes on Longevity in *Drosophila melanogaster*," *Journal of Biological Chemistry* 278, no. 29 (2003): 26418–22; Robin J. Mockett et al., "Overexpression of Mn-Containing Superoxide Dismutase in Transgenic *Drosophila melanogaster*," *Archives of Biochemistry and Biophysics* 371, no. 2 (1999): 260–9.

6. Peter P. Zandi et al., "Reduced Risk of Alzheimer Disease in Users of Antioxidant Vitamin Supplements: The Cache County Study," *Archives of Neurology* 61, no. 1 (2004): 82–88.

7. Jose A. Luchsinger, Ming-Xin Tang, Steven Shea, and Richard Mayeux, "Antioxidant Vitamin Intake and Risk of Alzheimer Disease," *Archives of Neurology* 60, no. 2 (2003): 203–8.

8. Shelly L. Gray et al., "Antioxidant Vitamin Supplement Use and Risk of Dementia or Alzheimer's Disease in Older Adults." *Journal of the American Geriatrics Society* 56, no. 2 (2008): 291–95.

9. Jae Hee Kang et al., "A Randomized Trial of Vitamin E Supplementation and Cognitive Function in Women," *Archives of Internal Medicine* 166, no. 22 (2006): 2462–68.

10. Richard J. Kryscio et al., "Association of Antioxidant Supplement Use and Dementia in the Prevention of Alzheimer's Disease by Vitamin E and Selenium Trial (PREADViSE)," *JAMA Neurology* 74, no. 5 (2017): 567–73.

11. Mary Sano et al., "A Controlled Trial of Selegiline, Alpha-Tocopherol, or Both as Treatment for Alzheimer's Disease," *New England Journal of Medicine* 336, no. 17 (1997): 1216–22; Maurice W. Dysken et al., "Effect of Vitamin E and Memantine on Functional Decline in Alzheimer Disease: The TEAM-AD VA Cooperative Randomized Trial," *JAMA* 311, no. 1 (2014): 33–44; Ronald C. Petersen et al., "Vitamin E and Donepezil for the Treatment of Mild Cognitive Impairment," *New England Journal of Medicine* 352, no. 23 (2005): 2379–88; Douglas R. Galasko et al., "Antioxidants for Alzheimer Disease:

A Randomized Clinical Trial with Cerebrospinal Fluid Biomarker Measures," *Archives of Neurology* 69, no. 7 (2012): 836–41.

12. Martha Clare Morris et al., "Dietary Intake of Antioxidant Nutrients and the Risk of Incident Alzheimer Disease in a Biracial Community Study," *JAMA* 287, no. 24 (2002): 3230–7.

13. Marianne J. Engelhart et al., "Dietary Intake of Antioxidants and Risk of Alzheimer Disease," *JAMA* 287, no. 24 (2002): 3223–29.

14. Engelhart et al., "Dietary Intake of Antioxidants."

15. Claudine Manach et al., "Polyphenols: Food Sources and Bioavailability," *American Journal of Clinical Nutrition* 79, no. 5 (2004): 727–47.

16. Manach et al., "Polyphenols."

17. James A. Joseph et al., "Reversals of Age-Related Declines in Neuronal Signal Transduction, Cognitive, and Motor Behavioral Deficits with Blueberry, Spinach, or Strawberry Dietary Supplementation," *Journal of Neuroscience: The Official Journal of the Society for Neuroscience* 19, no. 18 (1999): 8114–21; Barbara Shukitt-Hale, Vivian Cheng, and James A. Joseph, "Effects of Blackberries on Motor and Cognitive Function in Aged Rats," *Nutritional Neuroscience* 12, no. 3 (2009): 135–40.

18. James A. Joseph et al., "Blueberry Supplementation Enhances Signaling and Prevents Behavioral Deficits in an Alzheimer Disease Model," *Nutritional Neuroscience* 6, no. 3 (2003): 153–62.

19. Nurses' Health Study, "History," accessed April 1, 2020, https://www.nurseshealthstudy.org/about-nhs/history.

20. Nurses' Health Study, "History."

21. Elizabeth E. Devore, Jae Hee Kang, Monique M. B. Breteler, and Francine Grodstein, "Dietary Intakes of Berries and Flavonoids in Relation to Cognitive Decline," *Annals of Neurology* 72, no. 1 (2012): 135–43.

22. Jae H. Kang, Alberto Ascherio, and Francine Grodstein, "Fruit and Vegetable Consumption and Cognitive Decline in Aging Women," *Annals of Neurology* 57, no. 5 (2005): 713–20.

23. Martha Clare Morris et al., "Associations of Vegetable and Fruit Consumption with Age-Related Cognitive Change," *Neurology* 67, no. 8 (2006): 1370–6.

24. Jean M. Orgogozo et al., "Wine Consumption and Dementia in the Elderly: A Prospective Community Study in the Bordeaux Area," *Revue Neurologique* 153, no. 3 (1997): 185–92.

25. Yasutake Tomata et al., "Green Tea Consumption and the Risk of Incident Dementia in Elderly Japanese: The Ohsaki Cohort 2006 Study," *American Journal of Geriatric Psychiatry* 24, no. 10 (2016): 881–89.

26. Martha Clare Morris et al., "MIND Diet Slows Cognitive Decline with Aging," *Alzheimers Dement* 11, no. 9 (2015): 1015–22; Martha Clare Morris

et al., "MIND Diet Associated with Reduced Incidence of Alzheimer's Disease," *Alzheimers Dement* 11, no. 9 (2015): 1007–14.

27. Morris et al., "MIND Diet Slows Cognitive Decline"; Morris et al., "MIND Diet Associated with Reduced Incidence of Alzheimer's Disease."

28. Robert Krikorian et al., "Blueberry Supplementation Improves Memory in Older Adults," *Journal of Agricultural and Food Chemistry* 58, no. 7 (2010): 3996–4000.

29. Anne Nilsson et al., "Effects of a Mixed Berry Beverage on Cognitive Functions and Cardiometabolic Risk Markers; A Randomized Cross-Over Study in Healthy Older Adults," *PLoS One* 12, no. 11 (2017): e0188173.

30. Robert Krikorian et al., "Concord Grape Juice Supplementation Improves Memory Function in Older Adults with Mild Cognitive Impairment," *British Journal of Nutrition* 103, no. 5 (2010): 730–4.

31. Robert Krikorian et al., "Concord Grape Juice Supplementation and Neurocognitive Function in Human Aging," *Journal of Agricultural and Food Chemistry* 60, no. 23 (2012): 5736–42.

32. Manach et al., "Polyphenols."

33. Manach et al., "Polyphenols."

34. Manach et al., "Polyphenols."

35. M. Serafini, A. Ghiselli, and A. Ferro-Luzzi, "In Vivo Antioxidant Effect of Green and Black Tea in Man," *European Journal of Clinical Nutrition* 50 (1996): 28–32.

36. Orgogozo et al., "Wine Consumption and Dementia."

18. BLOOD, HEART, AND BRAIN

1. Michael Lemonick, "Secrets of the Lost Tomb," *Time*, June 24, 2001, http://content.time.com/time/magazine/article/0,9171,134211,00.html.

2. Lemonick, "Secrets of the Lost Tomb."

3. Stephanie Fitzgerald, *Ramses II: Egyptian Pharaoh, Warrior, and Builder* (Mankato, MN: Compass Point Books, 2009).

4. Donald Stein, Simón Brailowsky, and Bruno Will, *Brain Repair* (Oxford: Oxford University Press, 1995).

5. Magdi M. Saba, Hector O. Ventura, Mohamed Saleh, and Mandeep R. Mehra, "Ancient Egyptian Medicine and the Concept of Heart Failure," *Journal of Cardiac Failure* 12, no. 6 (2006): 416–21.

6. Ahmed Okasha, "Mental Health in the Middle East: An Egyptian Perspective," *Clinical Psychology Review* 19, no. 8 (1999): 917–33.

7. Gustavo C. Román, "Vascular Dementia: Distinguishing Characteristics, Treatment, and Prevention," *Journal of the American Geriatrics Society* 51, no. 5 (2003): S296–S304.

8. Román, "Vascular Dementia"; Alzheimer's Association, "Vascular Dementia," August 2019, https://www.alz.org/media/Documents/alzheimers-dementia -vascular-dementia-ts.pdf.

9. Ken Nagata et al., "Clinical Diagnosis of Vascular Dementia," *Journal of the Neurological Sciences* 257, no. 1 (2007): 44–48.

10. Anouk G. W. van Norden et al., "Dementia: Alzheimer Pathology and Vascular Factors: From Mutually Exclusive to Interaction," *Biochimica et Biophysica Acta* 1822, no. 3 (2012): 340–9.

11. Konrad Maurer, Stephan Volk, and Hector Gerbaldo, "Auguste D and Alzheimer's Disease," *The Lancet* 349 (1997): 1546–49; Hans-Jürgen Möller and Manuel B. Graeber, "The Case Described by Alois Alzheimer in 1911," *European Archives of Psychiatry and Clinical Neuroscience* 248, no. 3 (1998): 111–22.

12. Ingemar Björkhem and Steve Meaney, "Brain Cholesterol: Long Secret Life Behind a Barrier," *Arteriosclerosis, Thrombosis, and Vascular Biology* 24, no. 5 (2004): 806–15.

13. Björkhem and Meaney, "Brain Cholesterol."

14. Chia-Chan Liu, Takahisa Kanekiyo, Huaxi Xu, and Guojun Bu, "Apolipoprotein E and Alzheimer Disease: Risk, Mechanisms, and Therapy," *Nature Reviews Neurology* 9, no. 2 (2013): 106–18.

15. Lorenzo M. Refolo et al., "Hypercholesterolemia Accelerates the Alzheimer's Amyloid Pathology in a Transgenic Mouse Model," *Neurobiology of Disease* 7, no. 4 (2000): 321–31.

16. Refolo et al., "Hypercholesterolemia"; Lorenzo M. Refolo et al., "A Cholesterol-Lowering Drug Reduces β-Amyloid Pathology in a Transgenic Mouse Model of Alzheimer's Disease," *Neurobiology of Disease* 8, no. 5 (2001): 890–9.

17. Miia Kivipelto et al., "Midlife Vascular Risk Factors and Alzheimer's Disease in Later Life: Longitudinal, Population Based Study," *BMJ* 322, no. 7300 (2001): 1447–51.

18. Alina Solomon et al., "Midlife Serum Cholesterol and Increased Risk of Alzheimer's and Vascular Dementia Three Decades Later," *Dementia and Geriatric Cognitive Disorders* 28, no. 1 (2009): 75–80.

19. Zaldy Sy Tan et al., "Plasma Total Cholesterol Level as a Risk Factor for Alzheimer Disease: The Framingham Study," *Archives of Internal Medicine* 163, no. 9 (2003): 1053–57.

20. Christiane Reitz et al., "Association of Higher Levels of High-Density Lipoprotein Cholesterol in Elderly Individuals and Lower Risk of Late-Onset Alzheimer Disease," *Archives of Neurology* 67, no. 12 (2010): 1491–97.

21. Ľubica Cibičková, "Statins and Their Influence on Brain Cholesterol," *Journal of Clinical Lipidology* 5, no. 5 (2011): 373–79.

22. Refolo et al., "Hypercholesterolemia."
23. Françoise Forette et al., "Prevention of Dementia in Randomised Double-Blind Placebo-Controlled Systolic Hypertension in Europe (Syst-Eur) Trial," *The Lancet* 352, no. 9137 (1998): 1347–51.
24. Lenore J. Launer et al., "Midlife Blood Pressure and Dementia: The Honolulu-Asia Aging Study," *Neurobiology of Aging* 21, no. 1 (2000): 49–55.
25. Jack C. de la Torre, "Vascular Risk Factors: A Ticking Time Bomb to Alzheimer's Disease," *American Journal of Alzheimer's Disease & Other Dementias* 28, no. 6 (2013): 551–59.
26. Berislav V. Zlokovic, "Neurovascular Pathways to Neurodegeneration in Alzheimer's Disease and Other Disorders," *Nature Reviews Neuroscience* 12, no. 12 (2011): 723–38.
27. de la Torre, "Vascular Risk Factors."
28. K. A. Johnson et al., "Preclinical Prediction of Alzheimer's Disease Using SPECT," *Neurology* 50, no. 6 (1998): 1563–71; Guido Rodriguez et al., "Hippocampal Perfusion in Mild Alzheimer's Disease," *Psychiatry Research* 100, no. 2 (2000): 65–74.
29. Hillary Dolan et al., "Atherosclerosis, Dementia, and Alzheimer Disease in the Baltimore Longitudinal Study of Aging Cohort," *Annals of Neurology* 68, no. 2 (2010): 231–40.
30. Weiying Dai et al., "Mild Cognitive Impairment and Alzheimer Disease: Patterns of Altered Cerebral Blood Flow at MR Imaging," *Radiology* 250, no. 3 (2009): 856–66; David C. Alsop et al., "Hippocampal Hyperperfusion in Alzheimer's Disease," *NeuroImage* 42, no. 4 (2008): 1267–74.
31. Dai et al., "Mild Cognitive Impairment."
32. Kivipelto, "Midlife Vascular Risk Factors."
33. Benjamin Wolozin et al., "Decreased Prevalence of Alzheimer Disease Associated with 3-hydroxy-3-methyglutaryl Coenzyme A Reductase Inhibitors," *Archives of Neurology* 57, no. 10 (2000): 1439–43.
34. H. Jick et al., "Statins and the Risk of Dementia," *The Lancet* 356, no. 9242 (2000): 1627–31.
35. Peter P. Zandi et al., "Do Statins Reduce Risk of Incident Dementia and Alzheimer Disease? The Cache County Study," *Archives of General Psychiatry* 62, no. 2 (2005): 217–24.
36. Zandi et al., "Do Statins Reduce Risk?"
37. Thomas D. Rea et al., "Statin Use and the Risk of Incident Dementia: The Cardiovascular Health Study," *Archives of Neurology* 62, no. 7 (2005): 1047–51; G. Li et al., "Statin Therapy and Risk of Dementia in the Elderly: A Community-Based Prospective Cohort Study," *Neurology* 63, no. 9 (2004): 1624–28.

38. The U.S. study collected only snapshot data, while the U.K. study did consider time, but it went *back* in time, tracing participants' statin use before their Alzheimer's diagnoses. Because formal Alzheimer's diagnoses may not happen until years after cognitive symptoms appear, physicians may already suspect cognitive impairment and be less willing to prescribe statins. See Li et al., "Statin Therapy."

39. M. Sano et al., "A Randomized, Double-Blind, Placebo-Controlled Trial of Simvastatin to Treat Alzheimer Disease," *Neurology* 77, no. 6 (2011): 556–63.

40. Heart Protection Study Collaborative Group, "MRC/BHF Heart Protection Study of Cholesterol Lowering with Simvastatin in 20,536 High-Risk Individuals: A Randomised Placebo-Controlled Trial," *The Lancet* 360, no. 9326 (2002): 7–22; Stella Trompet et al., "Pravastatin and Cognitive Function in the Elderly: Results of the PROSPER Study," *Journal of Neurology* 257, no. 1 (2010): 85–90.

41. U.S. Food and Drug Administration, "FDA Drug Safety Communication: Important Safety Label Changes to Cholesterol-Lowering Statin Drugs," February 28, 2012, https://www.fda.gov/Drugs/DrugSafety/ucm293101 .htm. The FDA's ruling is challenged by some, however; for instance, see Brian Ott et al., "Do Statins Impair Cognition? A Systematic Review and Meta-Analysis of Randomized Controlled Trials," *Journal of General Internal Medicine* 30, no. 3 (2015): 348–58.

42. U.S. Food and Drug Administration, "Drug Safety."

43. Ara S. Khachaturian et al., "Antihypertensive Medication Use and Incident Alzheimer Disease: The Cache County Study," *Archives of Neurology* 63, no. 5 (2006): 686–92.

44. B. A. in't Veld et al., "Antihypertensive Drugs and Incidence of Dementia: The Rotterdam Study," *Neurobiology of Aging* 22, no. 3 (2001): 407–12.

45. Whitney Wharton et al., "The Effects of Ramipril in Individuals at Risk for Alzheimer's Disease: Results of a Pilot Clinical Trial," *Journal of Alzheimer's Disease* 32, no. 1 (2012): 147–56.

46. T. Ohrui et al., "Effects of Brain-Penetrating ACE Inhibitors on Alzheimer Disease Progression," *Neurology* 63, no. 7 (2004): 1324–25.

47. For more details, see Clinicaltrials.gov.

19. A MISSED OPPORTUNITY

1. Christopher Rowland, "Pfizer Had Clues Its Blockbuster Drug Could Prevent Alzheimer's. Why Didn't It Tell the World?" *Washington Post*, June 4, 2019. Retrieved January 8, 2020, from https://www.washingtonpost.com

/business/economy/pfizer-had-clues-its-blockbuster-drug-could-prevent-alzheimers-why-didnt-it-tell-the-world/2019/06/04/9092e08a-7a61-11e9-8bb7-0fc796cf2ec0_story.html.

2. Alex Philippidis, "Top 15 Best-Selling Drugs of 2018: Sales for Most Treatments Grow Year-Over-Year Despite Concerns over Rising Prices," *Genetic Engineering & Biotechnology News* 39, no. 4 (2019): 16–17. Enbrel is sold in the United States and Canada by Amgen, and elsewhere by Pfizer.

3. Rowland, "Pfizer Had Clues."

4. Rowland, "Pfizer Had Clues," comments.

5. Being Patient, "Pfizer Knew Its Drug Could Prevent Alzheimer's. They Did Nothing About It, Says *Post*," June 5, 2019, https://www.beingpatient.com/pfizer-coverup-enbrel-etanercept-alzheimers/.

6. Tsuyoshi Ishii, Seiji Haga, and Fujio Shimizu, "Identification of Components of Immunoglobulins in Senile Plaques by Means of Fluorescent Antibody Technique," *Acta Neuropathologica* 32, no. 2 (1975): 157–62.

7. Joseph Rogers et al., "Inflammation and Alzheimer's Disease Pathogenesis," *Neurobiology of Aging* 17, no. 5 (1996): 681–86.

8. Ehab E. Tuppo and Hugo R. Arias, "The Role of Inflammation in Alzheimer's Disease," *International Journal of Biochemistry and Cell Biology* 37, no. 2 (2005): 289–305.

9. Rogers et al., "Inflammation and Alzheimer's"; Tuppo and Arias, "The Role of Inflammation."

10. Tony Wyss-Coray and Joseph Rogers, "Inflammation in Alzheimer Disease—A Brief Review of the Basic Science and Clinical Literature," *Cold Spring Harbor Perspectives in Medicine* 2, no. 1 (2012): a006346; Frank L. Heppner, Richard M. Ransohoff, and Burkhard Becher, "Immune Attack: The Role of Inflammation in Alzheimer Disease," *Nature Reviews Neuroscience* 16, no. 6 (2015): 358–72.

11. Tuppo and Arias, "The Role of Inflammation."

12. Rogers et al., "Inflammation and Alzheimer's"; Heppner et al., "Immune Attack."

13. Rita Guerreiro et al., "TREM2 Variants in Alzheimer's Disease," *New England Journal of Medicine* 368, no. 2 (2013): 117–27.

14. Taylor R. Jay, Victoria E. von Saucken, and Gary E. Landreth, "TREM2 in Neurodegenerative Diseases," *Molecular Neurodegeneration* 12, no. 56 (2017): 1–33.

15. Yaming Wang et al., "TREM2 Lipid Sensing Sustains the Microglial Response in an Alzheimer's Disease Model," *Cell* 160, no. 6 (2015): 1061–71; Taylor R. Jay et al., "TREM2 Deficiency Eliminates TREM2+ Inflammatory Macrophages and Ameliorates Pathology in Alzheimer's Disease Mouse Models," *Journal of Experimental Medicine* 212, no. 3 (2015): 287–95.

16. Jean-Charles Lambert et al., "Genome-Wide Association Study Identifies Variants at *CLU* and *CR1* Associated with Alzheimer's Disease," *Nature Genetics* 41, no. 10 (2009): 1094–99.

17. Helen Crehan et al., "Complement Receptor 1 (CR1) and Alzheimer's Disease," *Immunobiology* 217, no. 2 (2012): 244–50.

18. Supakanya Wongrakpanich, Amaraporn Wongrakpanich, Katie Melhado, and Janani Rangaswami, "A Comprehensive Review of Non-Steroidal Anti-Inflammatory Drug Use in the Elderly," *Aging and Disease* 9, no. 1 (2018): 143–50.

19. Wyss-Coray and Rogers, "Inflammation in Alzheimer Disease."

20. Bas A. in't Veld et al., "Nonsteroidal Antiinflammatory Drugs and the Risk of Alzheimer's Disease," *New England Journal of Medicine* 345, no. 21 (2001): 1515–21.

21. The list of NSAIDs does not include aspirin.

22. Peter P. Zandi et al., "Reduced Incidence of AD with NSAID But Not H_2 Receptor Antagonists: The Cache County Study," *Neurology* 59, no. 6 (2002): 880–6.

23. S. A. Reines et al., "Rofecoxib: No Effect on Alzheimer's Disease in a 1-Year, Randomized, Blinded, Controlled Study," *Neurology* 62, no. 1 (2004): 66–71.

24. Patrizio Pasqualetti et al., "A Randomized Controlled Study on Effects of Ibuprofen on Cognitive Progression of Alzheimer's Disease," *Aging Clinical and Experimental Research* 21, no. 2 (2009): 102–10.

25. ADAPT Research Group, "Cardiovascular and Cerebrovascular Events in the Randomized, Controlled Alzheimer's Disease Anti-Inflammatory Prevention Trial (ADAPT)," *PLoS Clinical Trials* 1, no. 7 (2006).

26. ADAPT Research Group, "Cardiovascular and Cerebrovascular Events."

27. ADAPT Research Group, "Naproxen and Celecoxib Do Not Prevent AD in Early Results from a Randomized Controlled Trial," *Neurology* 68, no. 21 (2007): 1800–8.

28. John C. Breitner et al., "Extended Results of the Alzheimer's Disease Anti-Inflammatory Prevention Trial," *Alzheimer's & Dementia: The Journal of the Alzheimer's Association* 7, no. 4 (2011): 402–11.

29. Pierre-François Meyer et al., "INTREPAD: A Randomized Trial of Naproxen to Slow Progress of Presymptomatic Alzheimer Disease," *Neurology* 92, no. 18 (2019): e2070–e2080.

30. The participants had to be older than sixty, or fifty-five if their age was within fifteen years of their youngest-affected relative's age of onset.

31. Meyer et al., "INTREPAD," e2079.

32. Pat McCaffrey, "Closing the Book on NSAIDs for Alzheimer's Prevention," Alzforum, April 12, 2019, https://www.alzforum.org/news/research-news/closing-book-nsaids-alzheimers-prevention.

33. McCaffrey, "Closing the Book."
34. Joseph Butchart et al., "Etanercept in Alzheimer Disease: A Randomized, Placebo-Controlled, Double-Blind, Phase 2 Trial," *Neurology* 84, no. 21 (2015): 2161–68.
35. Richard Chou et al., "Treatment for Rheumatoid Arthritis and Risk of Alzheimer's Disease: A Nested Case-Control Analysis," *CNS Drugs* 30, no. 11 (2016): 1111–20.
36. ADAPT Research Group, "Naproxen and Celecoxib Do Not Prevent AD"; Paul T. Jantzen et al., "Microglial Activation and Beta-Amyloid Deposit Reduction Caused by a Nitric Oxide–Releasing Nonsteroidal Anti-Inflammatory Drug in Amyloid Precursor Protein Plus Presenilin-1 Transgenic Mice," *The Journal of Neuroscience: The Official Journal of the Society for Neuroscience* 22, no. 6 (2002): 2246–54.
37. Breitner et al., "Extended Results."
38. Green Valley, "Green Valley Announces NMPA Approval of Oligomannate for Mild to Moderate Alzheimer's Disease," November 2, 2019, https://www.greenvalleypharma.com/En/Index/pageView/catid/48/id/28.html.
39. Xinyi Wang et al., "Sodium Oligomannate Therapeutically Remodels Gut Microbiota and Suppresses Gut Bacterial Amino Acids-Shaped Neuroinflammation to Inhibit Alzheimer's Disease Progression," *Cell Research* 29, no. 10 (2019): 787–803.
40. Wang et al., "Sodium Oligomannate."
41. Jessica Shugart, "China Approves Seaweed Sugar as First New Alzheimer's Drug in 17 Years," Alzforum, November 7, 2019, https://www.alzforum.org/news/research-news/china-approves-seaweed-sugar-first-new-alzheimers-drug-17-years.
42. Shugart, "China Approves Seaweed Sugar."
43. Bloomberg, "Chinese Alzheimer's Drug Gets U.S. Approval for Stateside Trial," April 26, 2020, https://www.bloomberg.com/news/articles/2020-04-26/chinese-alzheimer-s-drug-gets-u-s-approval-for-stateside-trial.

20. PARADIGM SHIFT(?)

1. Thorlakur Jonsson et al., "A Mutation in APP Protects Against Alzheimer's Disease and Age-Related Cognitive Decline," *Nature* 487, no. 7409 (2012): 96–99; Ewen Callaway, "Gene Mutation Defends Against Alzheimer's Disease," *Nature* 487, no. 7406 (2012): 153.
2. Madolyn Bowman Rogers, "Protective APP Mutation Found—Supports Amyloid Hypothesis," Alzforum, July 13, 2012, https://www.alzforum.org

/news/research-news/protective-app-mutation-found-supports-amyloid
-hypothesis.
3. William F. Goure, Grant A. Krafft, Jasna Jerecic, and Franz Hefti, "Targeting the Proper Amyloid-Beta Neuronal Toxins: A Path Forward for Alzheimer's Disease Immunotherapeutics," *Alzheimer's Research & Therapy* 6, no. 42 (2014): 1–15; Kirsten Viola and William Klein, "Amyloid β Oligomers in Alzheimer's Disease Pathogenesis, Treatment, and Diagnosis," *Acta Neuropathologica: Pathology and Mechanisms of Neurological Disease* 129, no. 2 (2015): 183–206.
4. Goure et al., "Targeting the Proper Amyloid-Beta Neuronal Toxins."
5. William I. Rosenblum, "Why Alzheimer Trials Fail: Removing Soluble Oligomeric Beta Amyloid Is Essential, Inconsistent, and Difficult," *Neurobiology of Aging* 35, no. 5 (2014): 969–74.
6. James A. R. Nicoll et al., "Aβ Species Removal After $A\beta_{42}$ Immunization," *Journal of Neuropathology and Experimental Neurology* 65, no. 11 (2006): 1040–48.
7. Ksenia V. Kastanenka et al., "Immunotherapy with Aducanumab Restores Calcium Homeostasis in Tg2576 Mice," *Journal of Neuroscience: The Official Journal of the Society for Neuroscience* 36, no. 50 (2016): 12549–58.
8. James Ferrero et al., "First-in-Human, Double-Blind, Placebo-Controlled, Single-Dose Escalation Study of Aducanumab (BIIB037) in Mild-to-Moderate Alzheimer's Disease," *Alzheimer's & Dementia: Translational Research & Clinical Interventions* 2, no. 3 (2016): 169–76.
9. Jeff Sevigny et al., "The Antibody Aducanumab Reduces Aβ Plaques in Alzheimer's Disease," *Nature* 537, no. 7618 (2016): 50–56; Gabrielle Strobel, "Biogen Antibody Buoyed by Phase 1 Data and Hungry Investors," Alzforum, March 25, 2015, https://www.alzforum.org/news/conference-coverage/biogen-antibody-buoyed-phase-1-data-and-hungry-investors.
10. Strobel, "Biogen Antibody Buoyed."
11. Sevigny et al., "The Antibody Aducanumab," 54 (emphasis mine).
12. Sevigny et al., "The Antibody Aducanumab."
13. Biogen, "Biogen and Eisai to Discontinue Phase 3 Engage and Emerge Trials of Aducanumab in Alzheimer's Disease," March 21, 2019, http://investors.biogen.com/news-releases/news-release-details/biogen-and-eisai-discontinue-phase-3-engage-and-emerge-trials.
14. Madolyn Bowman Rogers, " 'Reports of My Death Are Greatly Exaggerated.' Signed, Aducanumab," Alzforum, October 24, 2019, https://www.alzforum.org/news/research-news/reports-my-death-are-greatly-exaggerated-signed-aducanumab.
15. ClinicalTrials.gov, identifier: NCT04241068.

16. Kazuma Murakami, "Conformation-Specific Antibodies to Target Amyloid β Oligomers and Their Application to Immunotherapy for Alzheimer's Disease," *Bioscience, Biotechnology, and Biochemistry* 78, no. 8 (2014): 1293–305; Heinz Hillen et al., "Generation and Therapeutic Efficacy of Highly Oligomer-Specific β-Amyloid Antibodies," *Journal of Neuroscience* 30, no. 31 (2010): 10369–79; P. J. Shughrue et al., "Anti-ADDL Antibodies Differentially Block Oligomer Binding to Hippocampal Neurons," *Neurobiology of Aging* 31, no. 2 (2010): 189–202.

17. Eric Karran, Marc Mercken, and Bart De Strooper, "The Amyloid Cascade Hypothesis for Alzheimer's Disease: An Appraisal for the Development of Therapeutics," *Nature Reviews Drug Discovery* 10, no. 9 (2011): 698–712.

18. Clifford R. Jack et al., "Serial PIB and MRI in Normal, Mild Cognitive Impairment and Alzheimer's Disease: Implications for Sequence of Pathological Events in Alzheimer's Disease," *Brain* 132, no. 5 (2009): 1355–65; Sylvain E. Lesné et al., "Brain Amyloid-β Oligomers in Ageing and Alzheimer's Disease," *Brain* 136, no. 5 (2013): 1383–98; Alaina Baker-Nigh et al., "Neuronal Amyloid-β Accumulation Within Cholinergic Basal Forebrain in Ageing and Alzheimer's Disease," *Brain* 138, no. 6 (2015): 1722–37.

19. Karran et al., "Amyloid Cascade Hypothesis."

20. Kenneth S. Kosik and Francisco Lopera, "Genetic Testing Must Recognize Impact of Bad News on Recipient," *Nature* 454, no. 7201 (2008): 159.

21. Ken Alltucker, "Solving the Alzheimer's Mystery. Part Three: The Questions," *Azcentral*, February 21, 2015, http://www.azcentral.com/story/news /local/best-reads/2015/02/22/alzheimers-research-colombia-questions -part-three/23787199/.

22. Howard Mackey et al., "Exploratory Analyses of Cognitive Effects of Crenezumab in a Mild Alzheimer's Disease Subpopulation of a Randomized, Double-Blind, Placebo-Controlled, Parallel-Group Phase 2 Study (ABBY)," *Alzheimer's & Dementia* 12, no. 7 (2016): P610.

23. Susanne Ostrowitzki et al., "Mechanism of Amyloid Removal in Patients with Alzheimer Disease Treated with Gantenerumab," *Archives of Neurology* 69, no. 2 (2012): 198–207.

24. Alzforum, "Gantenerumab." Last updated May 7, 2020, https://www.alzforum .org/therapeutics/gantenerumab.

25. Johannes Streffer et al., "Pharmacodynamics of the Oral BACE Inhibitor JNJ-54861911 in Early Alzheimer's Disease," *Alzheimer's & Dementia: The Journal of the Alzheimer's Association* 12, no. 7 (2016): P199–P200; Hiroko Shimizu et al., "Pharmacocokinetic and Pharmacodynamic Study (54861911alz1006) with a BACE Inhibitor, JNJ-54861911, in Healthy Elderly Japanese Subjects," *Alzheimer's & Dementia* 12, no. 7 (2016) P612; Masayoshi

Takahashi et al., "A Pharmacodynamic Study (54861911alz1008) with a BACE Inhibitor, JNJ-54861911 in Japanese Asymptomatic Subjects at Risk for Alzheimer's Dementia," *Alzheimer's & Dementia* 12, no. 7 (2016): P608; Maarten Timmers et al., "Profiling the Dynamics of CSF and Plasma Aβ Reduction After Treatment with JNJ-54861911, a Potent Oral BACE Inhibitor," *Alzheimer's & Dementia: Translational Research & Clinical Interventions* 2, no. 3 (2016): 202–12.

26. Tamara Bhandari, "Investigational Drugs Didn't Slow Memory Loss, Cognitive Decline in Rare, Inherited Alzheimer's, Initial Analysis Indicates," Washington University School of Medicine, St. Louis, February 10, 2020, https://medicine.wustl.edu/news/alzheimers-diantu-trial-initial-results/.

27. Gabrielle Strobel, "In DIAN-TU, Gantenerumab Brings Down Tau. By a Lot. Open Extension Planned," Alzforum, April 20, 2020, https://www.alzforum.org/news/conference-coverage/dian-tu-gantenerumab-brings-down-tau-lot-open-extension-planned.

28. Strobel, "In DIAN-TU."

29. Diane B. Howieson, "Cognitive Skills and the Aging Brain: What to Expect," *Cerebrum: Dana Foundation*, December 1, 2015, https://www.dana.org/article/cognitive-skills-and-the-aging-brain-what-to-expect/.

30. M. Kerry O'Banion, Paul D. Coleman, and Linda M. Callahan, "Regional Neuronal Loss in Aging and Alzheimer's Disease: A Brief Review," *Seminars in Neuroscience* 6, no. 5 (1994): 307–14; Jorge Azpurua and Benjamin A. Eaton, "Neuronal Epigenetics and the Aging Synapse," *Frontiers in Cellular Neuroscience* 9 (2015): article 208; Anders M. Fjell et al., "What Is Normal in Normal Aging? Effects of Aging, Amyloid and Alzheimer's Disease on the Cerebral Cortex and the Hippocampus," *Progress in Neurobiology* 117 (2014): 20–40.

31. Bruno Dubois et al., "Research Criteria for the Diagnosis of Alzheimer's Disease: Revising the NINCDS–ADRDA Criteria," *Lancet Neurology* 6, no. 8 (2007): 734–46.

32. Reisa A. Sperling et al., "The A4 Study: Stopping AD Before Symptoms Begin?" *Science Translational Medicine* 6, no. 228 (2014): 228fs13.

33. M. X. Tang et al., "Incidence of AD in African-Americans, Caribbean Hispanics, and Caucasians in Northern Manhattan," *Neurology* 56, no. 1 (2001): 49–56; Marita Golden, "African Americans Are More Likely Than Whites to Develop Alzheimer's. Why?" *Washington Post*, June 1, 2017, https://www.washingtonpost.com/lifestyle/magazine/why-are-african-americans-so-much-more-likely-than-whites-to-develop-alzheimers/2017/05/31/9bfbcccc-3132-11e7-8674-437ddb6e813e_story.html.

34. Bengt Winblad et al., "Safety, Tolerability, and Antibody Response of Active Aβ Immunotherapy with CAD106 in Patients with Alzheimer's Disease:

Randomised, Double-Blind, Placebo-Controlled, First-in-Human Study," *The Lancet Neurology* 11, no. 7 (2012): 597–604; Martin R. Farlow et al., "Long-Term Treatment with Active Aβ Immunotherapy with CAD106 in Mild Alzheimer's Disease." *Alzheimer's Research & Therapy* 7, no. 1 (2015): 1–13.

35. Mike Ufer et al., "Results from a First-in-Human Study with the BACE Inhibitor CNP520," *Alzheimer's & Dementia* 12, no. 7 (2016): P200; ClinicalTrials.gov, Identifier: NCT02576639.

36. Novartis, "Novartis, Amgen and Banner Alzheimer's Institute Discontinue Clinical Program with BACE Inhibitor CNP520 for Alzheimer's Prevention," July 11, 2019, https://www.novartis.com/news/media-releases/novartis-amgen -and-banner-alzheimers-institute-discontinue-clinical-program-bace -inhibitor-cnp520-alzheimers-prevention.

21. AN ENRICHED LIFE

1. David A. Snowdon et al., "Linguistic Ability in Early Life and Cognitive Function and Alzheimer's Disease in Late Life: Findings from the Nun Study," *JAMA* 275, no. 7 (1996): 530.

2. Robert Katzman et al., "Clinical, Pathological, and Neurochemical Changes in Dementia: A Subgroup with Preserved Mental Status and Numerous Neocortical Plaques," *Annals of Neurology* 23, no. 2 (1988): 138–44.

3. Laura Fratiglioni and Hui-Xin Wang, "Brain Reserve Hypothesis in Dementia," *Journal of Alzheimer's Disease* 12, no. 1 (2007): 11–22.

4. Yaakov Stern, "What Is Cognitive Reserve? Theory and Research Application of the Reserve Concept," *Journal of the International Neuropsychological Society* 8, no. 3 (2002): 448–60.

5. Stern, "What Is Cognitive Reserve?" Some use the term *brain reserve* to specify this innate capacity.

6. Gerd Kempermann, Daniela Gast, and Fred H. Gage, "Neuroplasticity in Old Age: Sustained Fivefold Induction of Hippocampal Neurogenesis by Long-Term Environmental Enrichment," *Annals of Neurology* 52, no. 2 (2002): 135–43; Arne Buschler and Denise Manahan-Vaughan, "Brief Environmental Enrichment Elicits Metaplasticity of Hippocampal Synaptic Potentiation *in Vivo*," *Frontiers in Behavioral Neuroscience* 6, no. 85 (2012): 1–10.

7. Jennifer R. Cracchiolo et al., "Enhanced Cognitive Activity—Over and Above Social or Physical Activity—Is Required to Protect Alzheimer's Mice Against Cognitive Impairment, Reduce Aβ Deposition, and Increase Synaptic Immunoreactivity," *Neurobiology of Learning and Memory* 88, no. 3 (2007): 277–94; Gary W. Arendash et al., "Environmental Enrichment

Improves Cognition in Aged Alzheimer's Transgenic Mice Despite Stable Beta-Amyloid Deposition," *Neuroreport* 15, no. 11 (2004): 1751–54.

8. Fratiglioni and Wang, "Brain Reserve Hypothesis"; Xiangfei Meng and Carl D'Arcy, "Education and Dementia in the Context of the Cognitive Reserve Hypothesis: A Systematic Review with Meta-Analyses and Qualitative Analyses," *PLoS One* 7, no. 6 (2012): e38268.

9. Yaakov Stern et al., "Influence of Education and Occupation on the Incidence of Alzheimer's Disease," *JAMA* 271, no. 13 (1994): 1004–10; Fratiglioni and Wang, "Brain Reserve Hypothesis."

10. Margaret Gatz et al., "Education and the Risk of Alzheimer's Disease: Findings from the Study of Dementia in Swedish Twins," *The Journals of Gerontology: Psychological Sciences* 56B, no. 5 (2001): P292–P300.

11. Fratiglioni and Wang, "Brain Reserve Hypothesis."

12. Christine Sattler, Pablo Toro, Peter Schönknecht, and Johannes Schröder, "Cognitive Activity, Education and Socioeconomic Status as Preventive Factors for Mild Cognitive Impairment and Alzheimer's Disease," *Psychiatry Research* 196, no. 1 (2012): 90–95; Joe Verghese et al., "Leisure Activities and the Risk of Dementia in the Elderly," *New England Journal of Medicine* 348, no. 25 (2003): 2508–16.

13. R. S. Wilson et al., "Cognitive Activity and Incident AD in a Population-Based Sample of Older Persons," *Neurology* 59, no. 12 (2002): 1910–14.

14. Eleanor A. Maguire et al., "Navigation-Related Structural Change in the Hippocampi of Taxi Drivers," *Proceedings of the National Academy of Sciences of the United States of America* 97, no. 8 (2000): 4398–403.

15. Bogdan Draganski et al., "Neuroplasticity: Changes in Grey Matter Induced by Training," *Nature* 427, no. 6972 (2004): 311–12.

16. Michelle E. Kelly et al., "The Impact of Cognitive Training and Mental Stimulation on Cognitive and Everyday Functioning of Healthy Older Adults: A Systematic Review and Meta-Analysis," *Ageing Research Reviews* 15, no. 1 (2014): 28–43; Javier Olazarán et al., "Nonpharmacological Therapies in Alzheimer's Disease: A Systematic Review of Efficacy," *Dementia and Geriatric Cognitive Disorders* 30, no. 2 (2010): 161–78; Alex Bahar-Fuchs, Linda Clare, and Bob Woods, "Cognitive Training and Cognitive Rehabilitation for Mild to Moderate Alzheimer's Disease and Vascular Dementia," *Cochrane Database of Systematic Reviews*, no. 6 (2013): CD003260; J. D. Huntley et al., "Do Cognitive Interventions Improve General Cognition in Dementia? A Meta-Analysis and Meta-Regression," *BMJ Open* 5, no. 4 (2015), e005247; Bob Woods, Elisa Aguirre, Aimee E. Spector, and Martin Orrell, "Cognitive Stimulation to Improve Cognitive Functioning in People with Dementia," *Cochrane Database of Systematic Reviews*, no. 2 (2012): CD005562.

17. Chandramallika Basak, Walter R. Boot, Michelle W. Voss, and Arthur F. Kramer, "Can Training in a Real-Time Strategy Video Game Attenuate Cognitive Decline in Older Adults?" *Psychology and Aging* 23, no. 4 (2008): 765–77; Verena C. Buschert et al., "Effects of a Newly Developed Cognitive Intervention in Amnestic Mild Cognitive Impairment and Mild Alzheimer's Disease: A Pilot Study," *Journal of Alzheimer's Disease* 25, no. 4 (2011): 679–94; Cássio Bottino et al., "Cognitive Rehabilitation Combined with Drug Treatment in Alzheimer's Disease Patients: A Pilot Study," *Clinical Rehabilitation* 19, no. 8 (2005): 861–69; Deborah Cahn-Weiner, Paul F. Malloy, George W. Rebok, and Brian R. Ott, "Results of a Randomized Placebo-Controlled Study of Memory Training for Mildly Impaired Alzheimer's Disease Patients," *Applied Neuropsychology* 10, no. 4 (2003): 215–23.

18. Olazarán et al., "Nonpharmacological Therapies."

19. Huntley et al., "Do Cognitive Interventions Improve General Cognition?"; Olazarán et al., "Nonpharmacological Therapies."

20. Aaron Kandola, Joshua Hendrikse, Paul J. Lucassen, and Murat Yucel, "Aerobic Exercise as a Tool to Improve Hippocampal Plasticity and Function in Humans: Practical Implications for Mental Health Treatment," *Frontiers in Human Neuroscience* 10 (2016): article 373; Karlie A. Intlekofer and Carl W Cotman, "Exercise Counteracts Declining Hippocampal Function in Aging and Alzheimer's Disease," *Neurobiology of Disease* 57, no. C (2013): 47–55; Peter Rasmussen et al., "Evidence for a Release of Brain-Derived Neurotrophic Factor from the Brain During Exercise," *Experimental Physiology* 94, no. 10 (2009): 1062–69; Kirk I. Erickson et al., "Exercise Training Increases Size of Hippocampus and Improves Memory," *Proceedings of the National Academy of Sciences of the United States of America* 108, no. 7 (2011): 3017–22; Zaldy S. Tan et al., "Physical Activity, Brain Volume, and Dementia Risk: The Framingham Study," *Journals of Gerontology Series A: Biomedical Sciences and Medical Sciences* 72, no. 6 (2017): 789–95.

21. Alejandro Santos-Lozano et al., "Physical Activity and Alzheimer Disease: A Protective Association," *Mayo Clinic Proceedings* 91, no. 8 (2016): 999–1020.

22. Santos-Lozano et al., "Physical Activity."

23. Francis Langlois et al., "Benefits of Physical Exercise Training on Cognition and Quality of Life in Frail Older Adults," *The Journals of Gerontology Series B: Psychological Sciences and Social Sciences* 68, no. 3 (2012): 400–4; Anthea Vreugdenhil, John Cannell, Andrew Davies, and George Razay, "A Community-Based Exercise Programme to Improve Functional Ability in People with Alzheimer's Disease: A Randomized Controlled Trial," *Scandinavian Journal of Caring Sciences* 26, no. 1 (2012): 12–19.

24. Olivia C. Küster et al., "Cognitive Change Is More Positively Associated with an Active Lifestyle Than with Training Interventions in Older Adults at Risk

of Dementia: A Controlled Interventional Clinical Trial," *BMC Psychiatry* 16, no. 315 (2016):1–12; Kristine Hoffmann et al., "Moderate-to-High Intensity Physical Exercise in Patients with Alzheimer's Disease: A Randomized Controlled Trial," *Journal of Alzheimer's Disease* 50, no. 2 (2016): 443–53.

25. Patricia Heyn, Beatriz C. Abreu, and Kenneth J. Ottenbacher, "The Effects of Exercise Training on Elderly Persons with Cognitive Impairment and Dementia: A Meta-Analysis," *Archives of Physical Medicine and Rehabilitation* 85, no. 10 (2004): 1694–704.

26. Andreas Ströhle et al., "Drug and Exercise Treatment of Alzheimer Disease and Mild Cognitive Impairment: A Systematic Review and Meta-Analysis of Effects on Cognition in Randomized Controlled Trials," *The American Journal of Geriatric Psychiatry* 23, no. 12 (2015): 1234–49.

27. Jeremy Young, Maaike Angevaren, Jennifer Rusted, and Naji Tabet, "Aerobic Exercise to Improve Cognitive Function in Older People Without Known Cognitive Impairment," *Cochrane Database of Systematic Reviews*, no. 4 (2015): CD005381.

28. Carla M. Yuede et al., "Effects of Voluntary and Forced Exercise on Plaque Deposition, Hippocampal Volume, and Behavior in the Tg2576 Mouse Model of Alzheimer's Disease," *Neurobiology of Disease* 35, no. 3 (2009): 426–32.

29. Linda Teri et al., "Exercise Plus Behavioral Management in Patients with Alzheimer Disease: A Randomized Controlled Trial," *JAMA* 290, no. 15 (2003): 2015–22.

30. Teri et al., "Exercise Plus Behavioral Management."

31. Buschler and Manahan-Vaughan, "Brief Environmental Enrichment."

32. Vanessa Doulames, Sangmook Lee, and Thomas B. Shea, "Environmental Enrichment and Social Interaction Improve Cognitive Function and Decrease Reactive Oxidative Species in Normal Adult Mice," *International Journal of Neuroscience* 124, no. 5 (2014): 369–76.

33. Shari S. Bassuk, Thomas A. Glass, and Lisa F. Berkman, "Social Disengagement and Incident Cognitive Decline in Community-Dwelling Elderly Persons," *Annals of Internal Medicine* 131, no. 3 (1999): 165–73; L. L. Barnes et al., "Social Resources and Cognitive Decline in a Population of Older African Americans and Whites," *Neurology* 63, no. 12 (2004): 2322–26.

34. Laura Fratiglioni et al., "Influence of Social Network on Occurrence of Dementia: A Community-Based Longitudinal Study," *The Lancet* 355, no. 9212 (2000): 1315–19.

35. David A. Bennett et al., "The Effect of Social Networks on the Relation Between Alzheimer's Disease Pathology and Level of Cognitive Function in Old People: A Longitudinal Cohort Study," *Lancet Neurology* 5, no. 5 (2006): 406–12.

36. Hui-Xin Wang, Anita Karp, Bengt Winblad, and Laura Fratiglioni, "Late-Life Engagement in Social and Leisure Activities Is Associated with a

Decreased Risk of Dementia: A Longitudinal Study from the Kungsholmen Project," *American Journal of Epidemiology* 155, no. 12 (2002): 1081–87; Jisca S. Kuiper et al., "Social Relationships and Risk of Dementia: A Systematic Review and Meta-Analysis of Longitudinal Cohort Studies," *Ageing Research Reviews* 22 (2015): 39–57.

37. Barnes et al., "Social Resources and Cognitive Decline."
38. Kristine Yaffe and Tina Hoang, "Nonpharmacologic Treatment and Prevention Strategies for Dementia," *Continuum* 19, no. 2 (2013): 372–81.
39. Bennett et al., "Effect of Social Networks."
40. Kuiper et al., "Social Relationships."
41. Bassuk et al., "Social Disengagement"; Barnes et al., "Social Resources and Cognitive Decline."
42. For an exception where three months of relationship building and peer support improved participants' cognitive test scores, see Kaisu H. Pitkala et al., "Effects of Socially Stimulating Group Intervention on Lonely, Older People's Cognition: A Randomized, Controlled Trial," *American Journal of Geriatric Psychiatry: The Official Journal of the American Association for Geriatric Psychiatry* 19, no. 7 (2011): 654–63.
43. James A. Mortimer et al., "Changes in Brain Volume and Cognition in a Randomized Trial of Exercise and Social Interaction in a Community-Based Sample of Non-Demented Chinese Elders," *Journal of Alzheimer's Disease* 30, no. 4 (2012): 763.
44. Fratiglioni and Wang, "Brain Reserve Hypothesis"; Bassuk et al., "Social Disengagement"; Küster et al., "Cognitive Change."
45. Heather A. Lindstrom et al., "The Relationships Between Television Viewing in Midlife and the Development of Alzheimer's Disease in a Case-Control Study," *Brain and Cognition* 58, no. 2 (2005): 157–65.

EPILOGUE

1. Calvin Moh et al., "Cell Cycle Deregulation in the Neurons of Alzheimer's Disease," *Results and Problems in Cell Differentiation* 53 (2011): 565–76.
2. Moh et al., "Cell Cycle Deregulation."
3. Kie Honjo, Robert van Reekum, and Nicolaas P. L. G. Verhoeff, "Alzheimer's Disease and Infection: Do Infectious Agents Contribute to Progression of Alzheimer's Disease?" *Alzheimer's & Dementia: The Journal of the Alzheimer's Association* 5, no. 4 (2009): 348–60; Francis Mawanda and Robert Wallace, "Can Infections Cause Alzheimer's Disease?" *Epidemiologic Reviews* 35, no. 1 (2013): 161–80.

4. Michael Beekes, Achim Thomzig, Walter Schulz-Schaeffer, and Reinhard Burger, "Is There a Risk of Prion-Like Disease Transmission by Alzheimer- or Parkinson-Associated Protein Particles?" *Acta Neuropathologica* 128, no. 4 (2014): 463–76.

5. Centers for Disease Control and Prevention, "Creutzfeldt-Jakob Disease, Classic (CJD)," May 8, 2019, https://www.cdc.gov/prions/cjd/occurrence -transmission.html.

6. Stanley B. Prusiner, *Madness and Memory: The Discovery of Prions—A New Biological Principle of Disease* (New Haven, CT: Yale University Press, 2014); Atsushi Aoyagi et al., "Aβ and Tau Prion-Like Activities Decline with Longevity in the Alzheimer's Disease Human Brain," *Science Translational Medicine* 11, no. 490 (2019): eaat8462.

7. Anant K. Paravastu et al., "Seeded Growth of Beta-Amyloid Fibrils from Alzheimer's Brain-Derived Fibrils Produces a Distinct Fibril Structure," *Proceedings of the National Academy of Sciences of the United States of America* 106, no. 18 (2009): 7443–48.

8. Jan Stöhr et al., "Purified and Synthetic Alzheimer's Amyloid Beta (Aβ) Prions," *Proceedings of the National Academy of Sciences of the United States of America* 109, no. 27 (2012): 11025–30; R. Morales et al., "*De Novo* Induction of Amyloid-β Deposition *in Vivo*," *Molecular Psychiatry* 17, no. 12 (2012): 1347–53; Yvonne S. Eisele et al., "Peripherally Applied Aβ-Containing Inoculates Induce Cerebral β-Amyloidosis," *Science* 330, no. 6006 (2010): 980–82. The extracts were of both human and mouse origins.

9. Zane Jaunmuktane et al., "Evidence for Human Transmission of Amyloid-β Pathology and Cerebral Amyloid Angiopathy," *Nature* 525, no. 7568 (2015): 247–50.

10. Editorial, "Alzheimergate? When Miscommunication Met Sensationalism," *The Lancet* 386, no. 9999 (2015): 1109.

11. Dominic M. Walsh and Dennis J. Selkoe, "A Critical Appraisal of the Pathogenic Protein Spread Hypothesis of Neurodegeneration," *Nature Reviews Neuroscience* 17, no. 4 (2016): 251–60.

12. Beekes et al, "Is There a Risk of Prion-Like Disease Transmission?"

13. David J. Irwin et al., "Evaluation of Potential Infectivity of Alzheimer and Parkinson Disease Proteins in Recipients of Cadaver-Derived Human Growth Hormone," *JAMA Neurology* 70, no. 4 (2013): 462–68.

14. Jaap Goudsmit et al., "Evidence For and Against the Transmissibility of Alzheimer Disease," *Neurology* 30, no. 9 (1980): 945–50; Paul Brown et al., "Human Spongiform Encephalopathy: The National Institutes of Health Series of 300 Cases of Experimentally Transmitted Disease," *Annals of Neurology* 35, no. 5 (1994): 513–29.

BIBLIOGRAPHY

Abedini, Andisheh, Annette Plesner, Ping Cao, Zachary Ridgway, Jinghua Zhang, Ling-Hsien Tu, Chris T. Middleton, et al. "Time-Resolved Studies Define the Nature of Toxic IAPP Intermediates, Providing Insight for Anti-Amyloidosis Therapeutics." *eLife* 5 (2016): 1–28.

Accera. "Accera Announces Results of Its First Phase 3 Study in Mild-to-Moderate Alzheimer's Disease." Cerecin. February 28, 2017. http://www.cerecin.com /newsroom/accera-announces-results-of-its-first-phase-3-study.html.

ADAPT Research Group. "Cardiovascular and Cerebrovascular Events in the Randomized, Controlled Alzheimer's Disease Anti-Inflammatory Prevention Trial (ADAPT)." *PLoS Clinical Trials* 1, no. 7 (2006): e33.

——. "Naproxen and Celecoxib Do Not Prevent AD in Early Results from a Randomized Controlled Trial." *Neurology* 68, no. 21 (2007): 1800–8.

Adler, Brittany L., Mark Yarchoan, Hae Min Hwang, Natalia Louneva, Jeffrey A. Blair, Russell Palm, Mark A. Smith, et al. "Neuroprotective Effects of the Amylin Analogue Pramlintide on Alzheimer's Disease Pathogenesis and Cognition." *Neurobiology of Aging* 35, no. 4 (2014): 793–801.

Agnvall, Elizabeth. "New Science Sheds Light on the Cause of Alzheimer's Disease." AARP. January 24, 2012. http://www.aarp.org/health/conditions-treatments/info-05-2010/alzheimers_disease.html.

Ahlijanian, Michael K., Nestor X. Barrezueta, Robert D. Williams, Amy Jakowski, Kim P. Kowsz, Sheryl McCarthy, Timothy Coskran, et al. "Hyperphosphorylated Tau and Neurofilament and Cytoskeletal Disruptions in Mice

Overexpressing Human P25, an Activator of cdk5." *Proceedings of the National Academy of Sciences of the United States of America* 97, no. 6 (2000): 2910–15.

Ainsworth, Claire. "Sex Redefined: The Idea of 2 Sexes Is Overly Simplistic." *Scientific American*, October 22, 2018. https://www.scientificamerican.com/article/sex-redefined-the-idea-of-2-sexes-is-overly-simplistic1/.

Aisen, Paul S., Serge Gauthier, Steven H. Ferris, Daniel Saumier, Denis Haine, Denis Garceau, Anh Duong, et al. "Tramiprosate in Mild-to-Moderate Alzheimer's Disease—A Randomized, Double-Blind, Placebo-Controlled, Multi-Centre Study (the Alphase Study)." *Archives of Medical Science* 7, no. 1 (2011): 102–11.

Akoury, Elias, Marcus Pickhardt, Michal Gajda, Jacek Biernat, Eckhard Mandelkow, and Markus Zweckstetter. "Mechanistic Basis of Phenothiazine-Driven Inhibition of Tau Aggregation." *Angewandte Chemie International Edition* 52, no. 12 (2013): 3511–15.

Alectos. "Alectos Therapeutics Announces Achievement of Phase I Clinical Milestone in Merck Alzheimer's Collaboration." December 12, 2014. http://alectos.com/content/phase1-milestone-merck-alzheimers/.

Ali, Omar. "Genetics of Type 2 Diabetes." *World Journal of Diabetes* 4, no. 4 (2013): 114–23.

Allon Therapeutics. "Allon Announces PSP Clinical Trial Results." CISION. December 18, 2012. http://www.prnewswire.com/news-releases/allon-announces-psp-clinical-trial-results-183980141.html.

——. "Allon's Phase II Clinical Trial Shows Statistically Significant Efficacy on Human Cognition and Memory." BioSpace. February 27, 2008. https://www.biospace.com/article/releases/allon-therapeutics-inc-s-phase-ii-clinical-trial-shows-statistically-significant-efficacy-on-human-cognition-and-memory-/.

Alltucker, Ken. "Solving the Alzheimer's Mystery. Part Three: The Questions." Azcentral. February 21, 2015. http://www.azcentral.com/story/news/local/best-reads/2015/02/22/alzheimers-research-colombia-questions-part-three/23787199/.

Alsop, David C., Melynda Casement, Cedric de Bazelaire, Tamara Fong, and Daniel Z. Press. "Hippocampal Hyperperfusion in Alzheimer's Disease." *NeuroImage* 42, no. 4 (2008): 1267–74.

Altmann, Andre, Lu Tian, Victor W. Henderson, and Michael D. Greicius. "Sex Modifies the APOE-Related Risk of Developing Alzheimer Disease." *Annals of Neurology* 75, no. 4 (2014): 563–73.

Alzforum. "Gantenerumab." Last updated May 7, 2020. https://www.alzforum.org/therapeutics/gantenerumab.

——. "LY2886721." Alzforum. Accessed March 26, 2020. http://www.alzforum.org/therapeutics/ly2886721.

——. "Tau P301S (Line PS19)." Alzforum. Last updated April 13, 2018. http://www.alzforum.org/research-models/tau-p301s-line-ps19.

Alzheimer's Association. "2018 Alzheimer's Disease Facts and Figures." *Alzheimers Dement* 14, no. 3 (2018): 367–429.

——. "2019 Alzheimer's Disease Facts and Figures." *Alzheimers Dement* 15, no. 3 (2019): 321–87.

——. "IRS Information Returns: Form 990. Year Ended June 30, 2019." December 18, 2019. https://www.alz.org/media/Documents/form-990-fy-2019.pdf.

——. "Vascular Dementia." August 2019. https://www.alz.org/media/Documents/alzheimers-dementia-vascular-dementia-ts.pdf.

Anderson, Porter. "Dale Schenk: Alzheimer's Researcher." CNN. December 11, 2001. http://www.cnn.com/2001/CAREER/jobenvy/07/23/dale.schenk/.

Ankel-Simons, Friderun, and Jim Cummins. "Misconceptions About Mitochondria and Mammalian Fertilization: Implications for Theories on Human Evolution." *Proceedings of the National Academy of Sciences of the United States of America* 93, no. 24 (1996): 13859–63.

Aoyagi, Atsushi, Carlo Condello, Jan Stöhr, Weizhou Yue, Brianna M. Rivera, Joanne C. Lee, Amanda L. Woerman, et al. "Aβ and Tau Prion-Like Activities Decline with Longevity in the Alzheimer's Disease Human Brain." *Science Translational Medicine* 11, no. 490 (2019), eaat8462.

Arendash, Gary W., Marcos F. Garcia, David A. Costa, Jennifer R. Cracchiolo, Inge M. Wefes, and H. Potter. "Environmental Enrichment Improves Cognition in Aged Alzheimer's Transgenic Mice Despite Stable Beta-Amyloid Deposition." *Neuroreport* 15, no. 11 (2004): 1751–54.

Arias, Elizabeth. "United States Life Tables, 2011." *National Vital Statistics Reports* 64, no. 11 (2015): 1–62.

Arvanitakis, Z., J. A. Schneider, R. S. Wilson, Y. Li, S. E. Arnold, Z. Wang, and D. A. Bennett. "Diabetes Is Related to Cerebral Infarction But Not to AD Pathology in Older Persons." *Neurology* 67, no. 11 (2006): 1960–65.

AXON Neuroscience. "Axon Announces Positive Results from Phase II ADAMANT Trial for AADvac1 in Alzheimer's Disease." CISION. September 9, 2019. http://www.axon-neuroscience.eu/docs/press_release_Axon_announces_positive_result_9-9-2019.pdf.

——. "AXON Neuroscience's Vaccine to Halt Alzheimer's Finishes Phase 1 Clinical Trial." Businesswire. July 8, 2015. http://www.businesswire.com/news/home/20150708005060/en/AXON-Neuroscience%E2%80%99s-Vaccine-Halt-Alzheimer%E2%80%99s-Finishes-Phase.

Azpurua, Jorge, and Benjamin A. Eaton. "Neuronal Epigenetics and the Aging Synapse." *Frontiers in Cellular Neuroscience* 9 (2015): article 208.

Bahar-Fuchs, Alex, Linda Clare, and Bob Woods. "Cognitive Training and Cognitive Rehabilitation for Mild to Moderate Alzheimer's Disease and Vascular Dementia." *Cochrane Database of Systematic Reviews*, no. 6 (2013): CD003260.

Baker-Nigh, Alaina, Shahrooz Vahedi, Elena Goetz Davis, Sandra Weintraub, Eileen H. Bigio, William L. Klein, and Changiz Geula. "Neuronal Amyloid-β Accumulation Within Cholinergic Basal Forebrain in Ageing and Alzheimer's Disease." *Brain* 138, no. 6 (2015): 1722–37.

Ballatore, Carlo, Virginia M. Y. Lee, and John Q. Trojanowski. "Tau-Mediated Neurodegeneration in Alzheimer's Disease and Related Disorders." *Nature Reviews Neuroscience* 8, no. 9 (2007): 663–72.

Bard, Frédérique, Catherine Cannon, Robin Barbour, Burke Rae-Lyn, Dora Games, Henry Grajeda, Teresa Guido, et al. "Peripherally Administered Antibodies Against Amyloid β-Peptide Enter the Central Nervous System and Reduce Pathology in a Mouse Model of Alzheimer Disease." *Nature Medicine* 6, no. 8 (2000): 916–19.

Barini, Erica, Odetta Antico, Yingjun Zhao, Francesco Asta, Valter Tucci, Tiziano Catelani, Roberto Marotta, et al. "Metformin Promotes Tau Aggregation and Exacerbates Abnormal Behavior in a Mouse Model of Tauopathy." *Molecular Neurodegeneration* 11, no. 16 (2016): 1–20.

Barnes, L. L., C. F. Mendes de Leon, R. S. Wilson, J. L. Bienias, and D. A. Evans. "Social Resources and Cognitive Decline in a Population of Older African Americans and Whites." *Neurology* 63, no. 12 (2004): 2322–26.

Basak, Chandramallika, Walter R. Boot, Michelle W. Voss, and Arthur F. Kramer. "Can Training in a Real-Time Strategy Video Game Attenuate Cognitive Decline in Older Adults?" *Psychology and Aging* 23, no. 4 (2008): 765–77.

Bassuk, Shari S., Thomas A. Glass, and Lisa F. Berkman. "Social Disengagement and Incident Cognitive Decline in Community-Dwelling Elderly Persons." *Annals of Internal Medicine* 131, no. 3 (1999): 165–73.

Bateman, Randall J., Eric R. Siemers, Kwasi G. Mawuenyega, Guolin Wen, Karen R. Browning, Wendy C. Sigurdson, Kevin E. Yarasheski, et al. "A γ-Secretase Inhibitor Decreases Amyloid-β Production in the Central Nervous System." *Annals of Neurology* 66, no. 1 (2009): 48–54.

Bayer, Anthony J., R. Bullock, R. W. Jones, David Wilkinson, K. R. Paterson, L. Jenkins, S. B. Millais, and S. Donoghue. "Evaluation of the Safety and Immunogenicity of Synthetic Abeta42 (AN1792) in Patients with AD." *Neurology* 64, no. 1 (2005): 94–101.

Beach, Thomas. "The History of Alzheimer's Disease: Three Debates." *Journal of the History of Medicine and Allied Sciences* 42, no. 3 (1987): 327–49.

Beekes, Michael, Achim Thomzig, Walter Schulz-Schaeffer, and Reinhard Burger. "Is There a Risk of Prion-Like Disease Transmission by Alzheimer- or Parkinson-Associated Protein Particles?" *Acta Neuropathologica* 128, no. 4 (2014): 463–76.

Being Patient. "Pfizer Knew Its Drug Could Prevent Alzheimer's. They Did Nothing About It, Says *Post*." June 5, 2019. https://www.beingpatient.com/pfizer-coverup-enbrel-etanercept-alzheimers/.

Belluck, Pam. "Alzheimer's Stalks a Colombian Family." *New York Times*. June 1, 2010. https://www.nytimes.com/2010/06/02/health/02alzheimers.html.

——. "Eli Lilly's Experimental Alzheimer's Drug Fails in Large Trial." *New York Times*. November 23, 2016. https://www.nytimes.com/2016/11/23/health/eli -lillys-experimental-alzheimers-drug-failed-in-large-trial.html.

Belvedere, Matthew J. "Eli Lilly Shares Tank After Alzheimer's Drug Fails in Late-Stage Trial." CNBC. November 23, 2016. https://www.cnbc.com/2016 /11/23/eli-lilly-shares-tank-after-alzheimers-drug-fails-in-late-stage -trial.html.

Benedict, Christian, Manfred Hallschmid, Astrid Hatke, Bernd Schultes, Horst L. Fehm, Jan Born, and Werner Kern. "Intranasal Insulin Improves Memory in Humans." *Psychoneuroendocrinology* 29, no. 10 (2004): 1326–34.

Benilova, Iryna, Eric Karran, and Bart De Strooper. "The Toxic Aβ Oligomer and Alzheimer's Disease: An Emperor in Need of Clothes." *Nature Neuroscience* 15, no. 3 (2012): 349–57.

Bennett, David A., Julie A. Schneider, Yuxiao Tang, Steven E. Arnold, and Robert S. Wilson. "The Effect of Social Networks on the Relation Between Alzheimer's Disease Pathology and Level of Cognitive Function in Old People: A Longitudinal Cohort Study." *Lancet Neurology* 5, no. 5 (2006): 406–12.

Berrios, Germán Elías. "Alzheimer's Disease: A Conceptual History." *International Journal of Geriatric Psychiatry* 5, no. 6 (1990): 355–65.

Bhandari, Tamara. "Investigational Drugs Didn't Slow Memory Loss, Cognitive Decline in Rare, Inherited Alzheimer's, Initial Analysis Indicates." Washington University School of Medicine, St. Louis. February 10, 2020. https://medicine .wustl.edu/news/alzheimers-diantu-trial-initial-results/.

Biogen. "Biogen and Eisai to Discontinue Phase 3 Engage and Emerge Trials of Aducanumab in Alzheimer's Disease." March 21, 2019. http://investors .biogen.com/news-releases/news-release-details/biogen-and-eisai-discontinue -phase-3-engage-and-emerge-trials.

Bird, Thomas D. "Alzheimer Disease Overview." In *GeneReviews®*, ed. Margaret P. Adam, Holly H. Ardinger, Roberta A. Pagon, and Stephanie E. Wallace. Seattle: University of Washington, 2018

Bird, Thomas D., Ellen Nemens, D. Nochlin, S. M. Sumi, Ellen Wijsman, and Gerald D. Schellenberg. "Familial Alzheimer's Disease in Germans from Russia: A Model of Genetic Heterogeneity in Alzheimer's Disease." In *Heterogeneity of Alzheimer's Disease*, ed. Francois Boller, F. Forette, Zaven S. Khachaturian, Michel Poncet, and Yves Christen, 118–29. Berlin and Heidelberg: Springer Berlin Heidelberg, 1992.

Björkhem, Ingemar, and Steve Meaney. "Brain Cholesterol: Long Secret Life Behind a Barrier." *Arteriosclerosis, Thrombosis, and Vascular Biology* 24, no. 5 (2004): 806–15.

Black, Ronald S., Reisa A. Sperling, Beth Safirstein, Ruth N. Motter, Allan Pallay, Alice Nichols, and Michael Grundman. "A Single Ascending Dose Study of Bapineuzumab in Patients with Alzheimer Disease." *Alzheimer Disease and Associated Disorders* 24, no. 2 (2010): 198–203.

Blessed, Garry, Bernard E. Tomlinson, and Martin Roth. "The Association Between Quantitative Measures of Dementia and of Senile Change in the Cerebral Grey Matter of Elderly Subjects." *British Journal of Psychiatry: The Journal of Mental Science* 114, no. 512 (1968): 797–811.

Bondy, Stephen C. "Prolonged Exposure to Low Levels of Aluminum Leads to Changes Associated with Brain Aging and Neurodegeneration." *Toxicology* 315 (2014): 1–7.

Born, Jan, Tanja Lange, Werner Kern, Gerard P. McGregor, Ulrich Bickel, and Horst L. Fehm. "Sniffing Neuropeptides: A Transnasal Approach to the Human Brain." *Nature Neuroscience* 5, no. 6 (2002): 514–16.

Bottino, Cássio, Isabel A. M. Carvalho, Ana Maria M. A. Alvarez, Renata Avila, Patrícia R. Zukauskas, Sonia E. Z. Bustamante, Flávia C. Andrade, et al. "Cognitive Rehabilitation Combined with Drug Treatment in Alzheimer's Disease Patients: A Pilot Study." *Clinical Rehabilitation* 19, no. 8 (2005): 861–69.

Boxer, Adam L., Anthony E. Lang, Murray Grossman, David S. Knopman, Bruce L. Miller, Lon S. Schneider, Rachelle S. Doody, et al. "Davunetide in Patients with Progressive Supranuclear Palsy: A Randomised, Double-Blind, Placebo-Controlled Phase 2/3 Trial." *Lancet Neurology* 13, no. 7 (2014): 676–85.

Boyd, W. D., J. Graham-White, G. Blackwood, I. Glen, and J. McQueen. "Clinical Effects of Choline in Alzheimer Senile Dementia." *The Lancet* 310, no. 8040 (1977): 711.

Brackey, Jolene. *Creating Moments of Joy for the Person with Alzheimer's or Dementia.* West Lafayette, IN: Purdue University Press, 2007.

Breitner, John C., Laura D. Baker, Thomas J. Montine, Curtis L. Meinert, Constantine G. Lyketsos, Karen H. Ashe, Jason Brandt, et al. "Extended Results of the Alzheimer's Disease Anti-Inflammatory Prevention Trial." *Alzheimer's & Dementia: The Journal of the Alzheimer's Association* 7, no. 4 (2011): 402–11.

Brown, Paul, C. J. Gibbs, Pamela Rodgers-Johnson, David M. Asher, Michael P. Sulima, Alfred Bacote, Lev G. Goldfarb, and D. Carleton Gajdusek. "Human Spongiform Encephalopathy: The National Institutes of Health Series of 300 Cases of Experimentally Transmitted Disease." *Annals of Neurology* 35, no. 5 (1994): 513–29.

Buschert, Verena C., Uwe Friese, Stefan J. Teipel, Philine Schneider, Wibke Merensky, Dan Rujescu, Hans-Jürgen Möller, et al.. "Effects of a Newly Developed Cognitive Intervention in Amnestic Mild Cognitive Impairment and Mild Alzheimer's Disease: A Pilot Study." *Journal of Alzheimer's Disease* 25, no. 4 (2011): 679–94.

Buschler, Arne, and Denise Manahan-Vaughan. "Brief Environmental Enrichment Elicits Metaplasticity of Hippocampal Synaptic Potentiation *in Vivo*." *Frontiers in Behavioral Neuroscience* 6, no. 85 (2012): 1–10.

Butchart, Joseph, Laura Brook, Vivienne Hopkins, Jessica Teeling, Ursula Püntener, David Culliford, Richard Sharples, et al. "Etanercept in Alzheimer Disease: A Randomized, Placebo-Controlled, Double-Blind, Phase 2 Trial." *Neurology* 84, no. 21 (2015): 2161–68.

Cahn-Weiner, Deborah, Paul F. Malloy, George W. Rebok, and Brian R. Ott. "Results of a Randomized Placebo-Controlled Study of Memory Training for Mildly Impaired Alzheimer's Disease Patients." *Applied Neuropsychology* 10, no. 4 (2003): 215–23.

Callaway, Ewen. "Gene Mutation Defends Against Alzheimer's Disease." *Nature* 487, no. 7406 (2012): 153.

Cao, Dongfeng, Hailin Lu, Terry L. Lewis, and Ling Li. "Intake of Sucrose-Sweetened Water Induces Insulin Resistance and Exacerbates Memory Deficits and Amyloidosis in a Transgenic Mouse Model of Alzheimer Disease." *Journal of Biological Chemistry* 282, no. 50 (2007): 36275–82.

Cardoso, Susana, Só Correia, Renato X. Santos, Cristina Carvalho, Maria S. Santos, Catarina R. Oliveira, George Perry, et al. "Insulin Is a Two-Edged Knife on the Brain." *Journal of Alzheimer's Disease* 18, no. 3 (2009): 483–507.

Caselli, Richard J., Ann Saunders, Michael Lutz, Matthew Huentelman, Eric Reiman, and Allen Roses. "TOMM40, ApoE, and Age of Onset of Alzheimer's Disease." *Alzheimer's and Dementia* 6, no. 4 (2010): S202.

Centers for Disease Control and Prevention. "Creutzfeldt-Jakob Disease, Classic (CJD)." May 8, 2019. https://www.cdc.gov/prions/cjd/occurrence-transmission.html.

Cerecin. "Accera Closes New Investment Led by Asia's Leading Agribusiness Group, Wilmar, and Rebrands as Cerecin." October 4, 2018. http://www.cerecin.com/newsroom/accera-closes-new-investment-led-by-asia-leading-agribusiness-group.html.

Cerf, Emilie, Adelin Gustot, Erik Goormaghtigh, Jean-Marie Ruysschaert, and Vincent Raussens. "High Ability of Apolipoprotein E4 to Stabilize Amyloid-β Peptide Oligomers, the Pathological Entities Responsible for Alzheimer's Disease." *FASEB Journal: Official Publication of the Federation of American Societies for Experimental Biology* 25, no. 5 (2011): 1585–95.

Cervantes, Sebastián, Lluís Samaranch, Jos Vidal-Taboada, Isabel Lamet, María Jesús Bullido, Ana Frank-García, Francisco Coria, et al. "Genetic Variation in *APOE* Cluster Region and Alzheimer's Disease Risk." *Neurobiology of Aging* 32, no. 11 (2011): 2107.e7–2107.e17.

Chan, David C., and Eric A. Schon. "Eliminating Mitochondrial DNA from Sperm." *Developmental Cell* 22, no. 3 (2012): 469–70.

Chan, Kit Yee, Wei Wang, Jing Jing Wu, Li Liu, Evropi Theodoratou, Josip Car, Lefkos Middleton, et al. "Epidemiology of Alzheimer's Disease and Other Forms of Dementia in China, 1990–2010: A Systematic Review and Analysis." *The Lancet* 381, no. 9882 (2013): 2016–23.

Chen, Yaomin, Kun Zhou, Ruishan Wang, Yun Liu, Young-Don Kwak, Tao Ma, Robert C. Thompson, et al. "Antidiabetic Drug Metformin (Glucophage®) Increases Biogenesis of Alzheimer's Amyloid Peptides via Up-Regulating BACE1 Transcription." *Proceedings of the National Academy of Sciences of the United States of America* 106, no. 10 (2009): 3907–12.

Chilman-Blair, Kim, and John Taddeo. *What's Up with Ella?Medikidz Explain Type 1 Diabetes.* London: Medikidz Ltd., 2009.

Chou, Richard, Michael Kane, Sanjay Ghimire, Shiva Gautam, and Jiang Gui. "Treatment for Rheumatoid Arthritis and Risk of Alzheimer's Disease: A Nested Case-Control Analysis." *CNS Drugs* 30, no. 11 (2016): 1111–20.

Chu, Su Hee, Kathryn Roeder, Robert E. Ferrell, Bernie Devlin, Mary Demichele-Sweet, M. I. Kamboh, Oscar L. Lopez, and Robert A. Sweet. "*TOMM40* Poly-T Repeat Lengths, Age of Onset and Psychosis Risk in Alzheimer Disease." *Neurobiology of Aging* 32, no. 12 (2011): 2328.e1–2328.e9.

Cibičková, L'ubica. "Statins and Their Influence on Brain Cholesterol." *Journal of Clinical Lipidology* 5, no. 5 (2011): 373–79.

Citron, Martin, Tilman Oltersdorf, Christian Haass, Lisa McConlogue, Albert Y. Hung, Peter Seubert, Carmen Vigo-Pelfrey, Ivan Lieberburg, and Dennis J. Selkoe. "Mutation of the β-Amyloid Precursor Protein in Familial Alzheimer's Disease Increases β-Protein Production." *Nature* 360, no. 6405 (1992): 672–74.

City of Rochester. "History of Rochester." Accessed March 27, 2020. http://www .rochestermn.gov/about-the-city/history-of-rochester.

Claxton, Amy, Laura Baker, Angela Hanson, Brenna Cholerton, Emily Trittschuh, Amy Morgan, Maureen Callaghan, et al. "Long-Acting Intranasal Insulin Detemir Improves Working Memory for Adults with Mild Cognitive Impairment or Early-Stage Alzheimer's Dementia." *Journal of Alzheimer's Disease* 44, no. 3 (2015): 897–906.

Cleary, James, Dominic Walsh, J. J. Hofmeister, G. M. Shankar, M. A. Kuskowski, D. J. Selkoe, and K. H. Ashe. "Natural Oligomers of the Amyloid-β Protein Specifically Disrupt Cognitive Function." *Nature Neuroscience* 8, no. 1 (2004): 79–84.

Cohen, Edith L., and Richard J. Wurtman. "Brain Acetylcholine: Control by Dietary Choline." *Science* 191, no. 4227 (1976): 561–62.

——. "Brain Acetylcholine: Increase After Systemic Choline Administration." *Life Sciences* 16, no. 7 (1975): 1095–102.

Collin, Ludovic, Bernd Bohrmann, Ulrich Göpfert, Krisztina Oroszlan-Szovik, Laurence Ozmen, and Fiona Grüninger. "Neuronal Uptake of Tau/PS422

Antibody and Reduced Progression of Tau Pathology in a Mouse Model of Alzheimer's Disease." *Brain* 137, no. 10 (2014): 2834–46.

Congdon, Erin E., Jessica W. Wu, Natura Myeku, Yvette H. Figueroa, Mathieu Herman, Paul S. Marinec, Jason E. Gestwicki, et al. "Methylthioninium Chloride (Methylene Blue) Induces Autophagy and Attenuates Tauopathy in Vitro and in Vivo." *Autophagy* 8, no. 4 (2012): 609–22.

Cooper, Geoffrey. *The Cell: A Molecular Approach.* Sunderland, MA: Sinauer Associates, 2000.

Cordain, Loren, Michael R. Eades, and Mary D. Eades. "Hyperinsulinemic Diseases of Civilization: More Than Just Syndrome X." *Comparative Biochemistry and Physiology, Part A* 136, no. 1 (2003): 95–112.

Corder, E. H., A. M. Saunders, N. J. Risch, W. J. Strittmatter, D. E. Schmechel, P. C. Gaskell, J. B. Rimmler, et al. "Protective Effect of Apolipoprotein E Type 2 Allele for Late Onset Alzheimer Disease." *Nature Genetics* 7, no. 2 (1994): 180–4.

Corder, E. H., A. M. Saunders, W. J. Strittmatter, D. E. Schmechel, P. C. Gaskell, G. W. Small, A. D. Roses, J. L. Haines, and M. A. Petricak-Vance. "Gene Dose of Apolipoprotein E Type 4 Allele and the Risk of Alzheimer's Disease in Late Onset Families." *Science* 261, no. 5123 (1993): 921–23.

Coric, Vladimir, Christopher H. van Dyck, Stephen Salloway, Niels Andreasen, Mark Brody, Ralph W. Richter, Hilkka Soininen, et al. "Safety and Tolerability of the γ-Secretase Inhibitor Avagacestat in a Phase 2 Study of Mild to Moderate Alzheimer Disease." *Archives of Neurology* 69, no. 11 (2012): 1430–40.

Corneveaux, Jason J., Amanda J. Myers, April N. Allen, Jeremy J. Pruzin, Manuel Ramirez, Anzhelika Engel, Michael A. Nalls, et al. "Association of CR1, CLU and PICALM with Alzheimer's Disease in a Cohort of Clinically Characterized and Neuropathologically Verified Individuals." *Human Molecular Genetics* 19, no. 16 (2010): 3295–301.

Correia, Só, Renato X. Santos, Cristina Carvalho, Susana Cardoso, Emanuel Candeias, Maria S. Santos, Catarina R. Oliveira, and Paula I. Moreira. "Insulin Signaling, Glucose Metabolism and Mitochondria: Major Players in Alzheimer's Disease and Diabetes Interrelation." *Brain Research* 1441 (2012): 64–78.

Coskun, Pinar E., Joanne Wyrembak, Olga Derbereva, Goar Melkonian, Eric Doran, Ira T. Lott, Elizabeth Head, Carl W. Cotman, and Douglas C. Wallace. "Systemic Mitochondrial Dysfunction and the Etiology of Alzheimer's Disease and Down Syndrome Dementia." *Journal of Alzheimer's Disease* 20 Suppl 2 (2010): S293–S310.

Cracchiolo, Jennifer R., Takashi Mori, Stanley J. Nazian, Jun Tan, Huntington Potter, and Gary W. Arendash. "Enhanced Cognitive Activity—Over and Above Social or Physical Activity—Is Required to Protect Alzheimer's Mice Against Cognitive Impairment, Reduce Aβ Deposition, and Increase

Synaptic Immunoreactivity." *Neurobiology of Learning and Memory* 88, no. 3 (2007): 277–94.

Craft, Suzanne. "The Role of Metabolic Disorders in Alzheimer Disease and Vascular Dementia: Two Roads Converged." *Archives of Neurology* 66, no. 3 (2009): 300–5.

Craft, Suzanne, Sanjay Asthana, David G. Cook, Laura D. Baker, Monique Cherrier, Kristina Purganan, Colby Wait, et al. "Insulin Dose–Response Effects on Memory and Plasma Amyloid Precursor Protein in Alzheimer's Disease: Interactions with Apolipoprotein E Genotype." *Psychoneuroendocrinology* 28, no. 6 (2003): 809–22.

Craft, Suzanne, Laura D. Baker, Thomas J. Montine, Satoshi Minoshima, G. Stennis Watson, Amy Claxton, Matthew Arbuckle, et al. "Intranasal Insulin Therapy for Alzheimer Disease and Amnestic Mild Cognitive Impairment: A Pilot Clinical Trial." *Archives of Neurology* 69, no. 1 (2012): 29–38.

Craft, Suzanne, Amy Claxton, Laura D. Baker, Angela J. Hanson, Brenna Cholerton, Emily H. Trittschuh, Deborah Dahl, et al. "Effects of Regular and Long-Acting Insulin on Cognition and Alzheimer's Disease Biomarkers: A Pilot Clinical Trial." *Journal of Alzheimer's Disease* 57, no. 4 (2017): 1325–34.

Craft, Suzanne, John Newcomer, Stephen Kanne, Samuel Dagogo-Jack, Philip Cryer, Yvette Sheline, Joan Luby, Agbani Dagogo-Jack, and Amy Alderson. "Memory Improvement Following Induced Hyperinsulinemia in Alzheimer's Disease." *Neurobiology of Aging* 17, no. 1 (1996): 123–30.

Crapper, D. R., S. S. Krishnan, and A. J. Dalton. "Brain Aluminum Distribution in Alzheimer's Disease and Experimental Neurofibrillary Degeneration." *Science* 180, no. 4085 (1973): 511–13.

Crehan, Helen, Patrick Holton, Selina Wray, Jennifer Pocock, Rita Guerreiro, and John Hardy. "Complement Receptor 1 (CR1) and Alzheimer's Disease." *Immunobiology* 217, no. 2 (2012): 244–50.

Crismon, M. Lynn. "Tacrine: First Drug Approved for Alzheimer's Disease." *Annals of Pharmacotherapy* 28, no. 6 (1994): 744–51.

Cruchaga, Carlos, Petra Nowotny, John Kauwe, Perry Ridge, Kevin Mayo, Sarah Bertelsen, Anthony Hinrichs, et al. "Association and Expression Analyses with Single-Nucleotide Polymorphisms in *TOMM40* in Alzheimer Disease." *Archives of Neurology* 68, no. 8 (2011): 1013–19.

Cunnane, Stephen C., Alexandre Courchesne-Loyer, Camille Vandenberghe, Valérie St-Pierre, M. A. Fortier, Marie Hennebelle, Etienne Croteau, et al. "Can Ketones Help Rescue Brain Fuel Supply in Later Life? Implications for Cognitive Health During Aging and the Treatment of Alzheimer's Disease." *Frontiers in Molecular Neuroscience* 9 (2016): 1–21.

Cunnane, Stephen, Scott Nugent, Maggie Roy, Alexandre Courchesne-Loyer, Etienne Croteau, Sé Tremblay, Alex Castellano, et al. "Brain Fuel Metabolism, Aging, and Alzheimer's Disease." *Nutrition* 27, no. 1 (2011): 3–20.

Dacy, Matthew. "What's in a Name? The Story of 'Mayo Clinic.'" Mayo Clinic. February 9, 2009. https://sharing.mayoclinic.org/2009/02/09/whats-in-a-name-the -story-of-mayo-clinic/.

Dai, Weiying, Oscar L. Lopez, Owen T. Carmichael, James T. Becker, Lewis H. Kuller, and H. Michael Gach. "Mild Cognitive Impairment and Alzheimer Disease: Patterns of Altered Cerebral Blood Flow at MR Imaging." *Radiology* 250, no. 3 (2009): 856–66.

Davies, Peter, and A. J. Maloney. "Selective Loss of Central Cholinergic Neurons in Alzheimer's Disease." *The Lancet* 308, no. 8000 (1976): 1403.

De Felice, Fernanda G., Marcelo N. N. Vieira, Theresa R. Bomfim, Helena Decker, Pauline T. Velasco, Mary P. Lambert, Kirsten L. Viola, et al. "Protection of Synapses Against Alzheimer's-Linked Toxins: Insulin Signaling Prevents the Pathogenic Binding of Abeta Oligomers." *Proceedings of the National Academy of Sciences of the United States of America* 106, no. 6 (2009): 1971–76.

de la Monte, Suzanne M., Ming Tong, Nataniel Lester-Coll, Michael Plater, and Jack R. Wands. "Therapeutic Rescue of Neurodegeneration in Experimental Type 3 Diabetes: Relevance to Alzheimer's Disease." *Journal of Alzheimer's Disease* 10, no. 1 (2006): 89–109.

de la Rubia Ortí, José Enrique, María Pilar García-Pardo, Eraci Drehmer, David Sancho Cantus, Mariano Julián Rochina, Maria Asunción Aguilar Calpe, and Iván Hu Yang. "Improvement of Main Cognitive Functions in Patients with Alzheimer's Disease After Treatment with Coconut Oil Enriched Mediterranean Diet: A Pilot Study." *Journal of Alzheimer's Disease* 65, no. 2 (2018): 577–87.

de la Torre, Jack C. "Vascular Risk Factors: A Ticking Time Bomb to Alzheimer's Disease." *American Journal of Alzheimer's Disease & Other Dementias* 28, no. 6 (2013): 551–59.

Del Ser, Teodoro, Klaus C. Steinwachs, Hermann J. Gertz, María V. Andrés, Belén Gómez-Carrillo, Miguel Medina, Joan A. Vericat, et al. "Treatment of Alzheimer's Disease with the GSK-3 Inhibitor Tideglusib: A Pilot Study." *Journal of Alzheimer's Disease* 33, no. 1 (2013): 205–15.

De Strooper, Bart. "Lessons from a Failed γ-Secretase Alzheimer Trial." *Cell* 159, no. 4 (2014): 721–26.

De Strooper, Bart, and Wim Annaert. "Novel Research Horizons for Presenilins and γ-Secretases in Cell Biology and Disease." *Annual Review of Cell and Developmental Biology* 26 (2010): 235–60.

Demattos, Ronald B., Kelly R. Bales, David J. Cummins, Jean-Cosme Dodart, Steven M. Paul, and David M. Holtzman. "Peripheral Anti-Aβ Antibody Alters CNS and Plasma Aβ Clearance and Decreases Brain Aβ Burden in a Mouse Model of Alzheimer's Disease." *Proceedings of the National Academy of Sciences of the United States of America* 98, no. 15 (2001): 8850–5.

Devore, Elizabeth E., Jae Hee Kang, Monique M. B. Breteler, and Francine Grodstein. "Dietary Intakes of Berries and Flavonoids in Relation to Cognitive Decline." *Annals of Neurology* 72, no. 1 (2012): 135–43.

Dingledine, Raymond, and Chris McBain. "Glutamate and Aspartate." In *Basic Neurochemistry: Molecular, Cellular and Medical Aspects*, ed. George Siegle, Bernard Agranoff, R. Wayne Albers, Stephen Fisher, and Michael Uhler. Philadelphia: Lippincott-Raven, 1999.

Divry, Paul. "Etude Histochimique Des Plaques Séniles." *Journal Belge de Neurologie et de Psychiatrie* 27 (1927): 643–57.

Dodart, Jean-Cosme, Kelly Bales, Kimberley Gannon, Stephen Greene, Ronald Demattos, Chantal Mathis, Cynthia DeLong, et al. "Immunization Reverses Memory Deficits Without Reducing Brain Aβ Burden in Alzheimer's Disease Model." *Nature Neuroscience* 5, no. 5 (2002): 452–57.

Dolan, Hillary, Barbara Crain, Juan Troncoso, Susan M. Resnick, Alan B. Zonderman, and Richard J. Obrien. "Atherosclerosis, Dementia, and Alzheimer Disease in the Baltimore Longitudinal Study of Aging Cohort." *Annals of Neurology* 68, no. 2 (2010): 231–40.

Doody, Rachelle S., Rema Raman, Martin Farlow, Takeshi Iwatsubo, Bruno Vellas, Steven Joffe, Karl Kieburtz, et al. "A Phase 3 Trial of Semagacestat for Treatment of Alzheimer's Disease." *The New England Journal of Medicine* 369, no. 4 (2013): 341–50.

Doody, Rachelle S., Ronald G. Thomas, Martin Farlow, Takeshi Iwatsubo, Bruno Vellas, Steven Joffe, Karl Kieburtz, et al. "Phase 3 Trials of Solanezumab for Mild-to-Moderate Alzheimer's Disease." *New England Journal of Medicine* 370, no. 4 (2014): 311–21.

Doulames, Vanessa, Sangmook Lee, and Thomas B. Shea. "Environmental Enrichment and Social Interaction Improve Cognitive Function and Decrease Reactive Oxidative Species in Normal Adult Mice." *International Journal of Neuroscience* 124, no. 5 (2014): 369–76.

Draganski, Bogdan, Christian Gaser, Volker Busch, Gerhard Schuierer, Ulrich Bogdahn, and Arne May. "Neuroplasticity: Changes in Grey Matter Induced by Training." *Nature* 427, no. 6972 (2004): 311–12.

Duarte, Ana I., Paula I. Moreira, and Catarina R. Oliveira. "Insulin in Central Nervous System: More Than Just a Peripheral Hormone." *Journal of Aging Research* 2012 (2012): 1–21.

Dubois, Bruno, Howard H. Feldman, Claudia Jacova, Steven T. Dekosky, Pascale Barberger-Gateau, Jeffrey Cummings, André Delacourte, et al. "Research Criteria for the Diagnosis of Alzheimer's Disease: Revising the NINCDS–ADRDA Criteria." *Lancet Neurology* 6, no. 8 (2007): 734–46.

Dumanchin, Cécile, Agnès Camuzat, Dominique Campion, Patrice Verpillat, Didier Hannequin, Bruno Dubois, Pascale Saugier-Veber, et al. "Segregation of a Missense Mutation in the Microtubule-Associated Protein Tau Gene with Familial Frontotemporal Dementia and Parkinsonism." *Human Molecular Genetics* 7, no. 11 (1998): 1825–29.

Dysken, Maurice W., Mary Sano, Sanjay Asthana, Julia E. Vertrees, Muralidhar Pallaki, Maria Llorente, Susan Love, et al. "Effect of Vitamin E and Memantine on Functional Decline in Alzheimer Disease: The TEAM-AD VA Cooperative Randomized Trial." *JAMA* 311, no. 1 (2014): 33–44.

Dysken, M. W., P. Fovall, C. M. Harris, J. M. Davis, and A. Noronha. "Lecithin Administration in Alzheimer Dementia." *Neurology* 32, no. 10 (1982): 1203–4.

Ebbert, Mark T. W., Perry G. Ridge, Andrew R. Wilson, Aaron R. Sharp, Matthew Bailey, Maria C. Norton, Joann T. Tschanz, et al. "Population-Based Analysis of Alzheimer's Disease Risk Alleles Implicates Genetic Interactions." *Biological Psychiatry* 75, no. 9 (2014): 732–37.

Editorial. "Alzheimergate? When Miscommunication Met Sensationalism." *The Lancet* 386, no. 9999 (2015): 1109.

Eenfeldt, Andreas, "20 and 50 Grams of Carbs—How Much Food Is That?" Diet Doctor. February 14, 2020. https://www.dietdoctor.com/low-carb/20-50-how-much.

Eisele, Yvonne S., Ulrike Obermüller, Götz Heilbronner, Frank Baumann, Stephan A. Kaeser, Hartwig Wolburg, Lary C. Walker, et al. "Peripherally Applied Aβ-Containing Inoculates Induce Cerebral β-Amyloidosis." *Science* 330, no. 6006 (2010): 980–82.

Elder, Gregory A., Miguel A. Gama Sosa, and Rita De Gasperi. "Transgenic Mouse Models of Alzheimer's Disease." *Mount Sinai Journal of Medicine: A Journal of Translational and Personalized Medicine* 77, no. 1 (2010): 69–81.

Elgee, Neil J., Robert H. Williams, and Norman D. Lee. "Distribution and Degradation Studies with Insulin-I^{131}." *Journal of Clinical Investigation* 33, no. 9 (1954): 1252–60.

Eli Lilly. "Lilly Announces Top-Line Results of Solanezumab Phase 3 Clinical Trial." November 23, 2016. https://investor.lilly.com/news-releases/news-release -details/lilly-announces-top-line-results-solanezumab-phase-3-clinical ?ReleaseID=1000871.

——. "Lilly Halts Development of Semagacestat for Alzheimer's Disease Based on Preliminary Results of Phase III Clinical Trials." August 17, 2010. https://investor .lilly.com/news-releases/news-release-details/lilly-halts-development -semagacestat-alzheimers-disease-based?releaseid=499794.

——. "Lilly Reports Fourth-Quarter and Full-Year 2016 Results." January 31, 2017. https://investor.lilly.com/news-releases/news-release-details/lilly-reports -fourth-quarter-and-full-year-2016-results?ReleaseID=1009682.

——. "Update on Phase 3 Clinical Trials of Lanabecestat for Alzheimer's Disease." June 12, 2018. https://investor.lilly.com/news-releases/news-release-details /update-phase-3-clinical-trials-lanabecestat-alzheimers-disease.

Engelhart, Marianne J., Mirjam I. Geerlings, Annemieke Ruitenberg, John C. van Swieten, Albert Hofman, Jacqueline C. M. Witteman, and Monique M. B. Breteler. "Dietary Intake of Antioxidants and Risk of Alzheimer Disease." *JAMA* 287, no. 24 (2002): 3223–29.

Erasmus University Medical Center. "The Rotterdam Study." Erasmus University Rotterdam. Accessed March 27, 2020. http://www.epib.nl/research/ergo.htm.

Erickson, Kirk I., Michelle W. Voss, Ruchika Shaurya Prakash, Chandramallika Basak, Amanda Szabo, Laura Chaddock, Jennifer S. Kim, et al. "Exercise Training Increases Size of Hippocampus and Improves Memory." *Proceedings of the National Academy of Sciences of the United States of America* 108, no. 7 (2011): 3017–22.

Fagan, Tom. "In First Phase 3 Trial, the Tau Drug LMTM Did Not Work. Period." Alzforum. July 29, 2016. https://www.alzforum.org/news/conference-coverage /first-phase-3-trial-tau-drug-lmtm-did-not-work-period.

——. "Las Vegas: AD, Risk, ApoE—Tomm40 No Tomfoolery." Alzforum. November 15, 2009. https://www.alzforum.org/news/conference-coverage/las-vegas-ad-risk -apoe-tomm40-no-tomfoolery.

——. "Liver Tox Ends Janssen BACE Program." Alzforum. May 18, 2018. https:// www.alzforum.org/news/research-news/liver-tox-ends-janssen-bace-program.

——. "Tau Inhibitor Fails Again—Subgroup Analysis Irks Clinicians at CTAD." Alzforum. December 16, 2016. https://www.alzforum.org/news/conference -coverage/tau-inhibitor-fails-again-subgroup-analysis-irks-clinicians-ctad.

Fagan, Tom, and Gabrielle Strobel. "Dale Schenk, 59, Pioneer of Alzheimer's Immunotherapy." Alzforum. October 3, 2016. https://www.alzforum.org/news /community-news/dale-schenk-59-pioneer-alzheimers-immunotherapy.

Farber, N. B., J. W. Newcomer, and J. W. Olney. "The Glutamate Synapse in Neuropsychiatric Disorders. Focus on Schizophrenia and Alzheimer's Disease." *Progress in Brain Research* 116 (1998): 421–37.

Farlow, Martin R., Niels Andreasen, Marie-Emmanuelle Riviere, Igor Vostiar, Alessandra Vitaliti, Judit Sovago, Angelika Caputo, Bengt Winblad, and Ana Graf. "Long-Term Treatment with Active Aβ Immunotherapy with CAD106 in Mild Alzheimer's Disease." *Alzheimer's Research & Therapy* 7, no. 1 (2015): 1–13.

Farlow, Martin, Steven E. Arnold, Christopher H. van Dyck, Paul S. Aisen, B. J. Snider, Anton P. Porsteinsson, Stuart Friedrich, et al. "Safety and Biomarker Effects of Solanezumab in Patients with Alzheimer's Disease." *Alzheimer's & Dementia: The Journal of the Alzheimer's Association* 8, no. 4 (2012): 261–71.

Farrer, Lindsay A., L. A. Cupples, Jonathan L. Haines, Bradley Hyman, Walter A. Kukull, Richard Mayeux, Richard H. Myers, et al. "Effects of Age, Sex, and Ethnicity on the Association Between Apolipoprotein E Genotype and Alzheimer Disease: A Meta-Analysis." *JAMA* 278, no. 16 (1997): 1349–56.

Fatouros, Chronis, Ghulam Jeelani Pir, Jacek Biernat, Sandhya Padmanabhan Koushika, Eckhard Mandelkow, Eva-Maria Mandelkow, Enrico Schmidt, and Ralf Baumeister. "Inhibition of Tau Aggregation in a Novel *Caenorhabditis elegans* Model of Tauopathy Mitigates Proteotoxicity." *Human Molecular Genetics* 21, no. 16 (2012): 3587–603.

Ferrer, Isidre, Mercé Boada Rovira, María Luisa Sánchez Guerra, María Jesús Rey, and Frederic Costa-Jussá. "Neuropathology and Pathogenesis of Encephalitis Following Amyloid β Immunization in Alzheimer's Disease." *Brain Pathology* 14, no. 1 (2004): 11–20.

Ferrero, James, Leslie Williams, Heather Stella, Kate Leitermann, Alvydas Mikulskis, John O'Gorman, and Jeff Sevigny. "First-in-Human, Double-Blind, Placebo-Controlled, Single-Dose Escalation Study of Aducanumab (BIIB037) in Mild-to-Moderate Alzheimer's Disease." *Alzheimer's & Dementia: Translational Research & Clinical Interventions* 2, no. 3 (2016): 169–76.

Filipcik, Peter, Martin Cente, Gabriela Krajciova, Ivo Vanicky, and Michal Novak. "Cortical and Hippocampal Neurons from Truncated Tau Transgenic Rat Express Multiple Markers of Neurodegeneration." *Cellular and Molecular Neurobiology* 29, no. 6 (2009): 895–900.

Fisher, Lawrence. "Athena Neurosciences Makes Itself Heard in Fight Against Alzheimer's." *New York Times.* February 15, 1995. https://www.nytimes.com/1995/02/15/business/business-technology-athena-neurosciences-makes-itself-heard-fight-against.html.

Fitzgerald, Stephanie. *Ramses II: Egyptian Pharaoh, Warrior, and Builder.* Mankato, MN: Compass Point, 2009.

Fjell, Anders M., Linda McEvoy, Dominic Holland, Anders M. Dale, Kristine B. Walhovd, and Alzheimer's Disease Neuroimaging Initative. "What Is Normal in Normal Aging? Effects of Aging, Amyloid and Alzheimer's Disease on the Cerebral Cortex and the Hippocampus." *Progress in Neurobiology* 117 (2014): 20–40.

Fleisher, Adam S., Rema Raman, Eric R. Siemers, Lida Becerra, Christopher M. Clark, Robert A. Dean, Martin R. Farlow, et al. "Phase 2 Safety Trial Targeting Amyloid β Production with a γ-Secretase Inhibitor in Alzheimer Disease." *Archives of Neurology* 65, no. 8 (2008): 1031–38.

Forette, Françoise, Marie-Laure Seux, Jan A. Staessen, Lutgarde Thijs, Willem H. Birkenhäger, Marija-Ruta Babarskiene, Speranta Babeanu, et al. "Prevention of Dementia in Randomised Double-Blind Placebo-Controlled Systolic Hypertension in Europe (Syst-Eur) Trial." *The Lancet* 352, no. 9137 (1998): 1347–51.

Fox, Patrick. "From Senility to Alzheimer's Disease: The Rise of the Alzheimer's Disease Movement." *Milbank Quarterly* 67, no. 1 (1989): 58–102.

Fratiglioni, Laura, and Hui-Xin Wang. "Brain Reserve Hypothesis in Dementia." *Journal of Alzheimer's Disease* 12, no. 1 (2007): 11–22.

Fratiglioni, Laura, Hui-Xin Wang, Kjerstin Ericsson, Margaret Maytan, and Bengt Winblad. "Influence of Social Network on Occurrence of Dementia: A Community-Based Longitudinal Study." *The Lancet* 355, no. 9212 (2000): 1315–19.

Freeman, John, Pierangelo Veggiotti, Giovanni Lanzi, Anna Tagliabue, and Emilio Perucca. "The Ketogenic Diet: From Molecular Mechanisms to Clinical Effects." *Epilepsy Research* 68, (2006): 145–80.

Fuller, Solomon C. "Alzheimer's Disease (Senium Praecox) I: The Report of a Case and Review of Published Cases." *Journal of Nervous & Mental Disease* 39, no. 7 (1912): 440–55.

——. "Alzheimer's Disease (Senium Praecox) II: The Report of a Case and Review of Published Cases." *Journal of Nervous & Mental Disease* 39, no. 8 (1912): 536–57.

Galasko, Douglas R., Elaine Peskind, Christopher M. Clark, Joseph F. Quinn, John M. Ringman, Gregory A. Jicha, Carl Cotman, et al. "Antioxidants for Alzheimer Disease: A Randomized Clinical Trial with Cerebrospinal Fluid Biomarker Measures." *Archives of Neurology* 69, no. 7 (2012): 836–41.

Games, Dora, David Adams, Ree Alessandrini, Robin Barbour, Patricia Borthelette, Catherine Blackwell, Tony Carr, et al. "Alzheimer-Type Neuropathology in Transgenic Mice Overexpressing V717F β-Amyloid Precursor Protein." *Nature* 373, no. 6514 (1995): 523–27.

Gasparini, R., D. Panatto, P. L. Lai, and D. Amicizia. "The 'Urban Myth' of the Association Between Neurological Disorders and Vaccinations." *Journal of Preventive Medicine and Hygiene* 56, no. 1 (2015), E1–E8.

Gatz, Margaret, Pia Svedberg, Nancy L. Pedersen, James A. Mortimer, Stig Berg, and Boo Johansson. "Education and the Risk of Alzheimer's Disease: Findings from the Study of Dementia in Swedish Twins." *The Journals of Gerontology: Psychological Sciences* 56B, no. 5 (2001): P292–P300.

Gauthier, Serge, Howard H. Feldman, Lon S. Schneider, Gordon K. Wilcock, Giovanni B. Frisoni, Jiri H. Hardlund, Hans J. Moebius, et al. "Efficacy and Safety of Tau-Aggregation Inhibitor Therapy in Patients with Mild or Moderate Alzheimer's Disease: A Randomised, Controlled, Double-Blind, Parallel-Arm, Phase 3 Trial." *The Lancet* 388, no. 10062 (2016): 2873–84.

Gejl, Michael, Albert Gjedde, Lærke Egefjord, Arne Møller, Søren B. Hansen, Kim Vang, Anders Rodell, et al. "In Alzheimer's Disease, Six-Month Treatment with GLP-1 Analogue Prevents Decline of Brain Glucose Metabolism: Randomized, Placebo-Controlled, Double-Blind Clinical Trial." *Frontiers in Aging Neuroscience* 8 (2016): 1–10.

Geldmacher, David S., Thomas Fritsch, Mckee J. McClendon, and Gary Landreth. "A Randomized Pilot Clinical Trial of the Safety of Pioglitazone in Treatment of Patients with Alzheimer Disease." *Archives of Neurology* 68, no. 1 (2011): 45–50.

Gemma, Carmelina, Jennifer Vila, Adam Bachstetter, and Paula Bickford. "Oxidative Stress and the Aging Brain: From Theory to Prevention." In *Brain Aging: Models, Methods, and Mechanisms*, ed. David Riddle, 353–74. Boca Raton, FL: CRC Press/Taylor & Francis, 2007.

Gervais, Francine, Julie Paquette, Céline Morissette, Pascale Krzywkowski, Mathilde Yu, Mounia Azzi, Diane Lacombe, et al. "Targeting Soluble Aβ Peptide with Tramiprosate for the Treatment of Brain Amyloidosis." *Neurobiology of Aging* 28, no. 4 (2007): 537–47.

Ghosh, Sangeeta, Nishant Patel, Douglas Rahn, Jenna McAllister, Sina Sadeghi, Geoffrey Horwitz, Diana Berry, Kai Xuan Wang, and Russell H. Swerdlow. "The Thiazolidinedione Pioglitazone Alters Mitochondrial Function in Human Neuron-Like Cells." *Molecular Pharmacology* 71, no. 6 (2007): 1695–702.

Gilman, S., M. Koller, R. S. Black, L. Jenkins, S. G. Griffith, N. C. Fox, L. Eisner, et al. "Clinical Effects of Aβ Immunization (AN1792) in Patients with AD in an Interrupted Trial." *Neurology* 64, no. 9 (2005): 1553–62.

Giuffrida, Maria Laura, Filippo Caraci, Bruno Pignataro, Sebastiano Cataldo, Paolo De Bona, Valeria Bruno, Gemma Molinaro, et al. "Beta-Amyloid Monomers Are Neuroprotective." *Journal of Neuroscience* 29, no. 34 (2009): 10582–87.

Glenner, George G., and Caine W. Wong. "Alzheimer's Disease: Initial Report of the Purification and Characterization of a Novel Cerebrovascular Amyloid Protein." *Biochemical and Biophysical Research Communications* 120, no. 3 (1984): 885–90.

Goate, Alison, Marie-Christine Chartier-Harlin, Mike Mullan, Jeremy Brown, Fiona Crawford, Liana Fidani, Luis Giuffra, et al. "Segregation of a Missense Mutation in the Amyloid Precursor Protein Gene with Familial Alzheimer's Disease." *Nature* 349, no. 6311 (1991): 704–6.

Goedert, Michel. "Oskar Fischer and the Study of Dementia." *Brain* 132, no. 4 (2009): 1102–11.

Gold, Michael, Claire Alderton, Marina Zvartau-Hind, Sally Egginton, Ann M. Saunders, Michael Irizarry, Suzanne Craft, et al. "Rosiglitazone Monotherapy in Mild-to-Moderate Alzheimer's Disease: Results from a Randomized, Double-Blind, Placebo-Controlled Phase III Study." *Dementia and Geriatric Cognitive Disorders* 30, no. 2 (2010): 131–46.

Goldbloom, Alton. "Some Observations on the Starvation Treatment of Epilepsy." *Canadian Medical Association Journal* 12, no. 8 (1922): 539–40.

Golden, Marita. "African Americans Are More Likely Than Whites to Develop Alzheimer's. Why?" *Washington Post*. June 1, 2017. https://www.washingtonpost.com/lifestyle/magazine/why-are-african-americans

-so-much-more-likely-than-whites-to-develop-alzheimers/2017/05/31
/9bfbcccc-3132-11e7-8674-437ddb6e813e_story.html.

Gong, Cheng-Xin, and Khalid Iqbal. "Hyperphosphorylation of Microtubule-Associated Protein Tau: A Promising Therapeutic Target for Alzheimer Disease." *Current Medicinal Chemistry* 15, no. 23 (2008): 2321–28.

Goudsmit, Jaap, Chuck H. Morrow, David M. Asher, Richard T. Yanagihara, Colin L. Masters, Clarence J. Gibbs, and D. Carleton Gajdusek. "Evidence For and Against the Transmissibility of Alzheimer Disease." *Neurology* 30, no. 9 (1980): 945–50.

Goure, William F., Grant A. Krafft, Jasna Jerecic, and Franz Hefti. "Targeting the Proper Amyloid-Beta Neuronal Toxins: A Path Forward for Alzheimer's Disease Immunotherapeutics." *Alzheimer's Research & Therapy* 6, no. 42 (2014): 1–15.

Graeber, Manuel B. "Alois Alzheimer (1864–1915)." International Brain Research Organization. 2003. http://ibro.org/wp-content/uploads/2018/07/Alzheimer-Alois-2003.pdf.

Gray, Michael W. "Mitochondrial Evolution." *Cold Spring Harbor Perspectives in Biology* 4, no. 9 (2012): a011403.

Gray, Sarah M., Rick I. Meijer, and Eugene J. Barrett. "Insulin Regulates Brain Function, but How Does It Get There?" *Diabetes* 63, no. 12 (2014): 3992–97.

Gray, Shelly L., Melissa L. Anderson, Paul K. Crane, John C. S. Breitner, Wayne McCormick, James D. Bowen, Linda Teri, and Eric Larson. "Antioxidant Vitamin Supplement Use and Risk of Dementia or Alzheimer's Disease in Older Adults." *Journal of the American Geriatrics Society* 56, no. 2 (2008): 291–95.

Greenamyre, J. Timothy, William F. Maragos, Roger L. Albin, John B. Penney, and Anne B. Young. "Glutamate Transmission and Toxicity in Alzheimer's Disease." *Progress in Neuropsychopharmacology & Biological Psychiatry* 12, no. 4 (1988): 421–30.

Greener, Mark. "Food for Thought: The Ketogenic Diet for Epilepsy." *Progress in Neurology and Psychiatry* 18, no. 3 (2014): 6–9.

Green Valley. "Green Valley Announces NMPA Approval of Oligomannate for Mild to Moderate Alzheimer's Disease." November 2, 2019. https://www.greenvalleypharma.com/En/Index/pageView/catid/48/id/28.html.

Growdon, John H. "Acetylcholine in AD: Expectations Meet Reality." In *Alzheimer: 100 Years and Beyond*, ed. Mathias Jucker, Konrad Beyreuther, Christian Haass, Roger M. Nitsch, and Yves Christen, 127–32. Heidelberg: Springer, 2006.

Grundke-Iqbal, Khalid Iqbal, Maureen Quinlan, Yunn-Chyn Tung, Masooma S. Zaidi, and Henryk M. Wisniewski. "Microtubule-Associated Protein Tau—A Component of Alzheimer Paired Helical Filaments." *Journal of Biological Chemistry* 261, no. 13 (1986): 6084–89.

Grundke-Iqbal, Khalid Iqbal, Yunn-Chyn Tung, Maureen Quinlan, Henryk M. Wisniewski, and Lester I. Binder. "Abnormal Phosphorylation of the Micro-tubule-Associated Protein τ (Tau) in Alzheimer Cytoskeletal Pathology." *Proceedings of the National Academy of Sciences of the United States of America* 83, no. 13 (1986): 4913–17.

Guénet, Jean-Louis, Annie Orth, and François Bonhomme. "Origins and Phylo-genetic Relationships of the Laboratory Mouse." In *The Laboratory Mouse*, ed. Hans Hedrich, 3–20. Cambridge, MA: Academic Press, 2012.

Guerreiro, Rita J., and John Hardy. "TOMM40 Association with Alzheimer Dis-ease: Tales of APOE and Linkage Disequilibrium." *Archives of Neurology* 69, no. 10 (2012): 1243–44.

Guerreiro, Rita, Aleksandra Wojtas, Jose Bras, Minerva Carrasquillo, Ekaterina Rogaeva, Elisa Majounie, Carlos Cruchaga, et al. "TREM2 Variants in Alzheim-er's Disease." *New England Journal of Medicine* 368, no. 2 (2013): 117–27.

Gupta, Amit, Bharti Bisht, and Chinmoy Sankar Dey. "Peripheral Insulin-Sensitizer Drug Metformin Ameliorates Neuronal Insulin Resistance and Alzheimer's-Like Changes." *Neuropharmacology* 60, no. 6 (2011): 910–20.

Gusdon, Aaron M., Jason Callio, Giovanna DiStefano, Robert M. O'Doherty, Bret H. Goodpaster, Paul M. Coen, and Charleen T. Chu. "Exercise Increases Mito-chondrial Complex I Activity and DRP1 Expression in the Brains of Aged Mice." *Experimental Gerontology* 90 (2017): 1–13.

Gusella, James F., Nancy S. Wexler, P. Michael Conneally, Susan L. Naylor, Mary Anne Anderson, Rudolph E. Tanzi, Paul C. Watkins, et al. "A Polymorphic DNA Marker Genetically Linked to Huntington's Disease." *Nature* 306, no. 5940 (1983): 234–38.

Hansen, Henrik H., Katrine Fabricius, Pernille Barkholt, Pernille Kongsbak-Wismann, Chantal Schlumberger, Jacob Jelsing, Dick Terwel, et al. "Long-Term Treatment with Liraglutide, a Glucagon-Like Peptide-1 (GLP-1) Receptor Agonist, Has No Effect on β-Amyloid Plaque Load in Two Transgenic APP/PS1 Mouse Models of Alzheimer's Disease." *PLoS One* 11, no. 7 (2016): E0158205.

Hansen, Lawrence A., Eliezer Masliah, Douglas Galasko, and Robert D. Terry. "Plaque-Only Alzheimer Disease Is Usually the Lewy Body Variant, and Vice Versa." *Journal of Neuropathology and Experimental Neurology* 52, no. 6 (1993): 648–54.

Hanyu, Haruo, Tomohiko Sato, Akihiro Kiuchi, Hirofumi Sakurai, and Toshihiko Iwamoto. "Pioglitazone Improved Cognition in a Pilot Study on Patients with Alzheimer's Disease and Mild Cognitive Impairment with Diabetes Mellitus." *Journal of the American Geriatrics Society* 57, no. 1 (2009): 177–79.

Hardy, John. "Alzheimer's Disease: The Amyloid Cascade Hypothesis: An Update and Reappraisal." *Journal of Alzheimer's Disease* 9, no. 3 (2006): 151–53.

Hardy, John, and David Allsop. "Amyloid Deposition as the Central Event in the Aetiology of Alzheimer's Disease." *Trends in Pharmacological Sciences* 12 (1991): 383–88.

Hardy, John A., and Gerald A. Higgins. "Alzheimer's Disease: The Amyloid Cascade Hypothesis." *Science* 256, no. 5054 (1992): 184–85.

Harrington, C., S. Sawchak, C. Chiang, J. Davies, C. Donovan, A. M. Saunders, M. Irizarry, et al. "Rosiglitazone Does Not Improve Cognition or Global Function When Used as Adjunctive Therapy to AChE Inhibitors in Mild-to-Moderate Alzheimer's Disease: Two Phase 3 Studies." *Current Alzheimer Research* 8, no. 5 (2011): 592–606.

Hashimoto, Tadafumi, Alberto Serrano-Pozo, Yukiko Hori, Kenneth W. Adams, Shuko Takeda, Adrian Olaf Banerji, Akinori Mitani, et al. "Apolipoprotein E, Especially Apolipoprotein E4, Increases the Oligomerization of Amyloid β Peptide." *Journal of Neuroscience* 32, no. 43 (2012): 15181–92.

Heart Protection Study Collaborative Group. "MRC/BHF Heart Protection Study of Cholesterol Lowering with Simvastatin in 20,536 High-Risk Individuals: A Randomised Placebo-Controlled Trial." *The Lancet* 360, no. 9326 (2002): 7–22.

Hefti, Franz, William F. Goure, Jasna Jerecic, Kent S. Iverson, Patricia A. Walicke, and Grant A. Krafft. "The Case for Soluble Aβ Oligomers as a Drug Target in Alzheimer's Disease." *Trends in Pharmacological Sciences* 34, no. 5 (2013): 261–66.

Henderson, Samuel T., Janet L. Vogel, Linda J. Barr, Fiona Garvin, Julie J. Jones, and Lauren C. Costantini. "Study of the Ketogenic Agent AC-1202 in Mild to Moderate Alzheimer's Disease: A Randomized, Double-Blind, Placebo-Controlled, Multicenter Trial." *Nutrition & Metabolism* 6, no. 31 (2009): 1–25.

Heppner, Frank L., Richard M. Ransohoff, and Burkhard Becher. "Immune Attack: The Role of Inflammation in Alzheimer Disease." *Nature Reviews Neuroscience* 16, no. 6 (2015): 358–72.

Hey, John, Susan Abushakra, Aidan Power, Jeremy Yu, Mark Versavel, and Martin Tolar. "Phase 1 Development of ALZ-801, a Novel Beta Amyloid Anti-Aggregation Prodrug of Tramiprosate with Improved Drug Properties, Supporting Bridging to the Phase 3 Program." *Alzheimer's & Dementia: The Journal of the Alzheimer's Association* 12, no. 7 (2016): P613.

Heyn, Patricia, Beatriz C. Abreu, and Kenneth J. Ottenbacher. "The Effects of Exercise Training on Elderly Persons with Cognitive Impairment and Dementia: A Meta-Analysis." *Archives of Physical Medicine and Rehabilitation* 85, no. 10 (2004): 1694–704.

Hieronymus, Laura, and Stacy Griffin. "Role of Amylin in Type 1 and Type 2 Diabetes." *The Diabetes Educator* 41, no. 1 (2015): 47S–56S.

Hillen, Heinz, Stefan Barghorn, Andreas Striebinger, Boris Labkovsky, Reinhold Müller, Volker Nimmrich, Marc W Nolte, et al. "Generation and Therapeutic

Efficacy of Highly Oligomer-Specific β-Amyloid Antibodies." *Journal of Neuroscience* 30, no. 31 (2010): 10369–79.

Hippocrates. "On the Sacred Disease." Classics Archive. Accessed March 30, 2020. http://classics.mit.edu/Hippocrates/sacred.html.

Hiscott, Rebecca. "At the Bench: John Hardy, PhD, on Unraveling the Genetics of Alzheimer's Disease and Attending the 'Oscars of Science.' " *Neurology Today* 16, no. 1 (2016): 21–22.

Hoffmann, Kristine, Nanna A. Sobol, Kristian S. Frederiksen, Nina Beyer, Asmus Vogel, Karsten Vestergaard, Hans Brændgaard, et al. "Moderate-to-High Intensity Physical Exercise in Patients with Alzheimer's Disease: A Randomized Controlled Trial." *Journal of Alzheimer's Disease* 50, no. 2 (2016): 443–53.

Höltta, Mikko, Oskar Hansson, Ulf Andreasson, Joakim Hertze, Lennart Minthon, Katarina Nägga, Niels Andreasen, Henrik Zetterberg, and Kaj Blennow. "Evaluating Amyloid-β Oligomers in Cerebrospinal Fluid as a Biomarker for Alzheimer's Disease." *PLoS One* 8, no. 6 (2013): E66381.

Holtzman, David M., John C. Morris, and Alison M. Goate. "Alzheimer's Disease: The Challenge of the Second Century." *Science Translational Medicine* 3, no. 77 (2011): 77sr1.

Honig, Lawrence S., Bruno Vellas, Michael Woodward, Mercè Boada, Roger Bullock, Michael Borrie, Klaus Hager, et al. "Trial of Solanezumab for Mild Dementia Due to Alzheimer's Disease." *New England Journal of Medicine* 378, no. 4 (2018): 321–30.

Honjo, Kie, Robert van Reekum, and Nicolaas P. L. G. Verhoeff. "Alzheimer's Disease and Infection: Do Infectious Agents Contribute to Progression of Alzheimer's Disease?" *Alzheimer's & Dementia: The Journal of the Alzheimer's Association* 5, no. 4 (2009): 348–60.

Howieson, Diane B. "Cognitive Skills and the Aging Brain: What to Expect." *Cerebrum: Dana Foundation.* December 1, 2015. https://www.dana.org/article /cognitive-skills-and-the-aging-brain-what-to-expect/.

Hoyer, S. "Brain Glucose and Energy Metabolism Abnormalities in Sporadic Alzheimer Disease. Causes and Consequences: An Update." *Experimental Gerontology* 35, no. 9 (2000): 1363–72.

Hoyer, S., R. Nitsch, and K. Oesterreich. "Predominant Abnormality in Cerebral Glucose Utilization in Late-Onset Dementia of the Alzheimer Type: A Cross-Sectional Comparison Against Advanced Late-Onset and Incipient Early-Onset Cases." *Journal of Neural Transmission—Parkinson's Disease and Dementia Section* 3, no. 1 (1991): 1–14.

Hsu, Chih-Cheng, Mark L Wahlqvist, Meei-Shyuan Lee, and Hsin-Ni Tsai. "Incidence of Dementia Is Increased in Type 2 Diabetes and Reduced by the Use of Sulfonylureas and Metformin." *Journal of Alzheimer's Disease* 24, no. 3 (2011): 485–93.

Hundal, Ripudaman S., Martin Krssak, Sylvie Dufour, Didier Laurent, Vincent Lebon, Visvanathan Chandramouli, Silvio E. Inzucchi, et al. "Mechanism by Which Metformin Reduces Glucose Production in Type 2 Diabetes." *Diabetes* 49, no. 12 (2000): 2063–69.

Huntley, J. D., R. L. Gould, K. Liu, M. Smith, and R. J. Howard. "Do Cognitive Interventions Improve General Cognition in Dementia? A Meta-Analysis and Meta-Regression." *BMJ Open* 5, no. 4 (2015), e005247.

Hutton, Mike, Corinne L. Lendon, Patrizia Rizzu, Matt Baker, Susanne Froelich, Henry Houlden, Stuart Pickering-Brown, et al. "Association of Missense and 5'-Splice-Site Mutations in Tau with the Inherited Dementia FTDP-17." *Nature* 393, no. 6686 (1998): 702–5.

Imfeld, Patrick, Michael Bodmer, Susan S. Jick, and Christoph R. Meier. "Metformin, Other Antidiabetic Drugs, and Risk of Alzheimer's Disease: A Population-Based Case-Control Study." *Journal of the American Geriatrics Society* 60, no. 5 (2012): 916–21.

Intlekofer, Karlie A., and Carl W Cotman. "Exercise Counteracts Declining Hippocampal Function in Aging and Alzheimer's Disease." *Neurobiology of Disease* 57, no. C (2013): 47–55.

in't Veld, Bas A., Annemieke Ruitenberg, Albert Hofman, Lenore J. Launer, Cornelia M. van Duijn, Theo Stijnen, Monique M. B. Breteler, and Bruno H. C. Stricker. "Nonsteroidal Antiinflammatory Drugs and the Risk of Alzheimer's Disease." *New England Journal of Medicine* 345, no. 21 (2001): 1515–21.

in't Veld, B. A., A. Ruitenberg, A. Hofman, B. H. C. Stricker, and M. M. B. Breteler. "Antihypertensive Drugs and Incidence of Dementia: The Rotterdam Study." *Neurobiology of Aging* 22, no. 3 (2001): 407–12.

Iqbal, Khalid, Fei Liu, Cheng-Xin Gong, and Inge Grundke-Iqbal. "Tau in Alzheimer Disease and Related Tauopathies." *Current Alzheimer Research* 7, no. 8 (2010): 656–64.

Irrcher, Isabella, Peter Adhihetty, Anna-Maria Joseph, Vladimir Ljubicic, and David Hood. "Regulation of Mitochondrial Biogenesis in Muscle by Endurance Exercise." *Sports Medicine* 33, no. 11 (2003): 783–93.

Irwin, David J., Joseph Y. Abrams, Lawrence B. Schonberger, Ellen Werber Leschek, James L. Mills, Virginia M. Y. Lee, and John Q. Trojanowski. "Evaluation of Potential Infectivity of Alzheimer and Parkinson Disease Proteins in Recipients of Cadaver-Derived Human Growth Hormone." *JAMA Neurology* 70, no. 4 (2013): 462–68.

Ishii, Tsuyoshi, Seiji Haga, and Fujio Shimizu. "Identification of Components of Immunoglobulins in Senile Plaques by Means of Fluorescent Antibody Technique." *Acta Neuropathologica* 32, no. 2 (1975): 157–62.

Isik, Ahmet Turan, Pinar Soysal, Adnan Yay, and Cansu Usarel. "The Effects of Sitagliptin, a DPP-4 Inhibitor, on Cognitive Functions in Elderly Diabetic Patients With or Without Alzheimer's Disease." *Diabetes Research and Clinical Practice* 123 (2017): 192–98.

Ittner, Lars M., and Jürgen Götz. "Amyloid-β and Tau—A Toxic Pas de Deux in Alzheimer's Disease." *Nature Reviews Neuroscience* 12, no. 2 (2011): 67–72.

Jack, Clifford R., Val J. Lowe, Stephen D. Weigand, Heather J. Wiste, Matthew L. Senjem, David S. Knopman, Maria M. Shiung, et al. "Serial PIB and MRI in Normal, Mild Cognitive Impairment and Alzheimer's Disease: Implications for Sequence of Pathological Events in Alzheimer's Disease." *Brain* 132, no. 5 (2009): 1355–65.

Jackson, Kaleena, Gustavo A. Barisone, Elva Diaz, Lee-Way Jin, Charles Decarli, and Florin Despa. "Amylin Deposition in the Brain: A Second Amyloid in Alzheimer Disease?" *Annals of Neurology* 74, no. 4 (2013): 517–26.

Janson, Juliette, Thomas Laedtke, Joseph E. Parisi, Peter O'Brien, Ronald C. Petersen, and Peter C. Butler. "Increased Risk of Type 2 Diabetes in Alzheimer Disease." *Diabetes* 53, no. 2 (2004): 474–81.

Jantzen, Paul T., Karen E. Connor, Giovanni Dicarlo, Gary L. Wenk, John L. Wallace, Amyn M. Rojiani, Domenico Coppola, Dave Morgan, and Marcia N. Gordon. "Microglial Activation and Beta-Amyloid Deposit Reduction Caused by a Nitric Oxide–Releasing Nonsteroidal Anti-Inflammatory Drug in Amyloid Precursor Protein Plus Presenilin-1 Transgenic Mice." *Journal of Neuroscience* 22, no. 6 (2002): 2246–54.

Janus, Christopher, Jacqueline Pearson, JoAnne McLaurin, Paul M. Mathews, Ying Jiang, Stephen D. Schmidt, M. Azhar Chishti, et al. "Aβ Peptide Immunization Reduces Behavioural Impairment and Plaques in a Model of Alzheimer's Disease." *Nature* 408, no. 6815 (2000): 979–82.

Jaunmuktane, Zane, Simon Mead, Matthew Ellis, Jonathan D. F. Wadsworth, Andrew J. Nicoll, Joanna Kenny, Francesca Launchbury, et al. "Evidence for Human Transmission of Amyloid-β Pathology and Cerebral Amyloid Angiopathy." *Nature* 525, no. 7568 (2015): 247–50.

Jay, Taylor R., Crystal M. Miller, Paul J. Cheng, Leah C. Graham, Shane Bemiller, Margaret L. Broihier, Guixiang Xu, et al. "TREM2 Deficiency Eliminates TREM2[+] Inflammatory Macrophages and Ameliorates Pathology in Alzheimer's Disease Mouse Models." *Journal of Experimental Medicine* 212, no. 3 (2015): 287–95.

Jay, Taylor R., Victoria E. von Saucken, and Gary E. Landreth. "TREM2 in Neurodegenerative Diseases." *Molecular Neurodegeneration* 12, no. 56 (2017): 1–33.

Jick, H., G. L. Zornberg, S. S. Jick, S. Seshadri, and D. A. Drachman. "Statins and the Risk of Dementia." *The Lancet* 356, no. 9242 (2000): 1627–31.

Johnson, K. A., K. Jones, B. L. Holman, J. A. Becker, P. A. Spiers, A. Satlin, and M. S. Albert. "Preclinical Prediction of Alzheimer's Disease Using SPECT." *Neurology* 50, no. 6 (1998): 1563–71.

Johnson, Sterling C., Asenath La Rue, Bruce P. Hermann, Guofan Xu, Rebecca L. Koscik, Erin M. Jonaitis, Barbara B. Bendlin, et al. "The Effect of *TOMM40* Poly-T Length on Gray Matter Volume and Cognition in Middle-Aged Persons with *APOE* Ɛ3/Ɛ3 Genotype." *Alzheimer's & Dementia: The Journal of the Alzheimer's Association* 7, no. 4 (2011): 456–65.

Jonsson, Thorlakur, Jasvinder Atwal, Stacy Steinberg, Jon Snaedal, Palmi Jonsson, Sigurbjorn Bjornsson, Hreinn Stefansson, et al. "A Mutation in APP Protects Against Alzheimer's Disease and Age-Related Cognitive Decline." *Nature* 487, no. 7409 (2012): 96–99.

Joseph, James A., G. Arendash, M. Gordon, D. Diamond, B. Shukitt-Hale, D. Morgan, and N. A. Denisova. "Blueberry Supplementation Enhances Signaling and Prevents Behavioral Deficits in an Alzheimer Disease Model." *Nutritional Neuroscience* 6, no. 3 (2003): 153–62.

Joseph, James A., Barbara Shukitt-Hale, Natalia A. Denisova, Donna Bielinski, Antonio Martin, John J. McEwen, and Paula C. Bickford. "Reversals of Age-Related Declines in Neuronal Signal Transduction, Cognitive, and Motor Behavioral Deficits with Blueberry, Spinach, or Strawberry Dietary Supplementation." *Journal of Neuroscience* 19, no. 18 (1999): 8114–21.

Jucker, Mathias, Lary C. Walker, Lee J. Martin, Cheryl A. Kitt, Hynda K. Kleinman, Donald K. Ingram, Donald L. Price, et al. "Age-Associated Inclusions in Normal and Transgenic Mouse Brain." *Science* 255, no. 5050 (1992): 1443–45.

Jun, Gyungah, Badri N. Vardarajan, Jacqueline Buros, Chang-En Yu, Michele V. Hawk, Beth A. Dombroski, Paul K. Crane, et al. "Comprehensive Search for Alzheimer Disease Susceptibility Loci in the ApoE Region." *Archives of Neurology* 69, no. 10 (2012): 1270–79.

Kaiyla, Karl, Ronald Prigeon, Steven Kahn, Stephen Woods, and Michael Schwartz. "Obesity Induced by a High-Fat Diet Is Associated with Reduced Brain Insulin Transport in Dogs." *Diabetes* 49, no. 9 (2000): 1525–33.

Kandola, Aaron, Joshua Hendrikse, Paul J. Lucassen, and Murat Yucel. "Aerobic Exercise as a Tool to Improve Hippocampal Plasticity and Function in Humans: Practical Implications for Mental Health Treatment." *Frontiers in Human Neuroscience* 10 (2016): article 373.

Kanekiyo, Takahisa, Huaxi Xu, and Guojun Bu. "ApoE and Aβ in Alzheimer's Disease: Accidental Encounters or Partners?" *Neuron* 81, no. 4 (2014): 740–54.

Kang, Jae H., Alberto Ascherio, and Francine Grodstein. "Fruit and Vegetable Consumption and Cognitive Decline in Aging Women." *Annals of Neurology* 57, no. 5 (2005): 713–20.

Kang, Jae Hee, Nancy Cook, Joann Manson, Julie E. Buring, and Francine Grodstein. "A Randomized Trial of Vitamin E Supplementation and Cognitive Function in Women." *Archives of Internal Medicine* 166, no. 22 (2006): 2462–68.

Kang, Jie, Hans-Georg Lemaire, Axel Unterbeck, J. Michael Salbaum, Colin L. Masters, Karl-Heinz Grzeschik, Gerd Multhaup, Konrad Beyreuther, and Benno Müller-Hill. "The Precursor of Alzheimer's Disease Amyloid A4 Protein Resembles a Cell-Surface Receptor." *Nature* 325, no. 6106 (1987): 733–36.

Karran, Eric, and John Hardy. "A Critique of the Drug Discovery and Phase 3 Clinical Programs Targeting the Amyloid Hypothesis for Alzheimer Disease." *Annuals of Neurology* 76, no. 2 (2014): 185–205.

Karran, Eric, Marc Mercken, and Bart De Strooper. "The Amyloid Cascade Hypothesis for Alzheimer's Disease: An Appraisal for the Development of Therapeutics." *Nature Reviews Drug Discovery* 10, no. 9 (2011): 698–712.

Kastanenka, Ksenia V., Thierry Bussiere, Naomi Shakerdge, Fang Qian, Paul H. Weinreb, Ken Rhodes, and Brian J. Bacskai. "Immunotherapy with Aducanumab Restores Calcium Homeostasis in Tg2576 Mice." *Journal of Neuroscience* 36, no. 50 (2016): 12549–58.

Katzman, Robert. "Editorial: The Prevalence and Malignancy of Alzheimer Disease. A Major Killer." *Archives of Neurology* 33, no. 4 (1976): 217–18.

Katzman, Robert, and Katherine Bick. *Alzheimer Disease: The Changing View.* San Diego: Academic Press, 2000.

Katzman, Robert, Robert Terry, Richard Deteresa, Theodore Brown, Peter Davies, Paula Fuld, Xiong Renbing, and Arthur Peck. "Clinical, Pathological, and Neurochemical Changes in Dementia: A Subgroup with Preserved Mental Status and Numerous Neocortical Plaques." *Annals of Neurology* 23, no. 2 (1988): 138–44.

Kaufmann, Petra, Anne R. Pariser, and Christopher Austin. "From Scientific Discovery to Treatments for Rare Diseases—The View from the National Center for Advancing Translational Sciences—Office of Rare Diseases Research." *Orphanet Journal of Rare Diseases* 13, no. 196 (2018): 1–8.

Kawabata, Shigeki, Gerald Higgins, and Jon Gordon. "Amyloid Plaques, Neurofibrillary Tangles and Neuronal Loss in Brains of Transgenic Mice Overexpressing a C-Terminal Fragment of Human Amyloid Precursor Protein." *Nature* 354, no. 6353 (1991): 476–78.

Kelly, Michelle E., David Loughrey, Brian A. Lawlor, Ian H. Robertson, Cathal Walsh, and Sabina Brennan. "The Impact of Cognitive Training and Mental Stimulation on Cognitive and Everyday Functioning of Healthy Older Adults: A Systematic Review and Meta-Analysis." *Ageing Research Reviews* 15, no. 1 (2014): 28–43.

Kempermann, Gerd, Daniela Gast, and Fred H. Gage. "Neuroplasticity in Old Age: Sustained Fivefold Induction of Hippocampal Neurogenesis by Long-Term Environmental Enrichment." *Annals of Neurology* 52, no. 2 (2002): 135–43.

Khachaturian, Ara S., Peter P. Zandi, Constantine G. Lyketsos, Kathleen M. Hayden, Ingmar Skoog, Maria C. Norton, Joann T. Tschanz, et al. "Antihypertensive Medication Use and Incident Alzheimer Disease: The Cache County Study." *Archives of Neurology* 63, no. 5 (2006): 686–92.

Khachaturian, Zaven S. "A Chapter in the Development on Alzheimer's Disease Research." In *Alzheimer: 100 Years and Beyond*, ed. Mathias Jucker, Konrad Beyreuther, Christian Haass, Roger M. Nitsch, and Yves Christen, 63–86. Heidelberg: Springer, 2006.

Khan, Shaharyar, Rafal Smigrodzki, and Russell Swerdlow. "Cell and Animal Models of mtDNA Biology: Progress and Prospects." *American Journal of Physiology* 292, no. 2 (2007): C658–C669.

Kickstein, Eva, Sybille Krauss, Paul Thornhill, Désirée Rutschow, Raphael Zeller, John Sharkey, Ritchie Williamson, et al. "Biguanide Metformin Acts on Tau Phosphorylation via mTOR/Protein Phosphatase 2A (PP2A) Signaling." *Proceedings of the National Academy of Sciences of the United States of America* 107, no. 50 (2010): 21830–5.

Kim, Jungsu, Jacob M. Basak, and David M. Holtzman. "The Role of Apolipoprotein E in Alzheimer's Disease." *Neuron* 63, no. 3 (2009): 287–303.

Kivipelto, Miia, Eeva-Liisa Helkala, Mikko P. Laakso, Tuomo Hanninen, Merja Hallikainen, Kari Alhainen, Hilkka Soininen, Jaakko Tuomilehto, and Aulikki Nissien. "Midlife Vascular Risk Factors and Alzheimer's Disease in Later Life: Longitudinal, Population Based Study." *BMJ* 322, no. 7300 (2001): 1447–51.

Kivipelto, Miia, Tiia Ngandu, Laura Fratiglioni, Matti Viitanen, Ingemar Kåreholt, Bengt Winblad, Eeva-Liisa Helkala, et al. "Obesity and Vascular Risk Factors at Midlife and the Risk of Dementia and Alzheimer Disease." *Archives of Neurology* 62, no. 10 (2005): 1556–60.

Klatzo, Igor, Henryk Wisniewski, and Eugene Streicher. "Experimental Production of Neurofibrillary Degeneration. I. Light Microscopic Observations." *Journal of Neuropathology and Experimental Neurology* 24 (1965): 187–99.

Klein, William L. "Synaptotoxic Amyloid-β Oligomers: A Molecular Basis for the Cause, Diagnosis, and Treatment of Alzheimer's Disease?" *Journal of Alzheimer's Disease* 33 Suppl 1 (2013): S49–S65.

Klünemann, Hans H., Wolfgang Fronhöfer, Herbert Wurster, Wolfgang Fischer, Bernd Ibach, and Helmfried E. Klein. "Alzheimer's Second Patient: Johann F. and His Family." *Annals of Neurology* 52, no. 4 (2002): 520–3.

Knapton, Sarah. "First Drug to Slow Alzheimer's Disease Unveiled in Landmark Breakthrough." *The Telegraph*. July 22, 2015. https://www.telegraph.co.uk/news

/health/news/11755380/First-drug-to-slow-Alzheimers-Disease-unveiled -in-landmark-breakthrough.html.

Koch, Fred C. *The Volga Germans: In Russia and the Americas, from 1763 to the Present.* University Park: Pennsylvania State University Press, 1977.

Koenig, Aaron M., Dawn Mechanic-Hamilton, Sharon X. Xie, Martha F. Combs, Anne R. Cappola, Long Xie, John A. Detre, David A. Wolk, and Steven E. Arnold. "Effects of the Insulin Sensitizer Metformin in Alzheimer Disease: Pilot Data from a Randomized Placebo-Controlled Crossover Study." *Alzheimer Disease and Associated Disorders* 31, no. 2 (2017): 107–13.

Kolarova, Michala, Francisco García-Sierra, Ales Bartos, Jan Ricny, and Daniela Ripova. "Structure and Pathology of Tau Protein in Alzheimer Disease." *International Journal of Alzheimer's Disease* (2012): article 731526.

Kolata, Gina. "Landmark in Alzheimer Research: Breeding Mice with the Disease." *New York Times.* February 9, 1995. https://www.nytimes.com/1995/02/09/us /landmark-in-alzheimer-research-breeding-mice-with-the-disease.html.

Kontsekova, Eva, Norbert Zilka, Branislav Kovacech, Petr Novak, and Michal Novak. "First-in-Man Tau Vaccine Targeting Structural Determinants Essential for Pathological Tau-Tau Interaction Reduces Tau Oligomerisation and Neurofibrillary Degeneration in an Alzheimer's Disease Model." *Alzheimer's Research & Therapy* 6, no. 44 (2014): 1–12.

Kornelius, Edy, Chih-Li Lin, Hsiu-Han Chang, Hsin-Hua Li, Wen-Nung Huang, Yi-Sun Yang, Ying-Li Lu, Chiung-Huei Peng, and Chien-Ning Huang. "DPP-4 Inhibitor Linagliptin Attenuates Aβ-Induced Cytotoxicity Through Activation of AMPK in Neuronal Cells." *CNS Neuroscience & Therapeutics* 21, no. 7 (2015): 549–57.

Kosaraju, Jayasankar, R. M. Damian Holsinger, Lixia Guo, and Kin Tam. "Linagliptin, a Dipeptidyl Peptidase-4 Inhibitor, Mitigates Cognitive Deficits and Pathology in the 3xTg-AD Mouse Model of Alzheimer's Disease." *Molecular Neurobiology* 54, no. 8 (2017): 6074–84.

Kosaraju, Jayasankar, Rizwan Basha Khatwal, Anil Dubala, Chinni Santhi Vardhan, M. N. S. Kumar, and Basavan Duraiswamy. "Saxagliptin: A Dipeptidyl Peptidase-4 Inhibitor Ameliorates Streptozotocin-Induced Alzheimer's Disease." *Neuropharmacology* 72 (2013): 291–300.

Kosaraju, Jayasankar, Vishakantha Murthy, Rizwan Basha Khatwal, Anil Dubala, Santhivardhan Chinni, Satish Kumar Muthureddy Nataraj, and Duraiswamy Basavan. "Vildagliptin: An Anti-Diabetes Agent Ameliorates Cognitive Deficits and Pathology Observed in Streptozotocin-Induced Alzheimer's Disease." *Journal of Pharmacy and Pharmacology* 65, no. 12 (2013): 1773–84.

Kosik, Kenneth S., and Francisco Lopera. "Genetic Testing Must Recognize Impact of Bad News on Recipient." *Nature* 454, no. 7201 (2008): 158–59.

Kossoff, Eric, Zahava Turner, Sarah Doerrer, Mackenzie Cervenka, and Bobbie Henry. *The Ketogenic and Modified Atkins Diets: Treatments for Epilepsy and Other Disorders*. New York: Demos Medical Publishing, 2016.

Kowall, Neil W., M. Flint Beal, Jorge Busciglio, Lawrence K. Duffy, and Bruce A. Yankner. "An in Vivo Model for the Neurodegenerative Effects of Beta Amyloid and Protection by Substance P." *Proceedings of the National Academy of Sciences of the United States of America* 88, no. 16 (1991): 7247–51.

Krebs, Albin. "Rita Hayworth, Movie Legend, Dies." *New York Times*. May 16, 1987. https://www.nytimes.com/1987/05/16/obituaries/rita-hayworth-movie -legend-dies.html.

Krikorian, Robert, Erin L. Boespflug, David E. Fleck, Amanda L. Stein, Jolynne D. Wightman, Marcelle D. Shidler, and Sara Sadat-Hossieny. "Concord Grape Juice Supplementation and Neurocognitive Function in Human Aging." *Journal of Agricultural and Food Chemistry* 60, no. 23 (2012): 5736–42.

Krikorian, Robert, Tiffany A. Nash, Marcelle D. Shidler, Barbara Shukitt-Hale, and James A. Joseph. "Concord Grape Juice Supplementation Improves Memory Function in Older Adults with Mild Cognitive Impairment." *British Journal of Nutrition* 103, no. 5 (2010): 730–4.

Krikorian, Robert, Marcelle D. Shidler, Krista Dangelo, Sarah C. Couch, Stephen C. Benoit, and Deborah J. Clegg. "Dietary Ketosis Enhances Memory in Mild Cognitive Impairment." *Neurobiology of Aging* 33, no. 2 (2012): 425.e19–425.e27.

Krikorian, Robert, Marcelle D. Shidler, Tiffany A. Nash, Wilhelmina Kalt, Melinda Vinqvist-Tymchuk, Barbara Shukitt-Hale, and James A. Joseph. "Blueberry Supplementation Improves Memory in Older Adults." *Journal of Agricultural and Food Chemistry* 58, no. 7 (2010): 3996–4000.

Kryscio, Richard J., Erin L. Abner, Allison Caban-Holt, Mark Lovell, Phyllis Goodman, Amy K. Darke, Monica Yee, John Crowley, and Frederick A. Schmitt. "Association of Antioxidant Supplement Use and Dementia in the Prevention of Alzheimer's Disease by Vitamin E and Selenium Trial (PREADViSE)." *JAMA Neurology* 74, no. 5 (2017): 567–73.

Kuhn, Thomas. *The Structure of Scientific Revolutions*. 4th ed. Chicago and London: University of Chicago Press, 2012.

Kuiper, Jisca S., Marij Zuidersma, Richard C. Oude Voshaar, Sytse U. Zuidema, Edwin R. van den Heuvel, Ronald P. Stolk, and Nynke Smidt. "Social Relationships and Risk of Dementia: A Systematic Review and Meta-Analysis of Longitudinal Cohort Studies." *Ageing Research Reviews* 22 (2015): 39–57.

Kumar, Devendra, Ankit Ganeshpurkar, Dileep Kumar, Gyan Modi, Sanjeev Kumar Gupta, and Sushil Kumar Singh. "Secretase Inhibitors for the Treatment of Alzheimer's Disease: Long Road Ahead." *European Journal of Medicinal Chemistry* 148 (2018): 436–52.

Kuperstein, Inna, Kerensa Broersen, Iryna Benilova, Jef Rozenski, Wim Jonckheere, Maja Debulpaep, Annelies Vandersteen, et al. "Neurotoxicity of Alzheimer's Disease Aβ Peptides Is Induced by Small Changes in the Aβ_{42} to Aβ_{40} Ratio." *EMBO Journal* 29, no. 19 (2010): 3408–20.

Küster, Olivia C., Patrick Fissler, Daria Laptinskaya, Franka Thurm, Andrea Scharpf, Alexander Woll, Stephan Kolassa, et al. "Cognitive Change Is More Positively Associated with an Active Lifestyle Than with Training Interventions in Older Adults at Risk of Dementia: A Controlled Interventional Clinical Trial." *BMC Psychiatry* 16, no. 315 (2016):1–12.

Lambert, Jean-Charles, Simon Heath, Gael Even, Dominique Campion, Kristel Sleegers, Mikko Hiltunen, Onofre Combarros, et al. "Genome-Wide Association Study Identifies Variants at *CLU* and *CR1* Associated with Alzheimer's Disease." *Nature Genetics* 41, no. 10 (2009): 1094–99.

Landhuis, Esther. "Medical Foods—Fallback Option for Elusive AD Drug Status?" Alzforum. October 14, 2009. https://www.alzforum.org/news/research-news/medical-foods-fallback-option-elusive-ad-drug-status.

Landsberg, Judith, Brendan McDonald, Geoff Grime, and Frank Watt. "Microanalysis of Senile Plaques Using Nuclear Microscopy." *Journal of Geriatric Psychiatry and Neurology* 6, no. 2 (1993): 97–104.

Langlois, Francis, Thien Tuong Minh Vu, Kathleen Chassé, Gilles Dupuis, Marie-Jeanne Kergoat, and Louis Bherer. "Benefits of Physical Exercise Training on Cognition and Quality of Life in Frail Older Adults." *The Journals of Gerontology. Series B: Psychological Sciences and Social Sciences* 68, no. 3 (2012): 400–4.

Lathia, Justin D., Mark P. Mattson, and Aiwu Cheng. "Notch: From Neural Development to Neurological Disorders." *Journal of Neurochemistry* 107, no. 6 (2008): 1471–81.

Launer, Lenore J., G. W. Ross, Helen Petrovitch, Kamal Masaki, Dan Foley, Lon R. White, and Richard J. Havlik. "Midlife Blood Pressure and Dementia: The Honolulu–Asia Aging Study." *Neurobiology of Aging* 21, no. 1 (2000): 49–55.

Ledesma, M. Dolores, Pedro Bonay, Camilo Colaço, and Jesús Avila. "Analysis of Microtubule-Associated Protein Tau Glycation in Paired Helical Filaments." *Journal of Biological Chemistry* 269, no. 34 (1994): 21614–19.

Leibson, C. L., W. A. Rocca, V. A. Hanson, R. Cha, E. Kokmen, P. C. O'Brien, and P. J. Palumbo. "Risk of Dementia Among Persons with Diabetes Mellitus: A Population-Based Cohort Study." *American Journal of Epidemiology* 145, no. 4 (1997): 301–8.

Lemonick, Michael. "Secrets of the Lost Tomb." *Time*. June 24, 2001. http://content.time.com/time/magazine/article/0,9171,134211,00.html.

Leonard, William R., J. Josh Snodgrass, and Marcia L. Robertson. "Evolutionary Perspectives on Fat Ingestion and Metabolism in Humans." In *Fat Detection: Taste, Texture, and Post Ingestive Effects*, ed. Jean-Pierre Montmayeur and Johannes le Coutre, 3–18. Boca Raton: CRC Press, 2010.

Lerner, Barron H. "Rita Hayworth's Misdiagnosed Struggle." *Los Angeles Times*. November 20, 2006. https://www.latimes.com/archives/la-xpm-2006-nov-20-he-myturn20-story.html.

Lesné, Sylvain E., Mathew A. Sherman, Marianne Grant, Michael Kuskowski, Julie A. Schneider, David A. Bennett, and Karen H. Ashe. "Brain Amyloid-β Oligomers in Ageing and Alzheimer's Disease." *Brain* 136, no. 5 (2013): 1383–98.

Lester-Coll, Nataniel, Enrique J. Rivera, Stephanie J. Soscia, Kathryn Doiron, Jack R. Wands, and Suzanne M. de la Monte. "Intracerebral Streptozotocin Model of Type 3 Diabetes: Relevance to Sporadic Alzheimer's Disease." *Journal of Alzheimer's Disease* 9, no. 1 (2006): 13–33.

Levy-Lahad, Ephrat, Wilma Wasco, Parvoneh Poorkaj, Donna M. Romano, Junko Oshima, Warren H. Pettingell, Chang-En Yu, et al. "Candidate Gene for the Chromosome 1 Familial Alzheimer's Disease Locus." *Science* 269, no. 5226 (1995): 973–77.

Lezi, E., Jeffrey M Burns, and Russell H Swerdlow. "Effect of High-Intensity Exercise on Aged Mouse Brain Mitochondria, Neurogenesis, and Inflammation." *Neurobiology of Aging* 35, no. 11 (2014): 2574–83.

Li, G., R. Higdon, W. A. Kukull, E. Peskind, K. Van Valen Moore, D. Tsuang, G. van Belle, et al. "Statin Therapy and Risk of Dementia in the Elderly: A Community-Based Prospective Cohort Study." *Neurology* 63, no. 9 (2004): 1624–28.

Lidsky, Theodore. "Is the Aluminum Hypothesis Dead?" *Journal of Occupational and Environmental Medicine* 56 (2014): S73–S79.

Linder, Carol C., and Muriel T. Davisson. "Historical Foundations." In *The Laboratory Mouse*, ed. Hans Hedrich, 21–35. Cambridge, MA: Academic Press, 2012.

Lindstrom, Heather A., Thomas Fritsch, Grace Petot, Kathleen A. Smyth, Chien H. Chen, Sara M. Debanne, Alan J. Lerner, and Robert P. Friedland. "The Relationships Between Television Viewing in Midlife and the Development of Alzheimer's Disease in a Case-Control Study." *Brain and Cognition* 58, no. 2 (2005): 157–65.

Liu, Chia-Chan, Takahisa Kanekiyo, Huaxi Xu, and Guojun Bu. "Apolipoprotein E and Alzheimer Disease: Risk, Mechanisms and Therapy." *Nature Reviews Neurology* 9, no. 2 (2013): 106–18.

Liu, Mengying, Chen Bian, Jiqiang Zhang, and Feng Wen. "Apolipoprotein E Gene Polymorphism and Alzheimer's Disease in Chinese Population: A Meta-Analysis." *Scientific Reports* 4, no. 4383 (2014): 1–7.

Lock, Margaret. *The Alzheimer Conundrum: Entanglements of Dementia and Aging.* Princeton, NJ: Princeton University Press, 2013.

Lovestone, Simon, Mercè Boada, Bruno Dubois, Michael Hüll, Juha O. Rinne, Hans-Jürgen Huppertz, Miguel Calero, et al. "A Phase II Trial of Tideglusib in Alzheimer's Disease." *Journal of Alzheimer's Disease* 45, no. 1 (2015): 75–88.

Luchsinger, José Alejandro. "Adiposity, Hyperinsulinemia, Diabetes and Alzheimer's Disease: An Epidemiological Perspective." *European Journal of Pharmacology* 585, no. 1 (2008): 119–29.

Luchsinger, José A., Thania Perez, Helena Chang, Pankaj Mehta, Jason Steffener, Gnanavalli Pradabhan, Masanori Ichise, et al. "Metformin in Amnestic Mild Cognitive Impairment: Results of a Pilot Randomized Placebo Controlled Clinical Trial." *Journal of Alzheimer's Disease* 51, no. 2 (2016): 501–14.

Luchsinger, Jose A., Ming-Xin Tang, Steven Shea, and Richard Mayeux. "Antioxidant Vitamin Intake and Risk of Alzheimer Disease." *Archives of Neurology* 60, no. 2 (2003): 203–8.

MacDonald, Patrick E., Wasim El-Kholy, Michael J. Riedel, Anne Marie F. Salapatek, Peter E. Light, and Michael B. Wheeler. "The Multiple Actions of GLP-1 on the Process of Glucose-Stimulated Insulin Secretion." *Diabetes* 51 Suppl 3 (2002): S434–S442.

Mackey, Howard, William Cho, Michael Ward, Yuan Fang, Shehnaaz Suliman, Carole Ho, and Robert Paul. "Exploratory Analyses of Cognitive Effects of Crenezumab in a Mild Alzheimer's Disease Subpopulation of a Randomized, Double-Blind, Placebo-Controlled, Parallel-Group Phase 2 Study (ABBY)." *Alzheimer's & Dementia* 12, no. 7 (2016): P610.

Maguire, Eleanor A., David G. Gadian, Ingrid S. Johnsrude, Catriona D. Good, John Ashburner, Richard S. Frackowiak, and Christopher D. Frith. "Navigation-Related Structural Change in the Hippocampi of Taxi Drivers." *Proceedings of the National Academy of Sciences of the United States of America* 97, no. 8 (2000): 4398–403.

Manach, Claudine, Augustin Scalbert, Christine Morand, Christian Rémésy, and Liliana Jiménez. "Polyphenols: Food Sources and Bioavailability." *American Journal of Clinical Nutrition* 79, no. 5 (2004): 727–47.

Mancuso, Michelangelo, Daniele Orsucci, Gabiele Siciliano, and Luigi Murri. "Mitochondria, Mitochondrial DNA and Alzheimer's Disease. What Comes First?" *Current Alzheimer Research* 5, no. 5 (2008): 457–68.

Mandler, Markus, Walter Schmidt, and Frank Mattner. "Development of AFFITOPE Alzheimer Vaccines: Results of Phase I Studies with AD01 and AD02." *Alzheimer's & Dementia* 7, no. 4 (2011): S793.

Manning, Carol A., Michael E. Ragozzino, and Paul E. Gold. "Glucose Enhancement of Memory in Patients with Probable Senile Dementia of the Alzheimer's Type." *Neurobiology of Aging* 14, no. 6 (1993): 523–28.

Mannucci, Edoardo, Matteo Monami, Mauro Di Bari, Caterina Lamanna, Francesca Gori, Gian Franco Gensini, and Niccolò Marchionni. "Cardiac Safety Profile of Rosiglitazone: A Comprehensive Meta-Analysis of Randomized Clinical Trials." *International Journal of Cardiology* 143, no. 2 (2010): 135–40.

Marcus, David L., Christopher Thomas, Charles Rodriguez, Katherine Simberkoff, Jir S. Tsai, James A. Strafaci, and Michael L. Freedman. "Increased Peroxidation and Reduced Antioxidant Enzyme Activity in Alzheimer's Disease." *Experimental Neurology* 150, no. 1 (1998): 40–44.

Margolis, Richard U., and Norman Altszuler. "Insulin in the Cerebrospinal Fluid." *Nature* 215, no. 5108 (1967): 1375–76.

Marotte, Bertrand. "Neurochem Plummets as Clinical Trial Flops." *The Globe and Mail.* Last modified April 30, 2018. https://www.theglobeandmail.com/report -on-business/neurochem-plummets-as-clinical-trial-flops/article4098414/.

Maurer, Konrad, and Ulrike Maurer. *Alzheimer: The Life of a Physician and the Career of a Disease.* Translated by Neil Levi and Alistair Burns. New York: Columbia University Press, 2003.

Maurer, Konrad, Stephan Volk, and Hector Gerbaldo. "Auguste D and Alzheimer's Disease." *The Lancet* 349 (1997): 1546–49.

Mawanda, Francis, and Robert Wallace. "Can Infections Cause Alzheimer's Disease?" *Epidemiologic Reviews* 35, no. 1 (2013): 161–80.

May, Patrick C., Robert A. Dean, Stephen L. Lowe, Ferenc Martenyi, Scott M. Sheehan, Leonard N. Boggs, Scott A. Monk, et al. "Robust Central Reduction of Amyloid-β in Humans with an Orally Available, Non-Peptidic β-Secretase Inhibitor." *Journal of Neuroscience* 31, no. 46 (2011): 16507–16.

May, Patrick C., Brian A. Willis, Stephen L. Lowe, Robert A. Dean, Scott A. Monk, Patrick J. Cocke, James E. Audia, et al. "The Potent BACE1 Inhibitor LY2886721 Elicits Robust Central Aβ Pharmacodynamic Responses in Mice, Dogs, and Humans." *Journal of Neuroscience* 35, no. 3 (2015): 1199–210.

Maynard, Steven Douglas, and Jeff Gelblum. "Retrospective Case Studies of the Efficacy of Caprylic Triglyceride in Mild-to-Moderate Alzheimer's Disease." *Neuropsychiatric Disease and Treatment* 9 (2013): 1629–35.

McCaffrey, Pat. "Boston: Neuroprotective Peptide Inches Forward in Clinic." Alzforum. May 6, 2008. https://www.alzforum.org/news/conference-coverage /boston-neuroprotective-peptide-inches-forward-clinic.

——. "Closing the Book on NSAIDs for Alzheimer's Prevention." Alzforum. April 12, 2019. https://www.alzforum.org/news/research-news/closing-book-nsaids -alzheimers-prevention.

McClean, Paula L., Vadivel Parthsarathy, Emilie Faivre, and Christian Hölscher. "The Diabetes Drug Liraglutide Prevents Degenerative Processes in a Mouse Model of Alzheimer's Disease." *Journal of Neuroscience* 31, no. 17 (2011): 6587–94.

McNeill, Leila. "The History of Breeding Mice for Science Begins with a Woman in a Barn." *Smithsonian Magazine*. March 20, 2018. https://www.smithsonianmag.com/science-nature/history-breeding-mice-science-leads-back-woman-barn-180968441/.

Meng, Xiangfei, and Carl D'Arcy. "Education and Dementia in the Context of the Cognitive Reserve Hypothesis: A Systematic Review with Meta-Analyses and Qualitative Analyses." *PLoS One* 7, no. 6 (2012): e38268.

Meyer, Pierre-François, Jennifer Tremblay-Mercier, Jeannie Leoutsakos, Cécile Madjar, Marie-Élyse Lafaille-Maignan, Melissa Savard, Pedro Rosa-Neto, et al. "INTREPAD: A Randomized Trial of Naproxen to Slow Progress of Presymptomatic Alzheimer Disease." *Neurology* 92, no. 18 (2019): e2070–e2080.

Mietlicki-Baase, Elizabeth. "Amylin-Mediated Control of Glycemia, Energy Balance, and Cognition." *Physiology & Behavior* 162 (2016): 130–40.

Mockett, Robin J., William C. Orr, Jennifer J. Rahmandar, Judith J. Benes, Svetlana N. Radyuk, Vladimir I. Klichko, and Rajindar S. Sohal. "Overexpression of Mn-Containing Superoxide Dismutase in Transgenic *Drosophila melanogaster*." *Archives of Biochemistry and Biophysics* 371, no. 2 (1999): 260–9.

Moh, Calvin, Jacek Z. Kubiak, Vladan P. Bajic, Xiongwei Zhu, Mark A. Smith, and Hyoung-Gon Lee. "Cell Cycle Deregulation in the Neurons of Alzheimer's Disease." *Results and Problems in Cell Differentiation* 53 (2011): 565–76.

Möller, Hans-Jürgen, and Manuel B. Graeber. "The Case Described by Alois Alzheimer in 1911." *European Archives of Psychiatry and Clinical Neuroscience* 248, no. 3 (1998): 111–22.

Moore, Eileen M., Alastair G. Mander, David Ames, Mark A. Kotowicz, Ross P. Carne, Henry Brodaty, Michael Woodward, et al. "Increased Risk of Cognitive Impairment in Patients with Diabetes Is Associated with Metformin." *Diabetes Care* 36, no. 10 (2013): 2981–87.

Morales, R., C. Duran-Aniotz, J. Castilla, L. D. Estrada, and C. Soto. "*De Novo* Induction of Amyloid-β Deposition *in Vivo*." *Molecular Psychiatry* 17, no. 12 (2012): 1347–53.

Moreira, Paula I., Cristina Carvalho, Xiongwei Zhu, Mark A. Smith, and George Perry. "Mitochondrial Dysfunction Is a Trigger of Alzheimer's Disease Pathophysiology." *Biochimica et Biophysica Acta* 1802, no. 1 (2010): 2–10.

Morris, Martha Clare, Denis A. Evans, Julia L. Bienias, Christine C. Tangney, David A. Bennett, Neelum Aggarwal, Robert S. Wilson, and Paul A. Scherr. "Dietary Intake of Antioxidant Nutrients and the Risk of Incident Alzheimer Disease in a Biracial Community Study." *JAMA* 287, no. 24 (2002): 3230–37.

Morris, Martha Clare, Denis A. Evans, Christine C. Tangney, Julia L. Bienias, and Robert S. Wilson. "Associations of Vegetable and Fruit Consumption with Age-Related Cognitive Change." *Neurology* 67, no. 8 (2006): 1370–6.

Morris, Martha Clare, Christy C. Tangney, Yamin Wang, Frank Martin Sacks, Lisa L. Barnes, David William Bennett, and Neelum T. Aggarwal. "MIND Diet Slows Cognitive Decline with Aging." *Alzheimers Dement* 11, no. 9 (2015): 1015–22.

Morris, Martha Clare, Christy C. Tangney, Yamin Wang, Frank Martin Sacks, David William Bennett, and Neelum T. Aggarwal. "MIND Diet Associated with Reduced Incidence of Alzheimer's Disease." *Alzheimers Dement* 11, no. 9 (2015): 1007–14.

Morris, Meaghan, Sumihiro Maeda, Keith Vossel, and Lennart Mucke. "The Many Faces of Tau." *Neuron* 70, no. 3 (2011): 410–26.

Mortimer, James A., Ding Ding, Amy R. Borenstein, Charles Decarli, Qihao Guo, Yougui Wu, Qianhua Zhao, and Shugang Chu. "Changes in Brain Volume and Cognition in a Randomized Trial of Exercise and Social Interaction in a Community-Based Sample of Non-Demented Chinese Elders." *Journal of Alzheimer's Disease* 30, no. 4 (2012): 757–66.

Murakami, Kazuma. "Conformation-Specific Antibodies to Target Amyloid β Oligomers and Their Application to Immunotherapy for Alzheimer's Disease." *Bioscience, Biotechnology, and Biochemistry* 78, no. 8 (2014): 1293–305.

Murphy, M. Paul, and Harry LeVine III. "Alzheimer's Disease and the β-Amyloid Peptide." *Journal of Alzheimer's Disease* 19, no. 1 (2010): 311–23.

Nagata, Ken, Hirohiko Saito, Tomoyuki Ueno, Mika Sato, Taizen Nakase, Tetsuya Maeda, Yuichi Satoh, et al. "Clinical Diagnosis of Vascular Dementia." *Journal of the Neurological Sciences* 257, no. 1 (2007): 44–48.

National Human Genome Research Institute. *Why Mouse Matters.* National Institutes of Health. Last modified July 23, 2010. https://www.genome.gov/10001345/importance-of-mouse-genome.

National Institutes of Health. "Estimates of Funding for Various Research, Condition, and Disease Categories (RCDC)." February 24, 2020, from https://report.nih.gov/categorical_spending.aspx.

National Weather Service. "Rochester Tornado August 21 1883." Accessed March 31, 2020. https://www.weather.gov/arx/aug211883.

Navarro, Ana, and Alberto Boveris. "The Mitochondrial Energy Transduction System and the Aging Process." *American Journal of Physiology: Cell Physiology* 292, no. 2 (2007): C670–C686.

Nelson, Peter T., Heiko Braak, and William R. Markesbery. "Neuropathology and Cognitive Impairment in Alzheimer Disease: A Complex but Coherent Relationship." *Journal of Neuropathology and Experimental Neurology* 68, no. 1 (2009): 1–14.

Newport, Mary. *Alzheimer's Disease: What If There Was a Cure? The Story of Ketones.* Laguna Beach, CA: Basic Health Publications, 2013.

Ng, Tze Pin, Liang Feng, Keng Bee Yap, Tih Shih Lee, Chay Hoon Tan, and Bengt Winblad. "Long-Term Metformin Usage and Cognitive Function among Older Adults with Diabetes." *Journal of Alzheimer's Disease* 41, no. 1 (2014): 61–68.

"Nicholas Wood Interviews John Hardy." YouTube. February 19, 2013. https://www.youtube.com/watch?v=YZThB_M8DXw.

Nicolakakis, Nektaria, Tahar Aboulkassim, Brice Ongali, Clotilde Lecrux, Priscilla Fernandes, Pedro Rosa-Neto, Xin-Kang Tong, and Edith Hamel. "Complete Rescue of Cerebrovascular Function in Aged Alzheimer's Disease Transgenic Mice by Antioxidants and Pioglitazone, a Peroxisome Proliferator-Activated Receptor Gamma Agonist." *Journal of Neuroscience* 28, no. 37 (2008): 9287–96.

Nicoll, James A. R., Edward Barton, Delphine Boche, Jim W. Neal, Isidro Ferrer, Petrina Thompson, Christina Vlachouli, et al. "Aβ Species Removal After Aβ$_{42}$ Immunization." *Journal of Neuropathology and Experimental Neurology* 65, no. 11 (2006): 1040–8.

Nilsson, Anne, Ilkka Salo, Plaza Merichel, and Inger Björck. "Effects of a Mixed Berry Beverage on Cognitive Functions and Cardiometabolic Risk Markers; A Randomized Cross-Over Study in Healthy Older Adults." *PLoS One* 12, no. 11 (2017): e0188173.

Ninomiya, Toshiharu. "Japanese Legacy Cohort Studies: The Hisayama Study." *Journal of Epidemiology* 28, no. 11 (2018): 444–51.

Nissen, Steven E., and Kathy Wolski. "Effect of Rosiglitazone on the Risk of Myocardial Infarction and Death from Cardiovascular Causes." *New England Journal of Medicine* 356, no. 24 (2007): 2457–71.

——. "Rosiglitazone Revisited: An Updated Meta-Analysis of Risk for Myocardial Infarction and Cardiovascular Mortality." *Archives of Internal Medicine* 170, no. 14 (2010): 1191–201.

Novak, Petr, Reinhold Schmidt, Eva Kontsekova, Norbert Zilka, Branislav Kovacech, Rostislav Skrabana, Zuzana Vince-Kazmerova, et al. "Safety and Immunogenicity of the Tau Vaccine AADvac1 in Patients with Alzheimer's Disease: A Randomised, Double-Blind, Placebo-Controlled, Phase 1 Trial." *The Lancet Neurology* 16, no. 2 (2017): 123–34.

Novartis. "Novartis, Amgen and Banner Alzheimer's Institute Discontinue Clinical Program with BACE Inhibitor CNP520 for Alzheimer's Prevention." July 11, 2019. https://www.novartis.com/news/media-releases/novartis-amgen-and-banner-alzheimers-institute-discontinue-clinical-program-bace-inhibitor-cnp520-alzheimers-prevention.

Nurses' Health Study. "History." Accessed April 1, 2020. https://www.nurseshealth-study.org/about-nhs/history.

O'Banion, M. Kerry, Paul D. Coleman, and Linda M. Callahan. "Regional Neuronal Loss in Aging and Alzheimer's Disease: A Brief Review." *Seminars in Neuroscience* 6, no. 5 (1994): 307–14.

O'Brien, Claire. "Auguste D. and Alzheimer's Disease." *Science* 273, no. 5271 (1996): 28.

Ohrui, T., N. Tomita, T. Sato-Nakagawa, T. Matsui, M. Maruyama, K. Niwa, H. Arai, and H. Sasaki. "Effects of Brain-Penetrating ACE Inhibitors on Alzheimer Disease Progression." *Neurology* 63, no. 7 (2004): 1324–25.

Okasha, Ahmed. "Mental Health in the Middle East: An Egyptian Perspective." *Clinical Psychology Review* 19, no. 8 (1999): 917–33.

Olazarán, Javier, Barry Reisberg, Linda Clare, Isabel Cruz, Jordi Peña-Casanova, Teodoro del Ser, Bob Woods, et al. "Nonpharmacological Therapies in Alzheimer's Disease: A Systematic Review of Efficacy." *Dementia and Geriatric Cognitive Disorders* 30, no. 2 (2010): 161–78.

Orgogozo, Jean M., Jean F. Dartigues, Sylviane Lafont, Luc Letenneur, Daniel Commenges, Roger Salamon, Susanne Renaud, and Monique B. Breteler. "Wine Consumption and Dementia in the Elderly: A Prospective Community Study in the Bordeaux Area." *Revue Neurologique* 153, no. 3 (1997): 185–92.

Orgogozo, J. M., S. Gilman, J. F. Dartigues, B. Laurent, M. Puel, L. C. Kirby, P. Jouanny, et al. "Subacute Meningoencephalitis in a Subset of Patients with AD After Aβ42 Immunization." *Neurology* 61, no. 1 (2003): 46–54.

Orr, William C., and Rajindar S. Sohal. "The Effects of Catalase Gene Overexpression on Life Span and Resistance to Oxidative Stress in Transgenic *Drosophila melanogaster*." *Archives of Biochemistry and Biophysics* 297, no. 1 (1992): 35–41.

——. "Effects of Cu-Zn Superoxide Dismutase Overexpression on Life Span and Resistance to Oxidative Stress in Transgenic *Drosophila melanogaster*." *Archives of Biochemistry and Biophysics* 301, no. 1 (1993): 34–40.

——. "Extension of Life-Span by Overexpression of Superoxide Dismutase and Catalase in *Drosophila melanogaster*." *Science* 263, no. 5150 (1994): 1128–30.

Orr, William C., Robin J. Mockett, Judith J. Benes, and Rajindar S. Sohal. "Effects of Overexpression of Copper-Zinc and Manganese Superoxide Dismutases, Catalase, and Thioredoxin Reductase Genes on Longevity in *Drosophila melanogaster*." *Journal of Biological Chemistry* 278, no. 29 (2003): 26418–22.

Ostrowitzki, Susanne, Dennis Deptula, Lennart Thurfjell, Frederik Barkhof, Bernd Bohrmann, David J. Brooks, William E. Klunk, et al. "Mechanism of Amyloid Removal in Patients with Alzheimer Disease Treated with Gantenerumab." *Archives of Neurology* 69, no. 2 (2012): 198–207.

Ott, A., R. P. Stolk, F. van Harskamp, H. A. Pols, A. Hofman, and M. M. Breteler. "Diabetes Mellitus and the Risk of Dementia: The Rotterdam Study." *Neurology* 53, no. 9 (1999): 1937–42.

Ott, Brian, Lori Daiello, Issa Dahabreh, Beth Springate, Kimberly Bixby, Manjari Murali, and Thomas Trikalinos. "Do Statins Impair Cognition? A Systematic Review and Meta-Analysis of Randomized Controlled Trials." *Journal of General Internal Medicine* 30, no. 3 (2015): 348–58.

Page, Sean, and Tracey Fletcher. "Auguste D: One Hundred Years On: 'The Person' Not 'the Case'." *Dementia* 5, no. 4 (2006): 571–83.

Paholikova, Kristina, Barbara Salingova, Alena Opattova, Rostislav Skrabana, Petra Majerova, Norbert Zilka, Branislav Kovacech, et al. "N-Terminal Truncation of Microtubule Associated Protein Tau Dysregulates Its Cellular Localization." *Journal of Alzheimer's Disease* 43, no. 3 (2015): 915–26.

Paravastu, Anant K., Isam Qahwash, Richard D. Leapman, Stephen C. Meredith, and Robert Tycko. "Seeded Growth of Beta-Amyloid Fibrils from Alzheimer's Brain–Derived Fibrils Produces a Distinct Fibril Structure." *Proceedings of the National Academy of Sciences of the United States of America* 106, no. 18 (2009): 7443–48.

Pardridge, William M. "Receptor-Mediated Peptide Transport Through the Blood-Brain Barrier." *Endocrine Reviews* 7, no. 3 (1986): 314–30.

Pasqualetti, Patrizio, Cristina Bonomini, Gloria Forno, Luca Paulon, Elena Sinforiani, Camillo Marra, Orazio Zanetti, and Paolo Maria Rossini. "A Randomized Controlled Study on Effects of Ibuprofen on Cognitive Progression of Alzheimer's Disease." *Aging Clinical and Experimental Research* 21, no. 2 (2009): 102–10.

Pedersen, Ward A., Pamela J. McMillan, J. Jacob Kulstad, James B. Leverenz, Suzanne Craft, and Gleb R. Haynatzki. "Rosiglitazone Attenuates Learning and Memory Deficits in Tg2576 Alzheimer Mice." *Experimental Neurology* 199, no. 2 (2006): 265–73.

Pericak-Vance, M., J. L. Bebout, P. C. Gaskell, L. H. Yamaoka, W. Y. Hung, M. J. Alberts, A. P. Walker, et al. "Linkage Studies in Familial Alzheimer Disease: Evidence for Chromosome 19 Linkage." *American Journal of Human Genetics* 48, no. 6 (1991): 1034–50.

Perry, Elaine K., Robert H. Perry, Garry Blessed, and Bernard E. Tomlinson. "Necropsy Evidence of Central Cholinergic Deficits in Senile Dementia." *Lancet* 1, no. 8004 (1977): 189.

Petersen, Ronald C., Ronald G. Thomas, Michael Grundman, David Bennett, Rachelle Doody, Steven Ferris, Douglas Galasko, et al. "Vitamin E and Donepezil for the Treatment of Mild Cognitive Impairment." *New England Journal of Medicine* 352, no. 23 (2005): 2379–88.

Philippidis, Alex. "Top 15 Best-Selling Drugs of 2018: Sales for Most Treatments Grow Year-over-Year Despite Concerns over Rising Prices." *Genetic Engineering & Biotechnology News* 39, no. 4 (2019): 16–17.

Pierson, Ransdell. "Lilly's Drug for Alzheimer's Fails Big Trial; Shares Drop." Reuters. November 23, 2016. https://www.reuters.com/article/us-health -alzheimer-s-lilly/lillys-drug-for-alzheimers-fails-big-trial-shares-drop -idUSKBN13I146.

Pitkala, Kaisu H., Pirkko Routasalo, Hannu Kautiainen, Harri Sintonen, and Reijo S. Tilvis. "Effects of Socially Stimulating Group Intervention on Lonely, Older People's Cognition: A Randomized, Controlled Trial." *American Journal of*

Geriatric Psychiatry: Official Journal of the American Association for Geriatric Psychiatry 19, no. 7 (2011): 654–63.

Platt, Lauren R., Concepción F. Estívariz, and Roland W. Sutter. "Vaccine-Associated Paralytic Poliomyelitis: A Review of the Epidemiology and Estimation of the Global Burden." *Journal of Infectious Diseases* 210 Suppl 1 (2014): S380–S389.

Prince, Martin, Anders Wimo, Maëlenn Guerchet, Gemma-Claire Ali, Yu-Tzu Wu, Matthew Prina, and Alzheimer's Disease International. *World Alzheimer Report 2015: The Global Impact of Dementia: An Analysis of Prevalence, Incidence, Cost and Trends.* London: Alzheimer's Disease International, 2015.

Prusiner, Stanley B. *Madness and Memory: The Discovery of Prions—A New Biological Principle of Disease.* New Haven, CT: Yale University Press, 2014.

Qizilbash, Nawab, Anne Whitehead, Julian Higgins, Gordon Wilcock, Lon Schneider, and Martin Farlow. "Cholinesterase Inhibition for Alzheimer Disease: A Meta-Analysis of the Tacrine Trials." *JAMA* 280, no. 20 (1998): 1777–82.

Quraishe, S., C. M. Cowan, and A. Mudher. "NAP (Davunetide) Rescues Neuronal Dysfunction in a Drosophila Model of Tauopathy." *Molecular Psychiatry* 18, no. 7 (2013): 834–42.

Rajasekhar, K., Malabika Chakrabarti, and T. Govindaraju. "Function and Toxicity of Amyloid Beta and Recent Therapeutic Interventions Targeting Amyloid Beta in Alzheimer's Disease." *Chemical Communications* 51, no. 70 (2015): 13434–50.

Rasmussen, Peter, Patrice Brassard, Helle Adser, Martin V. Pedersen, Lotte Leick, Emma Hart, Niels H. Secher, Bente K. Pedersen, and Henriette Pilegaard. "Evidence for a Release of Brain-Derived Neurotrophic Factor from the Brain During Exercise." *Experimental Physiology* 94, no. 10 (2009): 1062–69.

Rea, Thomas D., John C. Breitner, Bruce M. Psaty, Annette L. Fitzpatrick, Oscar L. Lopez, Anne B. Newman, William R. Hazzard, et al. "Statin Use and the Risk of Incident Dementia: The Cardiovascular Health Study." *Archives of Neurology* 62, no. 7 (2005): 1047–51.

Refolo, Lorenzo M., Miguel A. Pappolla, John Lafrancois, Brian Malester, S. D. Schmidt, Tara Thomas-Bryant, G. Stephen Tint, et al. "A Cholesterol-Lowering Drug Reduces β-Amyloid Pathology in a Transgenic Mouse Model of Alzheimer's Disease." *Neurobiology of Disease* 8, no. 5 (2001): 890–9.

Refolo, Lorenzo M., Miguel A. Pappolla, Brian Malester, John Lafrancois, Tara Bryant-Thomas, Rong Wang, G. Stephen Tint, Kumar Sambamurti, and Karen Duff. "Hypercholesterolemia Accelerates the Alzheimer's Amyloid Pathology in a Transgenic Mouse Model." *Neurobiology of Disease* 7, no. 4 (2000): 321–31.

Reger, Mark A., Samuel T. Henderson, Cathy Hale, Brenna Cholerton, Laura D. Baker, G. Stennis Watson, Karen Hyde, Darla Chapman, and Suzanne Craft. "Effects of β-Hydroxybutyrate on Cognition in Memory-Impaired Adults." *Neurobiology of Aging* 25, no. 3 (2004): 311–14.

Reger, Mark A., G. Stennis Watson, Pattie S. Green, Laura D. Baker, Brenna Cholerton, Mark A. Fishel, Stephen R. Plymate, et al. "Intranasal Insulin Administration Dose-Dependently Modulates Verbal Memory and Plasma Amyloid-β in Memory-Impaired Older Adults." *Journal of Alzheimer's Disease* 13, no. 3 (2008): 323-31.

Reger, Mark A., G. Stennis Watson, Pattie S. Green, Charles W. Wilkinson, Laura D. Baker, Brenna Cholerton, Mark A. Fishel, et al. "Intranasal Insulin Improves Cognition and Modulates β-Amyloid in Early AD." *Neurology* 70, no. 6 (2008): 440-8.

Reiman, Eric M., Kewei Chen, Gene E. Alexander, Richard J. Caselli, Daniel Bandy, David Osborne, Ann M. Saunders, and John Hardy. "Functional Brain Abnormalities in Young Adults at Genetic Risk for Late-Onset Alzheimer's Dementia." *Proceedings of the National Academy of Sciences of the United States of America* 101, no. 1 (2004): 284-89.

Reines, S. A., G. A. Block, J. C. Morris, G. Liu, M. L. Nessly, C. R. Lines, B. A. Norman, and C. C. Baranak. "Rofecoxib: No Effect on Alzheimer's Disease in a 1-Year, Randomized, Blinded, Controlled Study." *Neurology* 62, no. 1 (2004): 66-71.

Reisberg, Barry, Rachelle Doody, Albrecht Stöffler, Frederick Schmitt, Steven Ferris, and Hans Jörg Möbius. "Memantine in Moderate-to-Severe Alzheimer's Disease." *New England Journal of Medicine* 348, no. 14 (2003): 1333-41.

Reiswig, Gary. *The Thousand Mile Stare: One Family's Journey through the Struggle and Science of Alzheimer's.* Boston: Nicholas Brealey, 2010.

Reitz, Christiane, Ming-Xin Tang, Nicole Schupf, Jennifer Manly, Richard Mayeux, and José Luchsinger. "Association of Higher Levels of High-Density Lipoprotein Cholesterol in Elderly Individuals and Lower Risk of Late-Onset Alzheimer Disease." *Archives of Neurology* 67, no. 12 (2010): 1491-97.

Risner, M. E., A. M. Saunders, J. F. B. Altman, G. C. Ormandy, S. Craft, I. M. Foley, M. E. Zvartau-Hind, D. A. Hosford, and A. D. Roses. "Efficacy of Rosiglitazone in a Genetically Defined Population with Mild-to-moderate Alzheimer's Disease." *Pharmacogenomics Journal* 6, no. 4 (2006): 246-54.

Rivera, Enrique J., Alison Goldin, Noah Fulmer, Rose Tavares, Jack R. Wands, and Suzanne M. de la Monte. "Insulin and Insulin-Like Growth Factor Expression and Function Deteriorate with Progression of Alzheimer's Disease: Link to Brain Reductions in Acetylcholine." *Journal of Alzheimer's Disease* 8, no. 3 (2005): 247-68.

Rizzo, Maria Rosaria, Michelangela Barbieri, Virginia Boccardi, Edith Angellotti, Raffaele Marfella, and Giuseppe Paolisso. "Dipeptidyl Peptidase-4 Inhibitors Have Protective Effect on Cognitive Impairment in Aged Diabetic Patients with Mild Cognitive Impairment." *Journals of Gerontology Series A: Biomedical Sciences and Medical Sciences* 69, no. 9 (2014): 1122-31.

Roberts, Sam. "Allen Roses, Who Upset Common Wisdom on Cause of Alzheimer's, Dies at 73." *New York Times*. October 5, 2016. https://www.nytimes.com/2016/10/06/science/allen-roses-who-upset-common-wisdom-on-cause-of-alzheimers-dies-at-73.html.

Rodrigue, Karen, Kristen Kennedy, and Denise Park. "Beta-Amyloid Deposition and the Aging Brain." *Neuropsychology Review* 19, no. 4 (2009): 436–50.

Rodriguez, Guido, Paolo Vitali, Piero Calvini, Chiara Bordoni, Nicola Girtler, Gioconda Taddei, Giuliano Mariani, and Flavio Nobili. "Hippocampal Perfusion in Mild Alzheimer's Disease." *Psychiatry Research* 100, no. 2 (2000): 65–74.

Rogers, Joseph, Scott Webster, Lih-Fen Lue, Libuse Brachova, W. Harold Civin, Mark Emmerling, Brenda Shivers, Douglas Walker, and Patrick McGeer. "Inflammation and Alzheimer's Disease Pathogenesis." *Neurobiology of Aging* 17, no. 5 (1996): 681–86.

Rogers, Madolyn Bowman. "Immunotherapy II: Active Approaches Down, New Passive Crops Up." Alzforum. December 17, 2014. http://www.alzforum.org/news/conference-coverage/immunotherapy-ii-active-approaches-down-new-passive-crops.

——. "Large Study Questions Tomm40's Effect on AD Age of Onset." Alzforum. August 15, 2011. https://www.alzforum.org/news/research-news/large-study-questions-tomm40s-effect-ad-age-onset.

——. "Protective APP Mutation Found—Supports Amyloid Hypothesis." Alzforum. July 13, 2012. https://www.alzforum.org/news/research-news/protective-app-mutation-found-supports-amyloid-hypothesis.

Rogers, Sharon L., Rachelle S. Doody, Richard C. Mohs, and Lawrence T. Friedhoff. "Donepezil Improves Cognition and Global Function in Alzheimer Disease: A 15-Week, Double-Blind, Placebo-Controlled Study." *Archives of Internal Medicine* 158, no. 9 (1998): 1021–31.

Román, Gustavo C. "Vascular Dementia: Distinguishing Characteristics, Treatment, and Prevention." *Journal of the American Geriatrics Society* 51, no. 5 (2003): S296–S304.

Rondeau, Virginie. "A Review of Epidemiologic Studies on Aluminum and Silica in Relation to Alzheimer's Disease and Associated Disorders." *Reviews on Environmental Health* 17, no. 2 (2002): 107–22.

Rosenblum, William I. "Why Alzheimer Trials Fail: Removing Soluble Oligomeric Beta Amyloid Is Essential, Inconsistent, and Difficult." *Neurobiology of Aging* 35, no. 5 (2014): 969–74.

Roses, Allen D., M. W. Lutz, H. Amrine-Madsen, A. M. Saunders, D. G. Crenshaw, S. S. Sundseth, M. J. Huentelman, K. A. Welsh, and E. M. Reiman. "A TOMM40 Variable-Length Polymorphism Predicts the Age of Late-Onset Alzheimer's Disease." *Pharmacogenomics Journal* 10, no. 5 (2010): 375–84.

Rösler, Michael, Ravi Anand, Ana Cicin-Sain, Serge Gauthier, Yves Agid, Peter Dal-Bianco, Hannes B. Stähelin, Richard Hartman, and Marguirguis Gharabawi. "Efficacy and Safety of Rivastigmine in Patients with Alzheimer's Disease: International Randomised Controlled Trial." *BMJ* 318, no. 7184 (1999): 633–38.

Rowland, Christopher. "Pfizer Had Clues Its Blockbuster Drug Could Prevent Alzheimer's. Why Didn't It Tell the World?" *Washington Post.* June 4, 2019. Retrieved January 8, 2020, from https://www.washingtonpost.com/business /economy/pfizer-had-clues-its-blockbuster-drug-could-prevent-alzheimers -why-didnt-it-tell-the-world/2019/06/04/9092e08a-7a61-11e9-8bb7 -0fc796cf2ec0_story.html.

Saba, Magdi M., Hector O. Ventura, Mohamed Saleh, and Mandeep R. Mehra. "Ancient Egyptian Medicine and the Concept of Heart Failure." *Journal of Cardiac Failure* 12, no. 6 (2006): 416–21.

Sakono, Masafumi, and Tamotsu Zako. "Amyloid Oligomers: Formation and Toxicity of Aβ Oligomers." *FEBS Journal* 277 (2010): 1348–58.

Salkovic-Petrisic, M., and S. Hoyer. "Central Insulin Resistance as a Trigger for Sporadic Alzheimer-Like Pathology: An Experimental Approach." *Journal of Neural Transmission* Suppl 72 (2007): 217–33.

Salloway, Stephen, Reisa Sperling, Nick C. Fox, Kaj Blennow, William Klunk, Murray Raskind, Marwan Sabbagh, et al. "Two Phase 3 Trials of Bapineuzumab in Mild-to-Moderate Alzheimer's Disease." *New England Journal of Medicine* 370, no. 4 (2014): 322–33.

Salloway, Stephen, Reisa Sperling, S. Gilman, Nick C. Fox, Kaj Blennow, Murray Raskind, Marwan Sabbagh, et al. "A Phase 2 Multiple Ascending Dose Trial of Bapineuzumab in Mild to Moderate Alzheimer Disease." *Neurology* 73, no. 24 (2009): 2061–70.

Sano, M., K. L. Bell, D. Galasko, J. E. Galvin, R. G. Thomas, C. H. van Dyck, and P. S. Aisen. "A Randomized, Double-Blind, Placebo-Controlled Trial of Simvastatin to Treat Alzheimer Disease." *Neurology* 77, no. 6 (2011): 556–63.

Sano, Mary, Christopher Ernesto, Ronald G. Thomas, Melville R. Klauber, Kimberly Schafer, Michael Grundman, Peter Woodbury, et al. "A Controlled Trial of Selegiline, Alpha-Tocopherol, or Both as Treatment for Alzheimer's Disease." *New England Journal of Medicine* 336, no. 17 (1997): 1216–22.

Santacruz, K., J. Lewis, T. Spires, J. Paulson, L. Kotilinek, M. Ingelsson, A. Guimaraes, et al. "Tau Suppression in a Neurodegenerative Mouse Model Improves Memory Function." *Science* 309, no. 5733 (2005): 476–81.

Santos-Lozano, Alejandro, Helios Pareja-Galeano, Fabian Sanchis-Gomar, Miguel Quindós-Rubial, Carmen Fiuza-Luces, Carlos Cristi-Montero, Enzo Emanuele, Nuria Garatachea, and Alejandro Lucia. "Physical Activity and Alzheimer

Disease: A Protective Association." *Mayo Clinic Proceedings* 91, no. 8 (2016): 999–1020.

Sato, Tomohiko, Haruo Hanyu, Kentaro Hirao, Hidekazu Kanetaka, Hirofumi Sakurai, and Toshihiko Iwamoto. "Efficacy of PPAR-γ Agonist Pioglitazone in Mild Alzheimer Disease." *Neurobiology of Aging* 32, no. 9 (2011): 1626–33.

Sattler, Christine, Pablo Toro, Peter Schönknecht, and Johannes Schröder. "Cognitive Activity, Education and Socioeconomic Status as Preventive Factors for Mild Cognitive Impairment and Alzheimer's Disease." *Psychiatry Research* 196, no. 1 (2012): 90–95.

Saunders, A., W. Strittmatter, D. Schmechel, P. George-Hyslop, M. Pericak-Vance, S. Joo, B. Rosi, et al. "Association of Apolipoprotein E Allele ε-4 with Late-Onset Familial and Sporadic Alzheimer's Disease." *Neurology* 43, no. 8 (1993): 1467–72.

Schellenberg, Gerard D., Thomas D. Bird, Ellen M. Wijsman, Harry T. Orr, Leojean Anderson, Ellen Nemens, June A. White, et al. "Genetic Linkage Evidence for a Familial Alzheimer's Disease Locus on Chromosome 14." *Science* 258, no. 5082 (1992): 668–71.

Schenk, Dale, Robin Barbour, Whitney Dunn, Grace Gordon, Henry Grajeda, Teresa Guido, Kang Hu, et al. "Immunization with Amyloid-β Attenuates Alzheimer-Disease-Like Pathology in the PDAPP Mouse." *Nature* 400, no. 6740 (1999): 173–77.

Schenk, Dale, Michael Hagen, and Peter Seubert. "Current Progress in Beta-Amyloid Immunotherapy." *Current Opinion in Immunology* 16, no. 5 (2004): 599–606.

Schoonjans, Kristina, and Johan Auwerx. "Thiazolidinediones: An Update." *The Lancet* 355, no. 9208 (2000): 1008–10.

Schöll, Michael, Ove Almkvist, Karin Axelman, Elka Stefanova, Anders Wall, Eric Westman, Bengt Långström, et al. "Glucose Metabolism and PIB Binding in Carriers of a His163Tyr Presenilin 1 Mutation." *Neurobiology of Aging* 32, no. 8 (2011): 1388–99.

Selkoe, Dennis J. "The Molecular Pathology of Alzheimer's Disease." *Neuron* 6, no. 4 (1991): 487–98.

Serafini, M., A. Ghiselli, and A. Ferro-Luzzi. "In Vivo Antioxidant Effect of Green and Black Tea in Man." *European Journal of Clinical Nutrition* 50 (1996): 28–32.

Serrano-Pozo, Alberto, Matthew P. Frosch, Eliezer Masliah, and Bradley T. Hyman. "Neuropathological Alterations in Alzheimer Disease." *Cold Spring Harbor Perspectives in Medicine* 1, no. 1 (2011): a006189.

Sevigny, Jeff, Ping Chiao, Thierry Bussière, Paul H. Weinreb, Leslie Williams, Marcel Maier, Robert Dunstan, et al. "The Antibody Aducanumab Reduces Aβ Plaques in Alzheimer's Disease." *Nature* 537, no. 7618 (2016): 50–56.

Shah, Kaushik, Shanal DeSilva, and Thomas Abbruscato. "The Role of Glucose Transporters in Brain Disease: Diabetes and Alzheimer's Disease." *International Journal of Molecular Sciences* 13 (2012): 12629–55.

Sherrington, R., E. I. Rogaev, Y. Liang, E. A. Rogaeva, G. Levesque, M. Ikeda, H. Chi, et al. "Cloning of a Gene Bearing Missense Mutations in Early-Onset Familial Alzheimer's Disease." *Nature* 375, no. 6534 (1995): 754–60.

Shimizu, Hiroko, Luc Tritsmans, Tomoko Santoh, Ayako Shiraishi, Masayoshi Takahashi, Yushin Tominaga, and Johannes Streffer. "Pharmacocokinetic and Pharmacodynamic Study (54861911alz1006) with a BACE Inhibitor, JNJ-54861911, in Healthy Elderly Japanese Subjects." *Alzheimer's & Dementia* 12, no. 7(2016) P612.

Shugart, Jessica. "China Approves Seaweed Sugar as First New Alzheimer's Drug in 17 Years." Alzforum. November 7, 2019. https://www.alzforum.org/news/research-news/china-approves-seaweed-sugar-first-new-alzheimers-drug-17-years.

Shughrue, P. J., P. J. Acton, R. S. Breese, W. Q. Zhao, E. Chen-Dodson, R. W. Hepler, A. L. Wolfe, et al. "Anti-ADDL Antibodies Differentially Block Oligomer Binding to Hippocampal Neurons." *Neurobiology of Aging* 31, no. 2 (2010): 189–202.

Shukitt-Hale, Barbara, Vivian Cheng, and James A Joseph. "Effects of Blackberries on Motor and Cognitive Function in Aged Rats." *Nutritional Neuroscience* 12, no. 3 (2009): 135–40.

Siemers, Eric R., Stuart Friedrich, Robert A. Dean, Celedon R. Gonzales, Martin R. Farlow, Steven M. Paul, and Ronald B. Demattos. "Safety and Changes in Plasma and Cerebrospinal Fluid Amyloid Beta After a Single Administration of an Amyloid Beta Monoclonal Antibody in Subjects with Alzheimer Disease." *Clinical Neuropharmacology* 33, no. 2 (2010): 67–73.

Siemers, Eric R., J. F. Quinn, J. Kaye, Martin R. Farlow, A. Porsteinsson, P. Tariot, P. Zoulnouni, et al. "Effects of a Gamma-Secretase Inhibitor in a Randomized Study of Patients with Alzheimer Disease." *Neurology* 66, no. 4 (2006): 602–4.

Siemers, Eric R., Karen L. Sundell, Christopher Carlson, Michael Case, Gopalan Sethuraman, Hong Liu-Seifert, Sherie A. Dowsett, et al. "Phase 3 Solanezumab Trials: Secondary Outcomes in Mild Alzheimer's Disease Patients." *Alzheimer's & Dementia: The Journal of the Alzheimer's Association* 12, no. 2 (2016): 110–20.

Sigurdsson, Einar M. "Tau Immunotherapies for Alzheimer's Disease and Related Tauopathies: Progress and Potential Pitfalls." *Journal of Alzheimer's Disease* 64, no. S1 (2018): S555–S565.

Simons, John. "Lilly Goes Off Prozac: The Drugmaker Bounced Back from the Loss of Its Blockbuster, but the Recovery Had Costs." *Fortune.* June 28, 2004. https://archive.fortune.com/magazines/fortune/fortune_archive/2004/06/28/374398/index.htm.

Singh, P. P., M. Singh, and S. S. Mastana. "ApoE Distribution in World Populations with New Data from India and the UK." *Annals of Human Biology* 33, no. 3 (2006): 279–308.

Singh, Sonal, Yoon K. Loke, and Curt D. Furberg. "Long-Term Risk of Cardiovascular Events with Rosiglitazone: A Meta-Analysis." *JAMA* 298, no. 10 (2007): 1189–95.

Smith, C. M., M. Swash, A. N. Exton-Smith, M. J. Phillips, P. W. Overstall, M. E. Piper, and M. R. Bailey. "Choline Therapy in Alzheimer's Disease." *The Lancet* 312, no. 8084 (1978): 318.

Smyth, Chris. "Scientists Create the First Drug to Halt Alzheimer's." *Times* (London). July 28, 2016. https://www.thetimes.co.uk/article/scientists-create-the -first-drug-to-halt-alzheimers-xzlkvrkvp.

Snowdon, David A. "Aging and Alzheimer's Disease: Lessons from the Nun Study." *The Gerontologist* 37, no. 2 (1997): 150–56.

Snowdon, David A., Susan J. Kemper, James A. Mortimer, Lydia H. Greiner, David R. Wekstein, and William R. Markesbery. "Linguistic Ability in Early Life and Cognitive Function and Alzheimer's Disease in Late Life: Findings from the Nun Study." *JAMA* 275, no. 7 (1996): 528–32.

Solomon, Alina, Miia Kivipelto, Benjamin Wolozin, Jufen Zhou, and Rachel Whitmer, "Midlife Serum Cholesterol and Increased Risk of Alzheimer's and Vascular Dementia Three Decades Later," *Dementia and Geriatric Cognitive Disorders* 28, no. 1 (2009): 75–80.

Sperling, Reisa A., Dorene M. Rentz, Keith A. Johnson, Jason Karlawish, Michael Donohue, David P. Salmon, and Paul Aisen. "The A4 Study: Stopping AD Before Symptoms Begin?" *Science Translational Medicine* 6, no. 228 (2014): 228fs13.

Staton, Tracy. "Eli Lilly—10 Largest U.S. Patent Losses." FiercePharma. October 24, 2011. https://www.fiercepharma.com/special-report/eli-lilly-10-largest -u-s-patent-losses.

——. "GSK Settles Bulk of Avandia Suits for $460m." FiercePharma. July 14, 2010. https://www.fiercepharma.com/pharma/gsk-settles-bulk-of-avandia -suits-for-460m.

Steen, Eric, Benjamin M. Terry, Enrique J. Rivera, Jennifer L. Cannon, Thomas R. Neely, Rose Tavares, X. J. Xu, Jack R. Wands, and Suzanne M. de La Monte. "Impaired Insulin and Insulin-Like Growth Factor Expression and Signaling Mechanisms in Alzheimer's Disease—Is This Type 3 Diabetes?" *Journal of Alzheimer's Disease* 7, no. 1 (2005): 63–80.

Steensma, David P., Robert A. Kyle, and Marc A. Shampo. "Abbie Lathrop, the 'Mouse Woman of Granby': Rodent Fancier and Accidental Genetics Pioneer." *Mayo Foundation for Medical Education and Research* 85, no. 11 (2010): e83.

Stein, Donald, Simón Brailowsky, and Bruno Will. *Brain Repair*. Oxford: Oxford University Press, 1995.

Stelzmann, Rainulf A., H. Norman Schnitzlein, and F. Reed Murtagh. "An English Translation of Alzheimer's 1907 Paper, 'Über Eine Eigenartige Erkankung Der Hirnrinde.' " *Clinical Anatomy* 8, no. 6 (1995): 429–31.

Stern, Yaakov. "What Is Cognitive Reserve? Theory and Research Application of the Reserve Concept." *Journal of the International Neuropsychological Society* 8, no. 3 (2002): 448–60.

Stern, Yaakov, Barry Gurland, Thomas K. Tatemichi, Ming Xin Tang, David Wilder, and Richard Mayeux. "Influence of Education and Occupation on the Incidence of Alzheimer's Disease." *JAMA* 271, no. 13 (1994): 1004–10.

Streffer, Johannes, Anne Börjesson-Hanson, Bianca Van Broeck, Pascale Smekens, Maarten Timmers, Ina Tesseur, Kanaka Tatikola, et al. "Pharmacodynamics of the Oral BACE Inhibitor JNJ-54861911 in Early Alzheimer's Disease." *Alzheimer's & Dementia: The Journal of the Alzheimer's Association* 12, no. 7 (2016): P199–P200.

St George-Hyslop, Peter, Jonathan Haines, E. Rogaev, M. Mortilla, G. Vaula, Margaret Pericak-Vance, Jean-Francois Foncin, et al. "Genetic Evidence for a Novel Familial Alzheimer's Disease Locus on Chromosome 14." *Nature Genetics* 2, no. 4 (1992): 330–4.

St George-Hyslop, Peter H., Rudolph E. Tanzi, Ronald J. Polinsky, Jonathan L. Haines, Linda Nee, Paul C. Watkins, Richard H. Myers, et al. "The Genetic Defect Causing Familial Alzheimer's Disease Maps on Chromosome 21." *Science* 235, no. 4791 (1987): 885–90.

Stöhr, Jan, Joel C. Watts, Zachary L. Mensinger, Abby Oehler, Sunny K. Grillo, Stephen J. Dearmond, Stanley B. Prusiner, and Kurt Giles. "Purified and Synthetic Alzheimer's Amyloid Beta (Aβ) Prions." *Proceedings of the National Academy of Sciences of the United States of America* 109, no. 27 (2012): 11025–30.

Strittmatter, Warren, Ann Saunders, Donald Schmechel, Margaret Pericak-Vance, J. Enghild, G. Salvesen, and A. Roses. "Apolipoprotein E: High-Avidity Binding to β-Amyloid and Increased Frequency of Type 4 Allele in Late-Onset Familial Alzheimer Disease." *Proceedings of the National Academy of Sciences of the United States of America* 90, no. 5 (1993): 1977–81.

Strobel, Gabrielle. "Biogen Antibody Buoyed by Phase 1 Data and Hungry Investors." Alzforum. March 25, 2015. https://www.alzforum.org/news/conference -coverage/biogen-antibody-buoyed-phase-1-data-and-hungry-investors.

Strum, Jay C., Ron Shehee, David Virley, Jill Richardson, Michael Mattie, Paula Selley, Sujoy Ghosh, et al. "Rosiglitazone Induces Mitochondrial Biogenesis in Mouse Brain." *Journal of Alzheimer's Disease* 11, no. 1 (2007): 45–51.

Ströhle, Andreas, Dietlinde K. Schmidt, Florian Schultz, Nina Fricke, Theresa Staden, Rainer Hellweg, Josef Priller, Michael A. Rapp, and Nina Rieckmann. "Drug and Exercise Treatment of Alzheimer Disease and Mild Cognitive Impairment: A Systematic Review and Meta-Analysis of Effects on Cognition in Randomized Controlled Trials." *American Journal of Geriatric Psychiatry* 23, no. 12 (2015): 1234–49.

Sun, Jingtao, and John Tower. "FLP Recombinase-Mediated Induction of Cu /Zn-Superoxide Dismutase Transgene Expression Can Extend the Life Span of Adult *Drosophila melanogaster* Flies." *Molecular and Cellular Biology* 19, no. 1 (1999): 216–28.

Sun, Jingtao, Donna Folk, Timothy J. Bradley, and John Tower. "Induced Overexpression of Mitochondrial Mn-Superoxide Dismutase Extends the Life Span of Adult *Drosophila melanogaster.*" *Genetics* 161, no. 2 (2002): 661–72.

Sünram-Lea, Sandra I., Jonathan K. Foster, Paula Durlach, and Catalina Perez. "The Effect of Retrograde and Anterograde Glucose Administration on Memory Performance in Healthy Young Adults." *Behavioural Brain Research* 134, no. 1 (2002): 505–16.

Swerdlow, Russell H., Jeffrey M. Burns, and Shaharyar M. Khan. "The Alzheimer's Disease Mitochondrial Cascade Hypothesis." *Journal of Alzheimer's Disease* 20 Suppl 2 (2010): 265–79.

——. "The Alzheimer's Disease Mitochondrial Cascade Hypothesis: Progress and Perspectives." *Biochimica et Biophysica Acta* 1842, no. 8 (2014): 1219–31.

Swiatek, Jeff. "Lean Years Behind It, Eli Lilly Sees Growth in New Drugs." IndyStar. May 31, 2015. https://www.indystar.com/story/money/2015/06/01/lean -years-behind-eli-lilly-sees-growth-new-drugs/28172457/.

Sydow, Astrid, Ann Van der Jeugd, Fang Zheng, Tariq Ahmed, Detlef Balschun, Olga Petrova, Dagmar Drexler, et al. "Tau-Induced Defects in Synaptic Plasticity, Learning, and Memory Are Reversible in Transgenic Mice After Switching Off the Toxic Tau Mutant." *The Journal of Neuroscience* 31, no. 7 (2011): 2511–25.

Szabados, Tamás, Csaba Dul, Katalin Majtényi, Judit Hargitai, Zoltán Pénzes, and Rudolf Urbanics. "A Chronic Alzheimer's Model Evoked by Mitochondrial Poison Sodium Azide for Pharmacological Investigations." *Behavioural Brain Research* 154, no. 1 (2004): 31–40.

Takahashi, Masayoshi, Luc Tritsmans, Hiroko Shimizu, Tomoko Santoh, Ayako Shiraishi, Yushin Tominaga, and Johannes Streffer. "A Pharmacodynamic Study (54861911alz1008) with a BACE Inhibitor, JNJ-54861911 in Japanese Asymptomatic Subjects at Risk for Alzheimer's Dementia." *Alzheimer's & Dementia* 12, no. 7 (2016): P608.

Takeda Pharmaceutical Company. "Takeda and Zinfandel Pharmaceuticals Discontinue TOMMORROW Trial Following Planned Futility Analysis." January 25, 2018. https://www.takeda.com/newsroom/newsreleases/2018/takeda-tommorrow -trial/.

Tan, Zaldy S., Nicole L. Spartano, Alexa S. Beiser, Charles DeCarli, Sanford H. Auerbach, Ramachandran S. Vasan, and Sudha Seshadri. "Physical Activity, Brain Volume, and Dementia Risk: The Framingham Study." *Journals of Gerontology Series A: Biomedical Sciences and Medical Sciences* 72, no. 6 (2017): 789–95.

Tan, Zaldy Sy, Sudha Seshadri, Alexa Beiser, Peter W. F. Wilson, Douglas P. Kiel, Michael Tocco, Ralph B. D'Agostino, and Philip A. Wolf. "Plasma Total Cholesterol Level as a Risk Factor for Alzheimer Disease: The Framingham Study." *Archives of Internal Medicine* 163, no. 9 (2003): 1053–57.

Tang, Ming-Xin, Yaakov Stern, Karen Marder, Karen Bell, Barry Gurland, Rafael Lantigua, Howard Andrews, et al. "The Apoe-4 Allele and the Risk of Alzheimer Disease Among African Americans, Whites, and Hispanics." *JAMA* 279, no. 10 (1998): 751–55.

Tang, M. X., P. Cross, H. Andrews, D. M. Jacobs, S. Small, K. Bell, C. Merchant, et al. "Incidence of AD in African-Americans, Caribbean Hispanics, and Caucasians in Northern Manhattan." *Neurology* 56, no. 1 (2001): 49–56.

Tanskanen, Maarit. " 'Amyloid'—Historical Aspects." In *Amyloidosis*, ed. Dali Feng, 3–24. London: IntechOpen, 2013.

Tanzi, Rudolph E., and Ann B. Parson. *Decoding Darkness: The Search for the Genetic Causes of Alzheimer's Disease*. Cambridge, MA: Perseus, 2000.

Teri, Linda, Laura E. Gibbons, Susan M. McCurry, Rebecca G. Logsdon, David M. Buchner, William E. Barlow, Walter A. Kukull, et al. "Exercise Plus Behavioral Management in Patients with Alzheimer Disease: A Randomized Controlled Trial." *JAMA* 290, no. 15 (2003): 2015–22.

Terry, Robert D. "Alzheimer's Disease at Mid-Century (1927–1977) and a Little More." In *Alzheimer: 100 Years and Beyond*, ed. Mathias Jucker, Konrad Beyreuther, Christian Haass, Roger M. Nitsch, and Yves Christen, 59–61. Heidelberg: Springer, 2006.

——. "Dementia. A Brief and Selective Review." *Archives of Neurology* 33, no. 1 (1976): 1–4.

Thal, L. J., W. Rosen, N. S. Sharpless, and H. Crystal. "Choline Chloride Fails to Improve Cognition in Alzheimer's Disease." *Neurobiology of Aging* 2, no. 3 (1981): 205–8.

Thangthaeng, Nopporn, Margaret Rutledge, Jessica M. Wong, Philip H. Vann, Michael J. Forster, and Nathalie Sumien. "Metformin Impairs Spatial Memory and Visual Acuity in Old Male Mice." *Aging and Disease* 8, no. 1 (2017): 17–30.

Timmers, Maarten, Bianca Van Broeck, Steven Ramael, John Slemmon, Katja De Waepenaert, Alberto Russu, Jennifer Bogert, et al. "Profiling the Dynamics of CSF and Plasma Aβ Reduction After Treatment with JNJ-54861911, a Potent Oral BACE Inhibitor." *Alzheimer's & Dementia: Translational Research & Clinical Interventions* 2, no. 3 (2016): 202–12.

Tolar, Martin, Bruno Vellas, Jeffrey Cummings, Anton Porsteinsson, Susan Abushakra, and John Hey. "Efficacy of Tramiprosate in APOE4 Heterozygous Patients with Mild to Moderate AD: Combined Sub-Group Analyses from Two Phase 3 Trials." *Neurobiology of Aging* 39 (2016): S22.

Tomata, Yasutake, Kemmyo Sugiyama, Yu Kaiho, Kenji Honkura, Takashi Watanabe, Shu Zhang, Yumi Sugawara, and Ichiro Tsuji. "Green Tea Consumption and the Risk of Incident Dementia in Elderly Japanese: The Ohsaki Cohort 2006 Study." *American Journal of Geriatric Psychiatry* 24, no. 10 (2016): 881–89.

Trompet, Stella, Peter Vliet, Anton Craen, Jelle Jolles, Brendan Buckley, Michael Murphy, Ian Ford, et al. "Pravastatin and Cognitive Function in the Elderly. Results of the PROSPER Study." *Journal of Neurology* 257, no. 1 (2010): 85–90.

Tuppo, Ehab E., and Hugo R. Arias. "The Role of Inflammation in Alzheimer's Disease." *International Journal of Biochemistry and Cell Biology* 37, no. 2 (2005): 289–305.

Tuttle, Katherine. "A 60-Year-Old Man with Type 2 Diabetes, Hypertension, Dyslipidemia, and Albuminuria." *Advanced Studies in Medicine* 5, no. 1A (2005): S34–S35.

Tzimopoulou, Sofia, Vincent J. Cunningham, Thomas E. Nichols, Graham Searle, Nick P. Bird, Prafull Mistry, Ian J. Dixon, et al. "A Multi-Center Randomized Proof-of-Concept Clinical Trial Applying [^{18}F]FDG-PET for Evaluation of Metabolic Therapy with Rosiglitazone XR in Mild to Moderate Alzheimer's Disease." *Journal of Alzheimer's Disease* 22, no. 4 (2010): 1241–56.

Ueshima, Hirotsugu. "Hisayama Study." University of Minnesota. Accessed April 3, 2020. http://www.epi.umn.edu/cvdepi/study-synopsis/hisayama-study/.

Ufer, Mike, Marie Rouzade-Dominguez, Gunilla Huledal, Nicole Pezous, Alexandre Avrameas, Olivier David, Sandrine Kretz, et al. "Results from a First-in-Human Study with the BACE Inhibitor CNP520." *Alzheimer's & Dementia* 12, no. 7 (2016): P200.

United Nations, Department of Economic and Social Affairs Population Division. *World Population Ageing.* New York: United Nations, 2015.

U.S. Food and Drug Administration. "The Drug Development Process Step 3: Clinical Research." January 4, 2018. https://www.fda.gov/ForPatients/Approvals /Drugs/ucm405622.htm.

——. "FDA Drug Safety Communication: Important Safety Label Changes to Cholesterol-Lowering Statin Drugs." February 28, 2012. https://www.fda.gov /Drugs/DrugSafety/ucm293101.htm.

——. "FDA Warning Letter to Accera, Inc." Case Watch. December 26, 2013. https://quackwatch.org/cases/fdawarning/prod/fda-warning-letters -about-products-2013/accera/.

——. "Frequently Asked Questions about Medical Foods." May 2016. https://www .fda.gov/downloads/food/guidanceregulation/guidancedocumentsregula -toryinformation/ucm500094.pdf.

Van Cauwenberghe, Caroline, Christine Van Broeckhoven, and Kristel Sleegers. "The Genetic Landscape of Alzheimer Disease: Clinical Implications and Perspectives." *Genetics in Medicine* 18, no. 5 (2016): 421–30.

van Norden, Anouk G. W., Ewoud J. van Dijk, Karlijn F. de Laat, Philip Scheltens, Marcel G. M. Olderikkert, and F. E. de Leeuw. "Dementia: Alzheimer Pathology and Vascular Factors: From Mutually Exclusive to Interaction." *Biochimica et Biophysica Acta* 1822, no. 3 (2012): 340–9.

Verghese, Joe, Richard B. Lipton, Mindy J. Katz, Charles B. Hall, Carol A. Derby, Gail Kuslansky, Anne F. Ambrose, Martin Sliwinski, and Herman Buschke. "Leisure Activities and the Risk of Dementia in the Elderly." *New England Journal of Medicine* 348, no. 25 (2003): 2508–16.

Viola, Kirsten, and William Klein. "Amyloid β Oligomers in Alzheimer's Disease Pathogenesis, Treatment, and Diagnosis." *Acta Neuropathologica: Pathology and Mechanisms of Neurological Disease* 129, no. 2 (2015): 183–206.

Vitek, Michael, Keshab Bhattacharya, J. Michael Glendening, Edward Stopa, Helen Vlassara, Richard Bucala, Kirk Manogue, and Anthony Cerami. "Advanced Glycation End Products Contribute to Amyloidosis in Alzheimer Disease." *Proceedings of the National Academy of Sciences of the United States of America* 91, no. 11 (1994): 4766–70.

Vreugdenhil, Anthea, John Cannell, Andrew Davies, and George Razay. "A Community-Based Exercise Programme to Improve Functional Ability in People with Alzheimer's Disease: A Randomized Controlled Trial." *Scandinavian Journal of Caring Sciences* 26, no. 1 (2012): 12–19.

Wakefield, A., S. Murch, A. Anthony, J. Linnell, D. M. Casson, M. Malik, M. Berelowitz, et al. "Ileal-Lymphoid-Nodular Hyperplasia, Non-Specific Colitis, and Pervasive Developmental Disorder in Children." *The Lancet* 351, no. 9103 (1998): 637–41.

Walsh, Dominic M., and Dennis J. Selkoe. "A Critical Appraisal of the Pathogenic Protein Spread Hypothesis of Neurodegeneration." *Nature Reviews Neuroscience* 17, no. 4 (2016): 251–60.

Walsh, Dominic M., Igor Klyubin, Julia V. Fadeeva, William K. Cullen, Roger Anwyl, Michael S. Wolfe, Michael J. Rowan, and Dennis J. Selkoe. "Naturally Secreted Oligomers of Amyloid β Protein Potently Inhibit Hippocampal Long-Term Potentiation in Vivo." *Nature* 416, no. 6880 (2002): 535–39.

Walsh, Fergus. "Alzheimer's Drug Solanezumab Could Slow Patients' Decline." BBC News. July 23, 2015. https://www.bbc.com/news/av/health-33618682/alzheimer-s-drug-solanezumab-could-slow-patients-decline.

Wang, Hui-Xin, Anita Karp, Bengt Winblad, and Laura Fratiglioni. "Late-Life Engagement in Social and Leisure Activities Is Associated with a Decreased Risk of Dementia: A Longitudinal Study from the Kungsholmen Project." *American Journal of Epidemiology* 155, no. 12 (2002): 1081–87.

Wang, Jing, Denis Gallagher, Loren M. Devito, Gonzalo I. Cancino, David Tsui, Ling He, Gordon M. Keller, et al. "Metformin Activates an Atypical PKC-CBP Pathway to Promote Neurogenesis and Enhance Spatial Memory Formation." *Cell Stem Cell* 11, no. 1 (2012): 23–35.

Wang, J., S. Xiong, C. Xie, W. R. Markesbery, and M. A. Lovell. "Increased Oxidative Damage in Nuclear and Mitochondrial DNA in Alzheimer's Disease." *Journal of Neurochemistry* 93, no. 4 (2005): 953–62.

Wang, Xinglong, Wenzhang Wang, Li Li, George Perry, Hyoung-Gon Lee, and Xiongwei Zhu. "Oxidative Stress and Mitochondrial Dysfunction in Alzheimer's Disease." *Biochimica et Biophysica Acta* 1842, no. 8 (2014): 1240–7.

Wang, Xinyi, Guangqiang Sun, Jing Zhang, Xun Huang, Tao Wang, Zuoquan Xie, Xingkun Chu, et al. "Sodium Oligomannate Therapeutically Remodels Gut Microbiota and Suppresses Gut Bacterial Amino Acids–Shaped Neuroinflammation to Inhibit Alzheimer's Disease Progression." *Cell Research* 29, no. 10 (2019): 787–803.

Wang, Yaming, Marina Cella, Kaitlin Mallinson, Jason D. Ulrich, Katherine L. Young, Michelle L. Robinette, Susan Gilfillan, et al. "TREM2 Lipid Sensing Sustains the Microglial Response in an Alzheimer's Disease Model." *Cell* 160, no. 6 (2015): 1061–71.

Watson, G. Stennis, and Suzanne Craft. "Modulation of Memory by Insulin and Glucose: Neuropsychological Observations in Alzheimer's Disease." *European Journal of Pharmacology* 490, no. 1 (2004): 97–113.

Watson, G. Stennis, Brenna A. Cholerton, Mark A. Reger, Laura D. Baker, Stephen R. Plymate, Sanjay Asthana, Mark A. Fishel, et al. "Preserved Cognition in Patients with Early Alzheimer Disease and Amnestic Mild Cognitive Impairment During Treatment with Rosiglitazone: A Preliminary Study." *American Journal of Geriatric Psychiatry* 13, no. 11 (2005): 950–8.

Weggen, Sascha, and Dirk Beher. "Molecular Consequences of Amyloid Precursor Protein and Presenilin Mutations Causing Autosomal-Dominant Alzheimer's Disease." *Alzheimer's Research & Therapy* 4, no. 9 (2012): 1–14.

Weingarten, Murray D., Arthur H. Lockwood, Shu-Yin Hwo, and Marc W. Kirschner. "A Protein Factor Essential for Microtubule Assembly." *Proceedings of the National Academy of Sciences of the United States of America* 72, no. 5 (1975): 1858–62.

Weintraub, S., M. M. Mesulam, R. Auty, R. Baratz, B. N. Cholakos, L. Kapust, B. Ransil, et al. "Lethicin in the Treatment of Alzheimer's Disease." *Archives of Neurology* 40, no. 8 (1983): 527–28.

Wharton, Whitney, James H. Stein, Claudia Korcarz, Jane Sachs, Sandra R. Olson, Henrik Zetterberg, Maritza Dowling, et al. "The Effects of Ramipril in Individuals at Risk for Alzheimer's Disease: Results of a Pilot Clinical Trial." *Journal of Alzheimer's Disease* 32, no. 1 (2012): 147–56.

Wheless, James W. "History of the Ketogenic Diet." *Epilepsia* 49 (2008): 3–5.

White, P., M. J. Goodhardt, J. P. Keet, C. R. Hiley, L. H. Carrasco, I. E. I. Williams, and D. M. Bowen. "Neocortical Cholinergic Neurons in Elderly People." *The Lancet* 309, no. 8013 (1977): 668–71.

Whitehouse, Peter. *The Myth of Alzheimer's: What You Aren't Being Told About Today's Most Dreaded Diagnosis*. New York: St. Martin's Press, 2008.

Whitehouse, Peter J., Donald L. Price, Robert G. Struble, Arthur W. Clark, Joseph T. Coyle, and Mahlon R. Delong. "Alzheimer's Disease and Senile Dementia: Loss of Neurons in the Basal Forebrain." *Science* 215, no. 4537 (1982): 1237–39.

Wiedemann, Nils, Ann E. Frazier, and Nikolaus Pfanner. "The Protein Import Machinery of Mitochondria." *The Journal of Biological Chemistry* 279, no. 15 (2004): 14473–76.

Wilcock, Gordon K., Serge Gauthier, Giovanni B. Frisoni, Jianping Jia, Jiri H. Hardlund, Hans J. Moebius, Peter Bentham, et al. "Potential of Low Dose Leuco-Methylthioninium Bis(Hydromethanesulphonate) (LMTM) Monotherapy for Treatment of Mild Alzheimer's Disease: Cohort Analysis as Modified Primary Outcome in a Phase III Clinical Trial." *Journal of Alzheimer's Disease* 61, no. 1 (2017): 435–57.

Wilcock, Gordon K., Sean Lilienfeld, and Els Gaens. "Efficacy and Safety of Galantamine in Patients with Mild to Moderate Alzheimer's Disease: Multicentre Randomised Controlled Trial." *BMJ* 321, no. 7274 (2000): 1445–49.

Wilson, R. S., D. A. Bennett, J. L. Bienias, N. T. Aggarwal, Mendes De Leon, M. C. Morris, J. A. Schneider, and D. A. Evans. "Cognitive Activity and Incident AD in a Population-Based Sample of Older Persons." *Neurology* 59, no. 12 (2002): 1910–4.

Wilson-Fritch, Leanne, Sarah Nicoloro, My Chouinard, Mitchell Lazar, Patricia C. Chui, John Leszyk, Juerg Straubhaar, Michael P. Czech, and Silvia Corvera. "Mitochondrial Remodeling in Adipose Tissue Associated with Obesity and Treatment with Rosiglitazone." *Journal of Clinical Investigation* 114, no. 9 (2004): 1281–89.

Winblad, Bengt, and N. Poritis. "Memantine in Severe Dementia: Results of the ⁹M-Best Study (Benefit and Efficacy in Severely Demented Patients During Treatment with Memantine)." *International Journal of Geriatric Psychiatry* 14, no. 2 (1999): 135–46.

Winblad, Bengt, Niels Andreasen, Lennart Minthon, Annette Floesser, Georges Imbert, Thomas Dumortier, R. P. Maguire, et al. "Safety, Tolerability, and Antibody Response of Active Aβ Immunotherapy with CAD106 in Patients with Alzheimer's Disease: Randomised, Double-Blind, Placebo-Controlled, First-in-Human Study." *The Lancet Neurology* 11, no. 7 (2012): 597–604.

Winblad, Bengt, Ana Graf, Marie-Emmanuelle Riviere, Niels Andreasen, and J. M. Ryan. "Active Immunotherapy Options for Alzheimer's Disease." *Alzheimer's Research & Therapy* 6, no. 7 (2014):1–12.

Wirak, D. O., R. Bayney, T. V. Ramabhadran, R. P. Fracasso, J. T. Hart, P. E. Hauer, P. Hsiau, et al. "Deposits of Amyloid β Protein in the Central Nervous System of Transgenic Mice." *Science* 253, no. 5017 (1991): 323–25.

Wischik, Claude M., Roger T. Staff, Damon J. Wischik, Peter Bentham, Alison D. Murray, John M. D. Storey, Karin A. Kook, and Charles R. Harrington. "Tau Aggregation Inhibitor Therapy: An Exploratory Phase 2 Study in Mild or Moderate Alzheimer's Disease." *Journal of Alzheimer's Disease* 44, no. 2 (2015): 705–20.

Wolf-Klein, Gisele, Felix A. Siverstone, Meryl S. Brod, Arnold Levy, Conn J. Foley, Val Termotto, and Joseph Breuer. "Are Alzheimer Patients Healthier?" *Journal of the American Geriatrics Society* 36, no. 3 (1988): 219–24.

Wolozin, Benjamin, Wendy Kellman, Paul Ruosseau, Gastone G. Celesia, and George Siegel, "Decreased Prevalence of Alzheimer Disease Associated with 3-hydroxy-3-methyglutaryl Coenzyme A Reductase Inhibitors," *Archives of Neurology* 57, no. 10 (2000): 1439–43.

Wongrakpanich, Supakanya, Amaraporn Wongrakpanich, Katie Melhado, and Janani Rangaswami. "A Comprehensive Review of Non-Steroidal Anti-Inflammatory Drug Use in the Elderly." *Aging and Disease* 9, no. 1 (2018): 143–50.

Woods, Bob, Elisa Aguirre, Aimee E. Spector, and Martin Orrell. "Cognitive Stimulation to Improve Cognitive Functioning in People with Dementia." *Cochrane Database of Systematic Reviews*, no. 2 (2012): CD005562.

World Health Organization. "The Top 10 Causes of Death." May 24, 2018. https://www.who.int/news-room/fact-sheets/detail/the-top-10-causes-of-death.

Wyss-Coray, Tony, and Joseph Rogers. "Inflammation in Alzheimer Disease—A Brief Review of the Basic Science and Clinical Literature." *Cold Spring Harbor Perspectives in Medicine* 2, no. 1 (2012): a006346.

Yaffe, Kristine, and Tina Hoang. "Nonpharmacologic Treatment and Prevention Strategies for Dementia." *Continuum* 19, no. 2 (2013): 372–81.

Yankner, Bruce A., Linda R. Dawes, Shannon Fisher, Lydia Villa-Komaroff, Mary Oster-Granite, and Rachael L. Neve. "Neurotoxicity of a Fragment of the Amyloid Precursor Associated with Alzheimer's Disease." *Science* 245, no. 4916 (1989): 417–20.

Yankner, Bruce A., Lawrence K. Duffy, and Daniel A. Kirschner. "Neurotrophic and Neurotoxic Effects of Amyloid β Protein: Reversal by Tachykinin Neuropeptides." *Science* 250, no. 4978 (1990): 279–82.

Yoshitake, T., Y. Kiyohara, I. Kato, T. Ohmura, H. Iwamoto, K. Nakayama, S. Ohmori, et al. "Incidence and Risk Factors of Vascular Dementia and Alzheimer's Disease in a Defined Elderly Japanese Population: The Hisayama Study." *Neurology* 45, no. 6 (1995): 1161–68.

Young, Jeremy, Maaike Angevaren, Jennifer Rusted, and Naji Tabet. "Aerobic Exercise to Improve Cognitive Function in Older People Without Known Cognitive Impairment." *Cochrane Database of Systematic Reviews*, no. 4 (2015): CD005381.

Yu, Chang-En, Elizabeth Marchani, Georg Nikisch, Ulrich Müller, Dagmar Nolte, Andreas Hertel, Ellen M. Wijsman, and Thomas D. Bird. "The N141I Mutation in PSEN2: Implications for the Quintessential Case of Alzheimer Disease." *Archives of Neurology* 67, no. 5 (2010): 631–33.

Yuede, Carla M., Scott D. Zimmerman, Hongxin Dong, Matthew J. Kling, Adam W. Bero, David M. Holtzman, Benjamin F. Timson, and John G. Csernansky. "Effects of Voluntary and Forced Exercise on Plaque Deposition, Hippocampal Volume, and Behavior in the Tg2576 Mouse Model of Alzheimer's Disease." *Neurobiology of Disease* 35, no. 3 (2009): 426–32.

Zakaib, Gwyneth Dickey. "In Surprise, Placebo, Not Aβ Vaccine, Said to Slow Alzheimer's." Alzforum. June 6, 2014. https://www.alzforum.org/news/research-news/surprise-placebo-not-av-vaccine-said-slow-alzheimers.

Zandi, Peter P., James C. Anthony, Kathleen M. Hayden, Kala Mehta, Lawrence Mayer, and John C. S. Breitner. "Reduced Incidence of AD with NSAID But Not H$_2$ Receptor Antagonists: The Cache County Study." *Neurology* 59, no. 6 (2002): 880–6.

Zandi, Peter P., James C. Anthony, Ara S. Khachaturian, Stephanie V. Stone, Deborah Gustafson, Joann T. Tschanz, Maria C. Norton, Kathleen Welsh-Bohmer, and John C. S. Breitner. "Reduced Risk of Alzheimer Disease in Users of Antioxidant Vitamin Supplements: The Cache County Study." *Archives of Neurology* 61, no. 1 (2004): 82–88.

Zandi, Peter P., D. L. Sparks, Ara S. Khachaturian, Joann Tschanz, Maria Norton, Martin Steinberg, Kathleen Welsh-Bohmer, and John C. S. Breitner. "Do Statins Reduce Risk of Incident Dementia and Alzheimer Disease? The Cache County Study." *Archives of General Psychiatry* 62, no. 2 (2005): 217–24.

Zhang, Bin, Jenna Carroll, John Q. Trojanowski, Yuemang Yao, Michiyo Iba, Justin S. Potuzak, Anne-Marie L. Hogan, et al. "The Microtubule-Stabilizing Agent, Epothilone D, Reduces Axonal Dysfunction, Neurotoxicity, Cognitive Deficits, and Alzheimer-Like Pathology in an Interventional Study with Aged Tau Transgenic Mice." *Journal of Neuroscience* 32, no. 11 (2012): 3601–11.

Zhu, H., X. Wang, M. Wallack, H. Li, I. Carreras, A. Dedeoglu, J. Y. Hur, et al. "Intraperitoneal Injection of the Pancreatic Peptide Amylin Potently Reduces Behavioral Impairment and Brain Amyloid Pathology in Murine Models of Alzheimer's Disease." *Molecular Psychiatry* 20, no. 2 (2014): 232–39.

Zilka, Norbert, Peter Filipcik, Peter Koson, Lubica Fialova, Rostislav Skrabana, Monika Zilkova, Gabriela Rolkova, Eva Kontsekova, and Michal Novak. "Truncated Tau from Sporadic Alzheimer's Disease Suffices to Drive Neurofibrillary Degeneration in Vivo." *FEBS Letters* 580, no. 15 (2006): 3582–88.

Zlokovic, Berislav V. "Neurovascular Pathways to Neurodegeneration in Alzheimer's Disease and Other Disorders." *Nature Reviews Neuroscience* 12, no. 12 (2011): 723–38.

INDEX

Page numbers in *italics* indicate figures or tables.

CPSIA information can be obtained
at www.ICGtesting.com
Printed in the USA
JSHW020045210722
28356JS00004B/8

9 780231 198714